D1198843

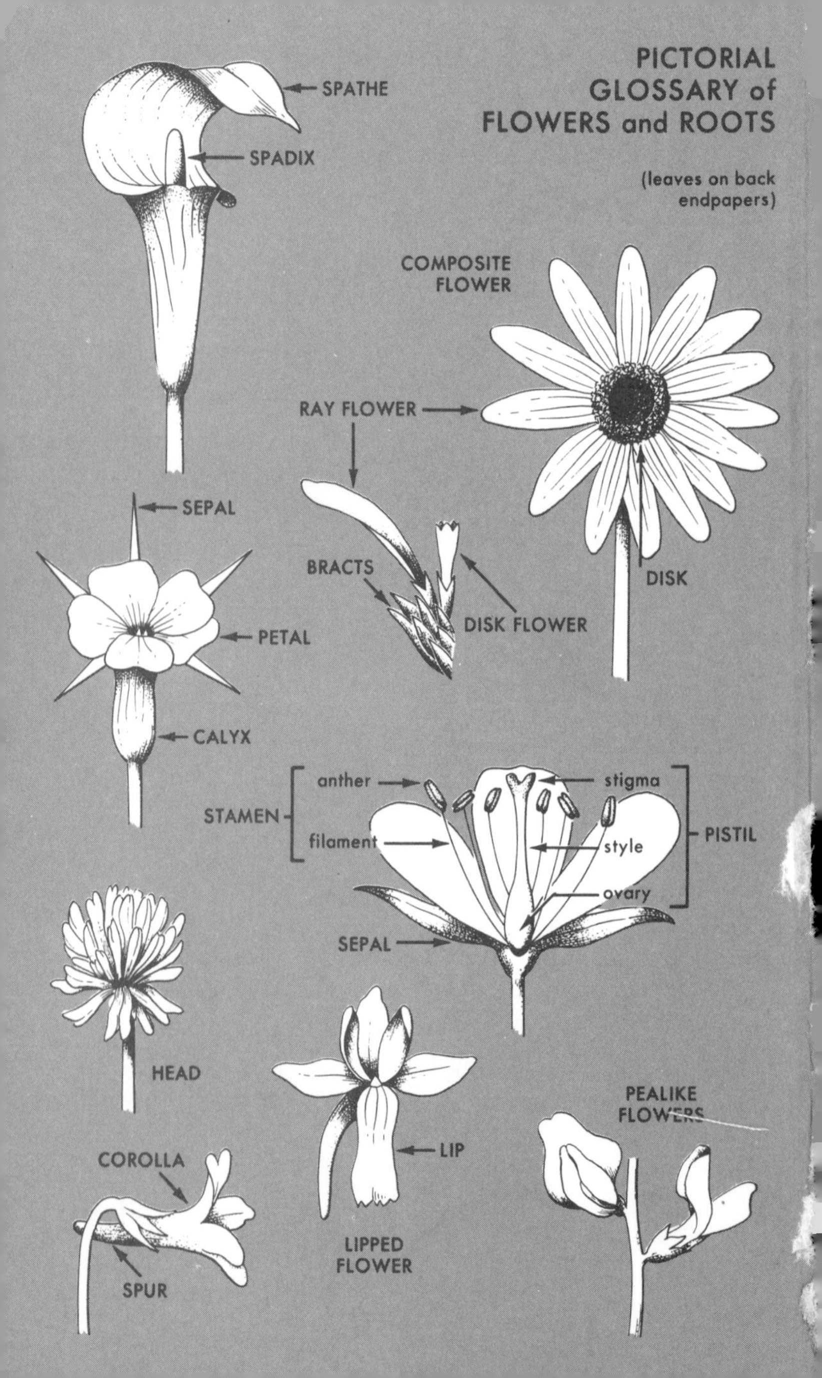

PICTORIAL GLOSSARY of FLOWERS and ROOTS
(leaves on back endpapers)
SPATHE
SPADIX
COMPOSITE FLOWER
RAY FLOWER
SEPAL
BRACTS
DISK
DISK FLOWER
PETAL
CALYX
anther
stigma
STAMEN
filament
style
PISTIL
ovary
SEPAL
HEAD
PEALIKE FLOWERS
COROLLA
LIP
LIPPED FLOWER
SPUR

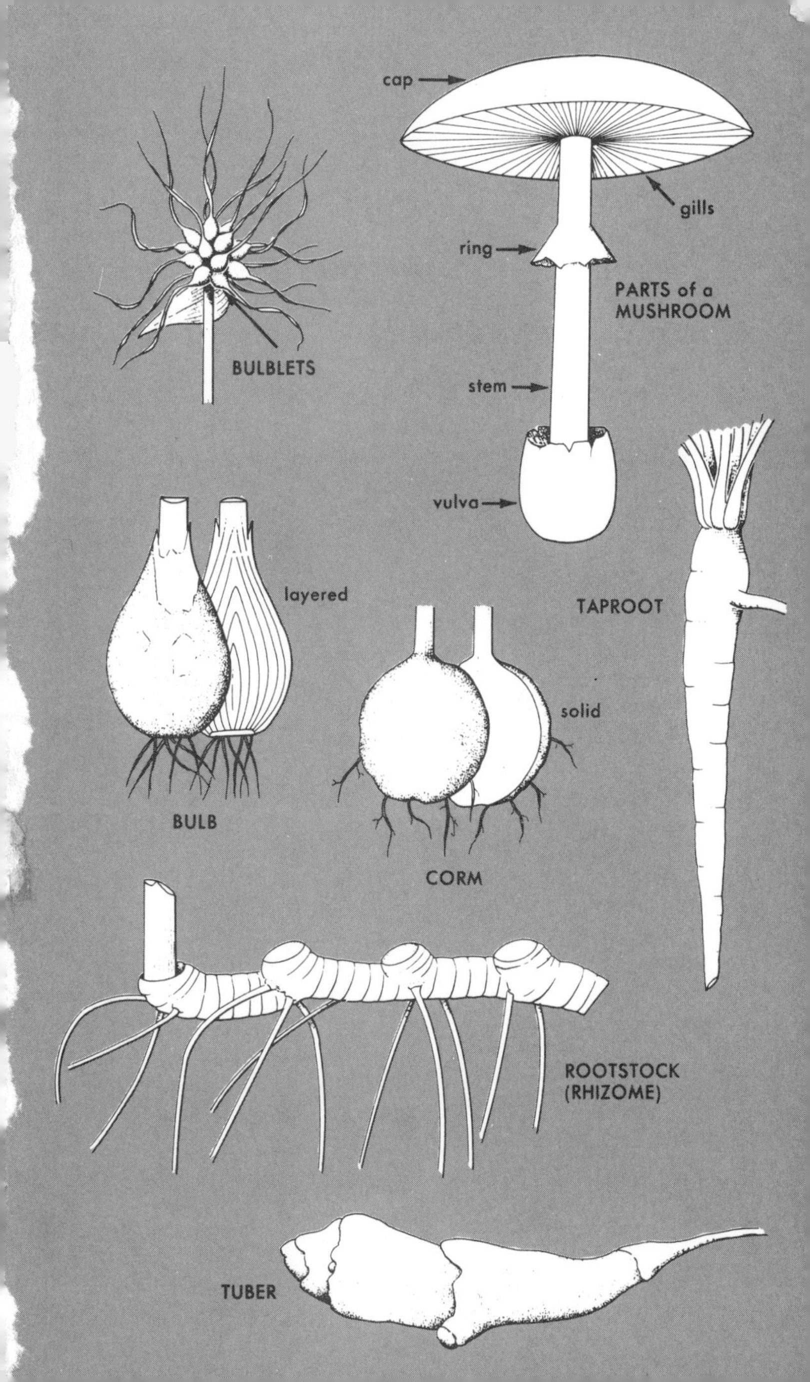

cap
gills
ring
PARTS of a MUSHROOM
BULBLETS
stem
vulva
layered
TAPROOT
solid
BULB
CORM
ROOTSTOCK (RHIZOME)
TUBER

THE PETERSON FIELD GUIDE SERIES®
Edited by Roger Tory Peterson

1. Birds (eastern) — *R.T. Peterson*
1A. Bird Songs (eastern) — *Cornell Laboratory of Ornithology*
2. Western Birds — *R.T. Peterson*
2A. Western Bird Songs — *Cornell Laboratory of Ornithology*
3. Shells of the Atlantic and Gulf Coasts, W. Indies — *Abbott & Morris*
4. Butterflies (eastern) — *Opler & Malikul*
5. Mammals — *Burt & Grossenheider*
6. Pacific Coast Shells (including Hawaii) — *Morris*
7. Rocks and Minerals — *Pough*
8. Birds of Britain and Europe — *Peterson, Mountfort, & Hollom*
9. Animal Tracks — *Murie*
10. Ferns (ne. and cen. N. America) — *Cobb*
11. Eastern Trees — *Petrides*
11A. Trees and Shrubs — *Petrides*
12. Reptiles and Amphibians (e. and cen. N. America) — *Conant & Collins*
13. Birds of Texas and Adjacent States — *R.T. Peterson*
14. Rocky Mt. Wildflowers — *Craighead, Craighead, & Davis*
15. Stars and Planets — *Pasachoff & Menzel*
16. Western Reptiles and Amphibians — *Stebbins*
17. Wildflowers (ne. and n.-cen. N. America) — *R.T. Peterson & McKenney*
18. Birds of the West Indies — *Bond*
19. Insects (America north of Mexico) — *Borror & White*
20. Mexican Birds — *R.T. Peterson & Chalif*
21. Birds' Nests (east of Mississippi River) — *Harrison*
22. Pacific States Wildflowers — *Niehaus & Ripper*
23. Edible Wild Plants (e. and cen. N. America) — *L. Peterson*
24. Atlantic Seashore — *Gosner*
25. Western Birds' Nests — *Harrison*
26. Atmosphere — *Schaefer & Day*
27. Coral Reefs (Caribbean and Florida) — *Kaplan*
28. Pacific Coast Fishes — *Eschmeyer, Herald, & Hammann*
29. Beetles — *White*
30. Moths — *Covell*
31. Southwestern and Texas Wildflowers — *Niehaus, Ripper, & Savage*
32. Atlantic Coast Fishes — *Robins, Ray, & Douglass*
33. Western Butterflies — *Tilden & Smith*
34. Mushrooms — *McKnight & McKnight*
35. Hawks — *Clark & Wheeler*
36. Southeastern and Caribbean Seashores — *Kaplan*
37. Ecology of Eastern Forests — *Kricher & Morrison*
38. Birding by Ear: Eastern and Central — *Walton & Lawson*
39. Advanced Birding — *Kaufman*
40. Medicinal Plants — *Foster & Duke*
41. Birding by Ear: Western — *Walton & Lawson*
42. Freshwater Fishes (N. America north of Mexico) — *Page & Burr*
43. Backyard Bird Song — *Walton & Lawson*
44. Western Trees — *Petrides*
46. Venomous Animals and Poisonous Plants — *Foster & Caras*
47. More Birding by Ear: Eastern and Central — *Walton & Lawson*
48. Geology — *Roberts & Hodsdon*
49. Warblers — *Dunn & Garrett*
50. California and Pacific Northwest Forests — *Kricher & Morrison*
51. Rocky Mountain and Southwest Forests — *Kricher & Morrison*

THE PETERSON FIELD GUIDE SERIES ®

A Field Guide To Venomous Animals And Poisonous Plants

North America
North of Mexico

STEVEN FOSTER
and
ROGER A. CARAS

Sponsored by the National Audubon Society, the National Wildlife Federation, and the Roger Tory Peterson Institute

HOUGHTON MIFFLIN COMPANY
BOSTON NEW YORK

Text copyright © 1994 by Steven Foster and Roger Caras
Editor's Note copyright © 1994 by Roger Tory Peterson
Line drawings of plants copyright © 1994 by Amy Eisenberg
Animal illustrations copyright © 1994 by Norman Arlott
Color photographs of plants copyright © 1994 by Steven Foster

Selected illustrations reproduced from *A Field Guide to Wildflowers of Northeastern and North-central North America* by Roger Tory Peterson and illustrated by Margaret McKenny, copyright © 1968 by Roger Tory Peterson and Margaret McKenny; *A Field Guide to Mushrooms of North America* by Kent H. McKnight and Vera B. McKnight, copyright © 1987 by Kent H. McKnight and Vera B. McKnight; *A Field Guide to Southwestern and Texas Wildflowers* by Theodore F. Niehaus and illustrated by Charles L. Ripper and Virginia Savage, illustrations copyright © 1984 by Charles L. Ripper and Virginia Savage; *A Field Guide to Edible Wild Plants of Eastern and Central North America* by Lee Allen Peterson, copyright © 1977 by Lee Peterson; *A Field Guide to Medicinal Plants* by Steven Foster and James A. Duke, illustrations copyright © 1990 by Jim Rose, photographs copyright © 1990 by Steven Foster; and *A Field Guide to Pacific States Wildflowers* by Theodore F. Niehaus and illustrated by Charles L. Ripper, copyright © 1976 by Theodore F. Niehaus and Charles L. Ripper.

All rights reserved.
For information about permission to reproduce selections from this book, write to Permissions, 215 Park Avenue South, New York, NY 10003.

PETERSON FIELD GUIDES and PETERSON FIELD GUIDE SERIES are registered trademarks of Houghton Mifflin Company.

Library of Congress Cataloging-in-Publication Data
Foster, Steven, 1957–
A field guide to venomous animals and poisonous plants of North America north of Mexico / by Steven Foster and Roger A. Caras.
p. cm. — (The Peterson field guide series ; 46)
"Sponsored by the National Audubon Society, the National Wildlife Federation, and the Roger Tory Peterson Institute."
Includes bibliographical references and index.
ISBN 0-395-51594-7 (cloth). — ISBN 0-395-93608-X (pbk.)
1.Poisonous animals—United States—Identification. 2. Poisonous animals—Canada—Identification. 3. Poisonous plants—United States—Identification. 4. Poisonous plants—Canada—Identification. I. Caras, Roger A. II. National Audubon Society.
III. National Wildlife Federation. IV. Roger Tory Peterson Institute.
V. Title. VI. Series.
QL100.F63 1994
574.6′5′0973—dc2094-1641

CIP

Printed in the United States of America

VB 14 13 12 11 10 9 8 7

EDITOR'S NOTE

In many years in the field, I have stayed out of trouble by staying alert. When birding in the South or Southwest, collecting minerals in Arizona, stargazing on a cool, clear prairie evening, or simply enjoying all aspects of nature on a camping trip in the West, you must be aware of the potential dangers in your environment, however remote they might be. In the South, if you enter the woods, you had better do so with "snake radar." The casual birder or wildflower enthusiast may spend years in the field without ever encountering a venomous snake. Nevertheless, you must be aware of their existence. Encounters with venomous animals are more likely to occur as the result of your entering their habitats than the animals entering yours.

No matter what takes you into the field, whether it be birding or rock collecting, no matter which Peterson Field Guide serves as your bible in the field, *A Field Guide to Venomous Animals and Poisonous Plants* is the second field guide to take with you.

The further south you go in the contiguous 48 states, the more likely you are to encounter a venomous animal or poisonous plant. The reason for this is simple. In warmer climates there is greater biological diversity—larger numbers of animal and plant species. In Massachusetts you are likely to encounter only one poisonous snake, the Timber Rattlesnake, and even this one will be seen only rarely, usually only if sought out. In Arkansas, on the other hand, you may encounter not only the timber rattler but the Western Pigmy Rattlesnake and the Western Diamondback Rattlesnake, plus the Texas Coral Snake, Western Cottonmouth, and several varieties of Copperheads.

In avoiding snakebite, prevention is the best medicine. Poisonous snakes are rarely aggressive toward humans. More than 30 percent of snakebites are from captives, and most bites are the result of aggravating the snake. In about 40 percent of bites from venomous snakes, the snake injects no venom. On average, there are about 10 fatalities per year from venomous snakebites. In a nation of 250 million people, that makes your chance of dying from snakebite about one in 25 million. Bee stings cause about 50 times more deaths each year. Rocky Mountain spotted fever results in about five times more fatalities. Still, poisonous snakes evoke a primeval human fear.

One of the best ways to avoid being bitten is simply to be attentive when you are in snake habitats. Learn how to identify venomous snakes, and keep a healthy distance from them. The same can be said for other poisonous animals such as scorpions and Gila monsters. Black Widow and Brown Recluse spiders, bees, wasps, and hornets may be more difficult to avoid. You are more likely to encounter them by accident, often because they have invaded your habitat.

Poisonous plants do not strike out at humans —unless you count an inadvertent brush with Poison Ivy. Generally, plants cause toxic reactions only if ingested, whether accidentally or intentionally. Again, the best way to avoid poisoning is prevention. One simply should not ingest a wild plant or mushroom unless it has been expertly identified.

A Field Guide to Venomous Animals and Poisonous Plants has been written to help keep you out of trouble with nature, whether in a remote wilderness or around the house.

Steven Foster and Roger Caras have provided the first survey of animals and plants dangerous to humans in North America (north of Mexico) in a field guide format. All of our rattlesnakes, Cottonmouths, Copperheads, and Coral Snakes are described and illustrated, along with rear-fanged lyre snakes, venomous insects, scorpions, spiders, and our only venomous mammal, the shrew. Norman Arlott has produced exquisite black and white artwork for the animals section. Color photographs of both venomous animals and poisonous plants have been provided. The book includes black and white illustrations and descriptions of over 300 native or introduced wildflowers, weeds, trees, shrubs, and mushrooms using the Peterson System, a practical approach based on visual impressions. Plant drawings have been adapted from my *Field Guide to Wildflowers of Northeastern and North-central North America* and from *A Field Guide to Edible Plants of Eastern and Central North America, A Field Guide to Pacific States Wildflowers, A Field Guide to Southwestern and Texas Wildflowers,* and *A Field Guide to Mushrooms.* More than 80 new pen-and-ink drawings were produced for this book by botanical artist Amy Eisenberg.

Obviously, one reason to learn to recognize venomous animals and poisonous plants is in order to avoid them. Perhaps a less obvious reason, however, is to gain greater appreciation for them. Venomous animals and poisonous plants represent the remarkable resiliency, adaptability, and diversity of nature. The biology of reptilians, insects, arachnids, plants, and mushrooms that are potentially dangerous to humans is endlessly fascinating. These organisms have developed complex and often unique defense mechanisms to help them survive in a dwindling natural environment. *A Field Guide to Venomous Animals and Poisonous Plants* serves to warn the reader of potential dangers in nature. However, it is sobering to remember at the same time that humans are often the greatest threat to venomous animals and poisonous plants. We should remember that their survival depends upon our attitudes and actions.

Roger Tory Peterson

CONTENTS

VENOMOUS ANIMALS

VENOMOUS ANIMALS

This book is designed to help you keep out of trouble in the field. But there is more than one reason for being able to recognize our venomous wildlife. The first, obviously, is so that you can avoid contact that could result in an injury. The second reason is also very important: to keep uninformed people from dashing about destroying harmless and often useful animals because they *might be* dangerous. If you know what to look for, there is no "might be" involved. An animal is either dangerous or harmless. In either case, the animal should be left in peace.

How many dangerous animals are there in North America? In order to answer that question, the word *dangerous* must be defined. The snake that accounts for most venomous bites in the United States, the Copperhead, is far from being the most venomous. Its bite is rarely lethal. In the United States you are unlikely to be bitten by any snake at all except in isolated localities. In those areas where venomous animals of one or several kinds congregate, there can be a real danger at certain times of the year. To go wandering along rocky ledges in some parts of Arizona, without boots and without paying attention to where you step, would be foolhardy. To worry much about snakebite north of the Great Lakes or in British Columbia (where rattlesnakes do, in fact, range) would be a waste of energy. The few thousand venomous snakebites that occur in the United States and Canada each year result in 9 to 15 deaths.

Spiders and scorpions are included in this book as well. We do have the world's most venomous spider, and although it is generally easy to avoid, a bite from a Black Widow spider can hardly be ignored. The large, hairy, so-called tarantulas of the American South and Southwest, however, are only mildly venomous, and their bite is not serious unless you are allergic to them; their bite is about as severe as a bee sting.

Try to put the presence of spiders into perspective. Every time you go into the field you are going to pass millions of spiders. A *single acre* of New England countryside, particularly pasture land, is home to about two and a quarter million spiders! Virtually all of them are venomous, but almost none of them seriously enough to bother a human being. Remember that all those millions of spiders are eating insects nearly all the time. They are a major control of the earth's insect population and so are essential to our well-being.

Snakebite

Venomous snakes have an envenomating apparatus that is designed and used primarily for getting food. These snakes cannot crawl rapidly, and they cannot constrict as nonvenomous snakes do. Instead, they have a chemical device—venom—for stopping the flight of the often very fast animals they need for food.

Snakes have no external ear openings (as lizards do) and are, by our definition, deaf. They are exquisitely equipped to detect our presence, however. First, because they are long, legless animals, they have extended contact with the ground. They feel vibrations, we are sure, at least those in the earth and perhaps some that are airborne. A snake's tongue, constantly darting in and out, captures small air samples and deposits them in openings in the roof of its mouth. These openings, known as Jacobson's organs, enable the snake to detect its prey chemically.

Pit vipers (rattlesnakes, Copperheads, and Cottonmouths) have an added detection system. Although it operates only to a distance of 12 to 18 inches, it is extremely sensitive, and enables the snake to direct its strike with great accuracy. Midway between the eye and the external nostril on each side of the snake's head is a two-chambered heat detection device. The snake can home in on its target by the minute differences in heat from one side of its head to the other.

Despite this impressive apparatus, snakes are fragile animals. They do not suffer trauma well; their bones are fragile, and they are susceptible to infection. They also have a low tolerance for temperature extremes. Optimum temperature for a snake is 72 degrees Fahrenheit. If it gets too much hotter than that, or if the snake is exposed for long to direct sunlight, the snake will die. If it gets too much colder than that, the snake will become moribund and be slow to react, and it will be easy prey for any predator seeking a meal.

Why snakes bite

The venomous snake does not want to bite you. The use of its food-getting device on anything as large and inedible as a human is for the secondary purpose of defense. And a snake will try not to bite in self-defense unless given what it considers to be no choice. A snake has nothing to gain in an encounter with a large, dangerous animal, and everything to lose.

As many as 25 percent of all bites by venomous species will be "dry," with no discharge of venom at all. Such a bite is analogous to a snap by a dog on a hot day. The dog—or the snake—doesn't attack, it gives a cranky, warning snap. Only when the snake is seriously disturbed, injured, or frightened will it discharge a full dose of venom. Snakes do vary in disposition from species to species and, we assume, from individual to individual.

The snake we may encounter in the field, then, is shy and nervous by nature because it is easily hurt and is preyed upon by many different kinds of animals. If you look, and listen, and know something of the habits of snakes, you can avoid trouble. Most snakes will try to hide or get away if they sense your approach, but if they feel cornered or endangered they may bite. If you do not approach a snake, the snake will be unlikely to approach you, except by accident.

How to avoid snakebite

The following information can help you avoid being bitten:

Note the ambient temperature. If it is below 70 degrees Fahrenheit, most snakes will be relatively inactive. If the temperature is below 60 degrees, they will be hidden and very still. If it is below 50 degrees, they probably will be unable to strike out at all. If the temperature is above 74 degrees, the snakes in the area are likely to be "hot" in more ways than one, and they may be irritable and more resentful of intrusion than at lower temperatures. The higher the temperature, the more easily upset many species are likely to be.

Watch where you put your hands and feet, especially if you are in "snaky" country. If you are on an isolated, rocky hillside in Arizona, or on the banks of a stream or on a riverine island or hammock in Florida or Georgia, watch where you walk, or you may sooner or later feel the searing jolt of a snakebite. Be careful about stepping over or poking under rotting logs, listen before crawling into a cave, use a flashlight, use common sense. Don't expose yourself to a snake that is trying to be as unobtrusive as possible.

Most snakebites occur on hands and feet, arms and legs, but pulling your body up level with a rocky ledge can bring your head, face, neck, and then your upper torso level to a surface on which a snake may be coiled. A bite in those areas of the body, which are closer to the heart and to the central nervous system, is even more dangerous than one on the hands and feet.

Do not attempt to handle venomous snakes unless you have a thoroughly professional reason for needing to do so, as well as a professional's skill. A large percentage of the snakebites in this country are suffered by people mishandling animals that should be left alone. Virtually *all* Coral Snake bites happen that way.

If a snake you know to be venomous invades territory where it simply cannot be tolerated (a school playground, a garage), employ someone who knows what they are doing to remove it. Call the zoo or the museum and see if they will help. They will probably simply move it to a place where its presence will be less resented.

If you really *must* do the task yourself, use a very long-handled shovel, and do not attempt to pick the snake up with your hands. Don't place it in a container; it will very quickly figure out how to escape. All snakes are escape artists. If you put a snake in a cardboard box on the back seat of your car with the intention of dropping it off in an empty field outside of town, the snake will probably be in the front seat with you long before you get to the town line. If it is absolutely essential to kill a snake—and it rarely is—a shovel swung with sufficient force can decapitate a snake. But remember that a thoroughly dead rattlesnake, even a severed head, can still bite in a reflexive movement. Use a long-handled shovel to move the snake, dead or alive.

Where the snakes are

The species account for each animal gives information on both habitats and geographical distribution under the heading **"Where found."** The Southwest, specifically Arizona, appears to be the epicenter of rattlesnake development. There are 17 forms in Arizona, 10 in Texas, and 10 in California. But by the time you get to the state of Washington or Massachusetts you are down to one species each. The really dangerous scorpions, all of the seriously venomous spiders, and the one venomous lizard in the United States also occur in the Southwest. The Southeast is not too far behind. The higher the latitude as you leave those areas, the less likely you are to encounter seriously venomous animals.

This field guide does not use maps. Snakes, because they are so heavily preyed upon by man, his animals, and his vehicles, are discontinuous in their distribution. There may be no more than two of three sites in several counties where a species of snake is likely to be found. To freckle what amounts to hundreds of square miles to indicate a few acres of rocky hillside is misleading. If you are rock hunting, bird watching, or otherwise outdoors and have reason to suspect you may be in snake country, ask people familiar with the area. Local museum people are probably the best sources, along with amateur herpetologists who work the area, but farmers and other outdoors people will also have thought the matter through.

How sick will snake venom make a person?

There are a great many factors involved in a venomous snakebite, and all play a role in determining the severity of the reaction.

Species. Clearly, this is of paramount importance. It is unlikely that anyone has ever died from the bite of some of the pigmy rattlesnakes. A good many people have been killed by the Mojave Rattlesnake and both the Eastern and Western diamondbacks.

Size. Although venomous snakes are born venomous (their venoms may, in fact, be more toxic, drop for drop, when they are small), a very small snake will deliver a relatively small dose of venom.

Mood. As noted earlier, a quarter of the bites by venomous snakes aren't venomous at all, simply because the snake saves its venom for the more important use of getting food. If a snake is mildly agitated, it may inject a small quantity of venom. If it is a hot day and you have intruded upon a thoroughly aroused snake, you may get a much larger dose.

Condition. An old or debilitated snake will probably produce less venom than one in its hunting prime.

Recent activity. A snake that has not hunted successfully or been in combat for several days may have more venom to deliver than one that has recently bitten something else. However, a snake rarely discharges all of its venom, so any snake is likely to have some venom available at all times.

Condition of fangs. A pit viper's long, hollow fangs, located at the front of the upper jaw, are designed for a swift, deep stab. A pit viper always has fangs in reserve, rather like cartridges in a pistol clip, and it takes only a few hours for a fang to be replaced. However, if a bite occurred very shortly after a snake broke one or both fangs on a tough target, obviously there would be a difference in the severity of the bite.

Site of the bite. As mentioned earlier, a bite around the face, head, or throat is potentially more serious than one on the big toe. The torso, generally, is a worse bite location than a limb.

Clothing. Obviously, the professional's heavy, multilayered snake boots and webbed snake chaps or leggings offer a great deal of protection, but even less dynamically designed clothing can help. I was once struck at by a fairly good-sized rattlesnake whose bite was deflected by nothing more protective than a pair of khakis. There were fang marks and venom stains on my pants, but none on me.

Foreign protein tolerance. Snake venom is a foreign protein, and people differ in their tolerance to foreign proteins. Some people are violently allergic to eggs or milk; some people are allergic to animal venom. A person who has a low tolerance for such protein may be made much more seriously ill by a snake bite.

Size and health of victim. A child who weighs under 75 pounds, or an older person with multiple health problems, may be seriously sickened by a bite or sting that would have only mild effects on a strapping, 200-pound 21-year-old.

First aid and ultimate medical management. How quickly a person receives first aid from someone who knows how to deal properly with snakebite will have a great deal to do with the outcome of the episode. And the quicker an envenomated victim is delivered from the first aid administrator into the hands of a fully equipped medical practitioner, the more effectively the damaging effects of the accident can be minimized.

What to do in case of snakebite

The treatment of snakebite, lacking the benefit of controlled clinical trials because of the obvious difficulties in designing such studies, is clearly an aspect of medicine where art reigns over science. *Don't* turn to an old Boy Scout manual, collecting dust on a book shelf, for advice on first aid treatment of snakebites. Current (and still evolving) medical consensus suggests that gone are the days where incision at the bite, suction, and application of ice and a tourniquet are considered rational measures for the management of snakebites. An incision may spread the venom into surrounding tissue. Tourniquets, unless applied by a medical professional, may cause unnecessary secondary complications, even gangrene, especially in the case of a nonlethal snakebite, like that of a Copperhead. Application of ice may cause needless "frostbite."

Initial symptoms of rattler bites often include intense, instantaneous burning pain, followed by swelling, discoloration of the skin, nausea, and vomiting. Pain may not develop until later. Blood blisters may also appear. Many venomous bites result only in moderate localized pain and swelling. More severe envenomation may involve symptoms of shock, such as faintness, cold sweat, weakness, and dizziness. Blood pressure and body temperature may drop, and the pulse may become rapid.

If a bite does occur, exertion should be kept to a minimum and hysteria avoided, on the part of both the victim and of companions who want to look, advise, and help. Many people believe that a snakebite will cause rapid, inevitable death, which in the vast majority of bites is simply not true. There is no animal in North America whose bite is inevitably fatal, folk tales to the contrary notwithstanding, and venoms are usually slow acting. Real problems are probably hours away. What you do and how quickly you do it, however, may determine how bad those problems may be.

If you are the victim, sit down, be calm, and let assistance come to you, if possible. If not, get help with a minimum of fuss and exertion. If you are with someone who is bitten by a venomous snake, get the victim to the closest hospital or medical facility as soon as possible. Immobilize the limb as much as possible with a splint or sling, and remove any restrictive clothing, jewelry, etc. Nonaspirin painkillers such as acetaminophen can be given to treat pain. Don't give aspirin; it reduces blood clotting and may aggravate bleeding. Likewise, don't give the victim a slug of whiskey or other alcohol. Alcohol is a vasodilator and may cause further unnecessary complications. Treatment for shock may be necessary, and if the patient vomits, the airway must be kept clear. If cardiac or respiratory arrest should occur, CPR (cardiopulmonary resuscitation) should be administered. If available, an intravenous line with isotonic fluid should be established before the victim is moved. If possible, the offending snake should be brought to the medical facility for proper identification. But remember, a dead snake can still inject venom.

About antivenins

If antivenin is available to treat an animal's bite, I have listed it in the species account. Although antivenins are the best medical weapon available against animal venoms, they should be administered only by a physician.

Antivenin #1 is effective against the venom of most North American snake species, including all of the rattlesnakes, Copperheads, and Cottonmouths. However, antivenin is not considered necessary in the majority of Copperhead bites. Antivenin #2 covers all the Coral Snakes in the United States except the Arizona Coral Snake. Although antivenins #1 and #3 can be purchased by prescription, the #2 serum is stocked only in poison control centers and hospitals in the nine southern states where the Coral Snakes it covers are found.

Antivenin #3 covers the Black Widow spider. It has no effect on the sting of the scorpion or other spiders.

A strong word of caution about the use of antivenins: All of the commercially prepared antivenins available in this country today are made from horse serum. *A person must be tested for horse serum sensitivity before antivenin is used.* Many people are more adversely affected by horse serum than by venom: as many as 75 percent of patients who receive antivenins develop "serum sickness." Even lethal venoms kill slowly in almost all cases. Horse serum can kill an allergic person in seconds. New antivenins, including those made with chicken egg–based preparations and sheep-derived products, are in development. Snakebite medical treatment in North America may see dramatic improvement in the next few years.

For detailed information on snakebite management, medical professionals are referred to the excellent treatment on "Snakebite Treatment and First Aid," pages 6–13 in Campbell and Lamar's *The Venomous Reptiles of Latin America* (see References).

Spiders and Scorpions

Since the Black Widow is truly the most venomous spider in the world and is capable of causing death, it deserves special attention. Deaths attributed to this arachnid are, fortunately, a rare occurrence—one survey recorded 55 deaths in 1,273 bites.

Like other web-spinning spiders, the Black Widow is nearly blind. It spins its web where flying insects are likely to collide with it, and by using the intimate contact it has with the strands of the trap it has laid (it does, after all, have eight feet in touch with the web), it dashes to the point of collision and injects its venom. If you blunder into the web, you may get injected.

Therefore, use caution anywhere a Black Widow may set up hunting webs, such as abandoned buildings, old tires, jars, and tin cans in abandoned dumps and wooded areas with good insect populations. Watch where you put your hands, feet, and other parts of your body. Carefully inspect outdoor privies, particularly those that are not in regular use. The latrine pit itself is ideal Black Widow habitat, and some very unpleasant bites have resulted from that fact.

Most scorpions have a venom far too mild to really bother a human being. A few in the American Southwest are more potent. Scorpions are cryptic in their habits. They come out at night, they hide under boards and rocks, and they pop up where they are least expected, including in the proverbial boot under the bed. If you are in scorpion country, watch where you put exposed parts of your body, especially hands and feet.

None of this information is intended to alarm or exaggerate the potential dangers of encounters with venomous animals. Such encounters, however, are threats to careless people in certain parts of the country. Our stress here is on alleviating the intensity of the threat. It is far more interesting to see a rattlesnake at a safe distance than to step on one; a Cottonmouth seen on a log, as you paddle down the river, is far more interesting when viewed from a dozen feet away. If you want to photograph venomous animals, use a telephoto lens and keep your distance. It is difficult to react properly when your face is glued to the back of a camera that you are trying to balance for exposure. And as you jockey for a good camera angle, watch where you step. Remember, you're in snake country.

—*Roger Caras*

MAMMALS

SHORTTAIL SHREW *Blarina brevicauda* **Pl. 1**

Large for a shrew, stocky build, 3–5 in. for head and body only. Tail an additional ½–1 in. *No visible external ears* (present, but extremely small), tiny eyes, short tail. Silver-gray to lead in color in northern and eastern parts of its range; more sooty further east and south. Extremely common in many areas, although all shrews can be difficult to find. Other shrews sharing range are likely to show brownish or reddish cast or have lighter underside more clearly defined.

Where found: Under vegetation, the more ground litter the better; woodlands, overgrown brush, marshes, and bogs. All of the eastern half of the United States (north into southern parts of corresponding Canadian provinces), west to cen. Texas, Okla., Kans., Neb., S.D., and N.D.

Comments: This shrew, like an unknown number of the insectivorous mammals, is truly venomous. The tiny animal's 32 black-tipped teeth create a wound into which a neurotoxic venom is introduced. It is a secretive animal that bites humans only if it is being handled, and its venom is too mild to be of concern.

Antivenin: None.

COMMON SHREW *Sorex cinereus*

Also called Masked Shrew. Head and body from 2–4 in., tail an additional 1–2 in. May vary according to part of range where found. In its northern range at high altitude, it is a grayish brown color, with markedly lighter underparts, a longish tail, and a very *pointed nose.* It is often quite difficult to make a positive identification unless the animal is captured and examined.

Where found: Mountainous areas, especially moist ones; swamps, marshes, tundra, mountains to tree line at least, forest floor and grassy areas. Likely to be most common shrew around. Its range is absolutely enormous: all of Canada, all of Alaska, all northern-tier states south into N.M., Ky. and N.C., in at least three southward fingers. All of New England and Great Lakes Region included. Absent in Ore. and Calif., Nev., probably Utah; not found in Ariz., Texas, or most of s.-cen. and se. states.

Comments: Almost certainly venomous, but this shrew's venom is of no real concern to an animal the size of a human being, and it poses no danger whatever if it is not handled.

Antivenin: None.

SHORTTAIL SHREW
COMMON SHREW

LIZARDS

RETICULATE GILA MONSTER **Pl. 1**
Heloderma suspectum suspectum

Almost impossible to confuse with any other kind of lizard: It is large, 12–16 in., and has a robust build, a sausagelike tail, a blunt wide head, and "beaded" scales. Black with pink, yellow, or orange in a blotched or banded pattern.

Where found: Primarily semiarid, rocky regions. Found in or near foothills of mountains, deserts, wooded areas, and near washes and streams. More commonly found near water and damp soil than most people realize. Will enter farmland. Extreme sw. N.M., se. and s.-cen. Ariz.; not found in n. Ariz., and missing in sw. corner of that state. Doubtful reports from extreme se. Calif. Ranges deep into Sonora, Mexico.

Comments: The Gila Monsters and their congenitors, the Mexican Beaded Lizards, are the world's only venomous lizards. Their venom is very potent, their bite is extremely tenacious, and they have been known to hang on and "chew in" their venom for 10 minutes or more. However, there has never been a report of an "attack" by an undisturbed wild specimen. All known bites (of which as many as 25 percent may have been fatal) were by captive specimens, usually those being teased. Protected in Ariz., N.M. Becoming increasingly rare.

Antivenin: None commercially available. Has been produced experimentally.

BANDED GILA MONSTER *Heloderma suspectum cinctum* **Pl. 1**

Essentially the same as Reticulate Gila Monster, except tail is strongly banded. This is the retention of a juvenile characteristic that is not found in the adult Reticulate form. On average, the Banded Gila Monster is slightly smaller than the Reticulate, 9–14 in. long.

Where found: Habits similar to those of Reticulate Gila Monster, but may be found on rocky ridges located in range. Not restricted to rocky areas, however, and certainly not to desert or Creosotebush areas, as is often thought. Slightly north and west of Reticulate form; w. Ariz., s. Nev., and extreme sw. Utah.

Comments: Dangerously venomous, but secretive. Attacks by unmolested free specimens unknown; bites uncommon. Increasingly rare, and should not be disturbed if encountered.

Antivenin: None.

MEXICAN BEADED LIZARD *Heloderma horridum*

Three subspecies of this lizard are found in Mexico exclusively. None ranges into the U.S.; the lizard is shown here only for comparison and because it is the only other venomous lizard in the world. Habitats and behavior of Mexican Beaded Lizards are similar to those of the Gila Monsters.

RETICULATE GILA MONSTER
BANDED GILA MONSTER
MEXICAN BEADED LIZARD

DIAMONDBACK RATTLESNAKES

EASTERN DIAMONDBACK RATTLESNAKE **Pl. 2**
Crotalus adamanteus

Largest of all North American rattlesnakes — perhaps rarely to 8 ft., average 2½–6 ft. Heavy, very *strong diamond markings* in all shades of brown to black outlined by a single row of cream to yellow scales. Ground color olive through browns to black. Markings are light and brilliant in specimens that have recently shed, but markings darken as shedding time approaches. Diagonal light lines on face are unique among rattlesnakes within range. Broad head. Rattle.

Where found: Found in many habitats. Most abundant in open prairie with saw palmetto clumps. Southern flatlands in palmetto or pine woods. Also in or near both fresh and salt water; strong swimmer. Takes advantage of all holes and cover, and frequently usurps burrows of other animals, especially gopher tortoises. Coastal lowlands. Northern limit apparently se. N.C.; south through coastal S.C., Ga., and Fla. throughout the Keys. West through s. Ala. and Miss. as far as the eastern edge of La.

Comments: Warning rattle *not* dependably given. Specimens vary greatly in disposition. A very large and seriously venomous snake. Potentially lethal. *Probably the most dangerous North American snake.*

Antivenin: #1

WESTERN DIAMONDBACK RATTLESNAKE *Crotalus atrox* **Pl. 2**

Second in size only to the Eastern Diamondback in North America. Average 2½–6 ft., perhaps rarely to 7 ft., although larger sizes now uncommon. Diamond pattern against dark brown to gray ground color less clearly defined than in Eastern Diamondback. Tail *ringed* in black to gray and white. Aggressive. When challenged, head is likely to be raised above coils. Rattle makes a loud and persistent buzzing. Some specimens have a reddish or yellowish cast to ground color. Light, diagonal facial stripes. Broad head. Rattle.

Where found: Lowlands to at least 7,000 ft. Deserts, canyons, rocky areas. Few undisturbed areas within its range are dependably free of this species. Pursues rodents and rabbits into ranching and farming areas, and will approach human habitats. Ark., Okla., almost all of Texas, s. two thirds of N.M. and Ariz., extreme se. Calif., ranges into Mexico. Rarely in extreme s. Nev., s. Colo., se. Kans., se. Mo.

Comments: A nervous and sometimes aggressive snake, seriously venomous and potentially lethal. Some areas may have dense populations.

Antivenin: #1

EASTERN
DIAMONDBACK
RATTLESNAKE

WESTERN
DIAMONDBACK
RATTLESNAKE

RATTLESNAKES: SIDEWINDERS

MOJAVE DESERT SIDEWINDER *Crotalus cerastes cerastes* **Pl. 3**
Small to medium-sized rattlesnake, 18–30 in., perhaps rarely 32 in. Sidewise locomotion leaves characteristic J-marks on loose sand. Has broad head, rattle, *hornlike protuberances* over eyes. First rattle segment may be brown in mature specimens. Generally pale on top, blending with habitat. Ground color may be cream to gray. Some have a pinkish cast.
Where found: A desert form is frequently found in areas with loose, blowing sand, but this sidewinder will also frequent rocky slopes, hardpan, and dunes with or without vegetation. Found near rodent burrows, where it hunts and lies in during the day. Se. Calif., s. Nev., w.-cen. Ariz., and extreme sw. tip of Utah.
Comments: Nocturnal; often hides from heat during day. May move without warning and can be surprisingly agile. May be nervous. Bites are characteristically not lethal.
Antivenin: #1

SONORAN SIDEWINDER *Crotalus cerastes cercobombus* **Pl. 3**
Medium-sized rattlesnake, 18 to approximately 30 in, rarely larger. Sidewise locomotion, broad head, rattle, *hornlike protuberances* over eyes. First rattle segment is black in adult rather than brown of Mojave subspecies. Light ground color tan to gray.
Where found: Habitats very similar to those of Mojave Desert Sidewinder. May range into hill country on rocky slopes or may be found on loose desert sand. Sw. Ariz. into Mexico.
Comments: Agile and unpredictable. Bites are generally not lethal.
Antivenin: #1

COLORADO DESERT SIDEWINDER *Crotalus cerastes laterorepens* **Pl. 3**
Medium-sized rattlesnake, 18–30 in. Characteristic sidewinder locomotion, broad head, rattle, *hornlike protuberances* over eyes. First rattle segment dark to black in mature specimens.
Where found: Arid country with or without vegetation. Sand dunes to hardpan to rocky slopes possibly to 6,000 ft., usually lower. Sw. Ariz., se. Calif. into Mexico on both sides of the Gulf of Calif.
Comments: The three sidewinders described on this page have overlapping ranges, each touching or nearly touching those of the other two. Agile, unpredictable. Bites are generally not lethal.
Antivenin: #1

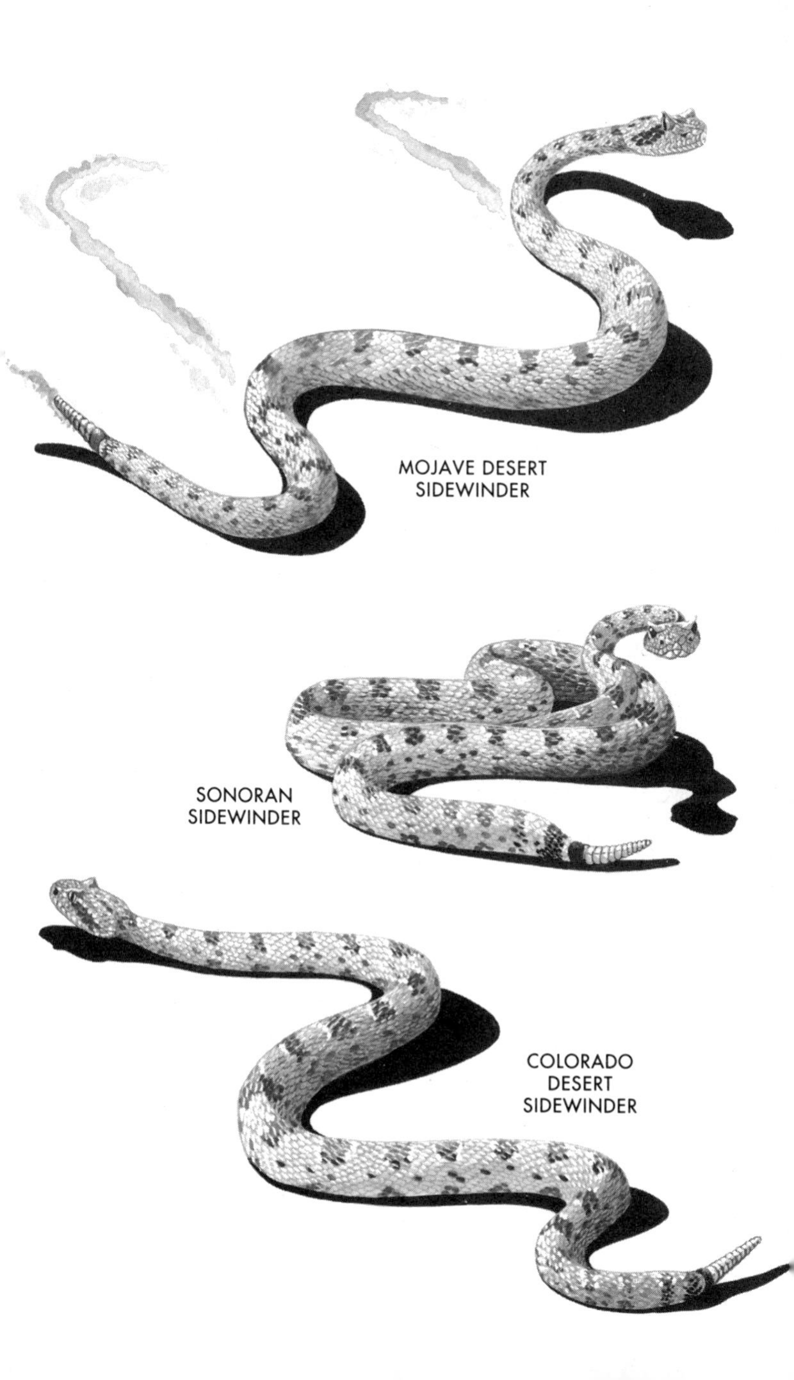
MOJAVE DESERT
SIDEWINDER
SONORAN
SIDEWINDER
COLORADO
DESERT
SIDEWINDER

TIMBER RATTLESNAKE *Crotalus horridus* **Pl. 3**

Average 3–5 ft., rarely to 6 ft. Frequently the only rattlesnake in the area. In *black phase,* completely black or with very dark stippling. In *yellow phase,* ground color is yellow, brown, or sometimes gray with very dark crossbands. Some specimens with rows of spots on back and sides. Variable colors and patterns. Southern variation ("canebrake rattlesnake") has a reddish brown stripe down the spine that divides the darker crossbands; ground color grayish brown to pinkish; dark facial stripe diagonally back from just under the eye. Tail usually much darker than rest of snake, and often velvety looking. Broad head. Rattle.

Where found: Rocky and well-wooded terrain. To 6,000 ft. in some areas. River bottoms out into low, flat areas. Southern lowlands: swamps, hammocks, cane thickets. Typically wooded hillsides with rocky outcroppings and natural caves and ledges. The Timber Rattlesnake is an adaptable species and one of the most widespread of all rattlesnakes. It is discontinuous but present in N.H., Mass., Vt., Conn., N.Y., N.J., Md., Pa., Va., N.C., S.C., n. Fla., Miss., La., Ga., Ala., Ky., W. Va., Tenn., Ohio, Ill., Ind., Mo., Mich., Ark., Okla., Texas, Kans., Neb., Wisc., Iowa. Extirpated in Me. and Del. Probably no longer in R.I. except accidentally. Occurs in Ont. in the region of Lake Erie and Lake Ontario.

Comments: Congregates into dense populations near winter denning sites, usually on elevated ground with caves and rocky crevices. This snake's disposition varies considerably, depending on the individual and its circumstances. The timber rattler is seriously venomous, but its bite is rarely lethal. Do not count on it to give a warning rattle or to retreat.

Antivenin: #1

TIMBER RATTLESNAKE

southern form
("canebrake rattlesnake")

RATTLESNAKES

MOTTLED ROCK RATTLESNAKE *Crotalus lepidus lepidus* **Pl. 2**

A small rattlesnake; average under 2 ft., occasionally a few inches longer. Highly variable ground color matches surroundings: gray, green, blue, brown, pink, and other earth tones. May be clear or flecked with darker tones. Pale brown to black crossbands may be darker toward tail. Dark stripe from eye to corner of mouth. Broad head. Rattle.

Where found: Chiefly mountains from 1,000 to almost 10,000 ft. Prefers rocky habitats: ledges, piles of boulders. Also seen near streams and in pine belts, frequently in rocky clearings. S.-cen. and sw. Texas, extreme s. N.M., into Mexico.

Comments: Rarely aggressive. Venom is powerful, but bites are rare; no human deaths recorded.

Antivenin: #1

BANDED ROCK RATTLESNAKE *Crotalus lepidus klauberi* **Pl. 2**

Average 1–2 ft., record slightly over 2½ ft. Green, blue-green, gray-green, or blue-gray ground color with brown to black crossbands widely spaced. No face stripe. Broad head. Rattle.

Where found: Ranges west of Mottled Rock Rattlesnake. Frequently in mountainous, rocky areas, arid to semiarid areas, streambeds. Extreme w. Texas, s. N.M, extreme se. Ariz., into Mexico.

Comments: Usually unaggressive. Venomous, but not lethal.

Antivenin: #1

SOUTHWESTERN SPECKLED RATTLESNAKE **Pl. 4**
Crotalus mitchellii pyrrhus

Average 2 to slightly over 4 ft. Ground color highly variable: white, gray, salmon, tan, yellow to orange-yellow. A handsome snake with speckled crossbands frequently pale but darker than ground color. May have diamond or hourglass markings or other geometric shapes. Broad head. Rattle.

Where found: Rocky areas, mountain slopes, down to coast and sea level. Occasionally found in open brushy and sandy areas. Southern Calif., extreme s. tip of Nev., extreme se. corner of Utah, w. Ariz., into Mexico.

Comments: Frequently stubborn. Nervous. Will hold ground and strike readily.

Antivenin: #1

PANAMINT RATTLESNAKE *Crotalus mitchellii stephensi* **Pl. 4**

Average 2–4 ft., some slightly larger. Ground color and markings are highly variable. Frequently with salt-and-pepper look. Broad head. Rattle.

Where found: Rocky, mountainous slopes from sea level to 8,000 ft. Sandy and brushy areas, pinyon-juniper woodlands. East-cen. Calif., sw. Nev.

Comments: Nervous. Strikes with little provocation.

Antivenin: #1

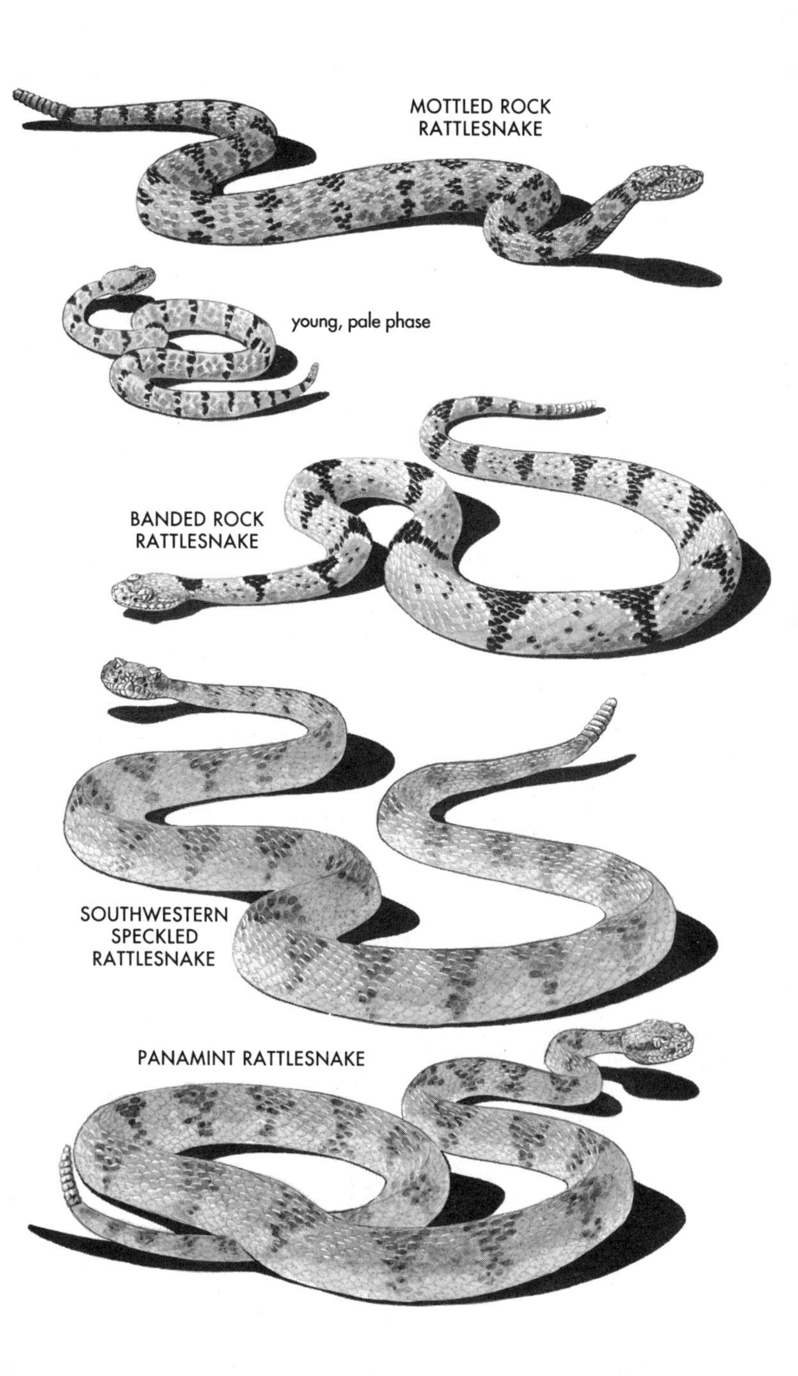
MOTTLED ROCK
RATTLESNAKE
young, pale phase
BANDED ROCK
RATTLESNAKE
SOUTHWESTERN
SPECKLED
RATTLESNAKE
PANAMINT RATTLESNAKE

NORTHERN BLACKTAIL RATTLESNAKE **Pl. 4**
Crotalus molossus molossus

A medium-sized rattler, average 2–4 ft., rarely slightly larger. Broad head, tail usually very dark to black, snout may also be black. Rattle. Ground color variable: yellow, gray, or olive to more pronounced green. Scales usually solid-colored.

Where found: Usually a highland species. Rocky slides and outcroppings, seldom on flatland or unvegetated deserts. Into wooded areas at altitudes to 9,000 ft. or slightly higher. Southern two thirds of Ariz. and N.M. to cen. Texas.

Comments: Both diurnal and nocturnal. Not particularly aggressive but agile and unpredictable.

Antivenin: #1

WESTERN TWIN-SPOTTED RATTLESNAKE *Crotalus pricei pricei* **Pl. 4**

Small; 12–24 in. Rattle, broad head, pale gray to gray-brown ground color with *two rows* of brown spots on back. Spots may be in pairs or alternate. Rarely, spots join across back. Brown bands on tail.

Where found: Mountains, 6,000–10,000 ft. Rocky slopes, coniferous and mixed woodlands, oak stands. Se. Ariz. into Mexico. Possibly extreme sw. corner of N.M.

Comments: Diurnal at higher altitudes. Generally but not inevitably mild mannered. Rattle makes a weak sound. Venom considered not lethal.

Antivenin: #1

RED DIAMOND RATTLESNAKE *Crotalus ruber* **Pl. 5**

Fairly large and heavy bodied, 30 in. to $5\frac{1}{2}$ ft. Broad head, rattle. Tan, pink, or reddish brown ground color, pale diamond pattern on back, *black and white ringed tail.*

Where found: Rocky slopes and brush, coastal and inland to desert. Extreme sw. Calif. down the Baja Peninsula. Usually restricted to lower altitudes.

Comments: May be found in cultivated areas. Generally but not reliably mild mannered. Venom powerful.

Antivenin: #1

MOJAVE RATTLESNAKE *Crotalus scutulatus* **Pl. 5**

Usually 24–50 in. Diamond pattern well defined; ground color usually greenish but may be olive, gray, brown, or yellow. Tail banded, with *darker bands narrower than light.* White face stripe to corner of mouth. Broad head, rattle.

Where found: Often high desert, but may be found in brush, grasslands, on mountain slopes, even in barren deserts. Seems to prefer scattered brush. S. Calif., Nev., extreme sw. Utah, w. and s. Ariz. to extreme sw. corner of N. M., sw. Texas.

Comments: An often short-tempered snake with strong neurotoxins in its venom. Potentially lethal.

Antivenin: #1

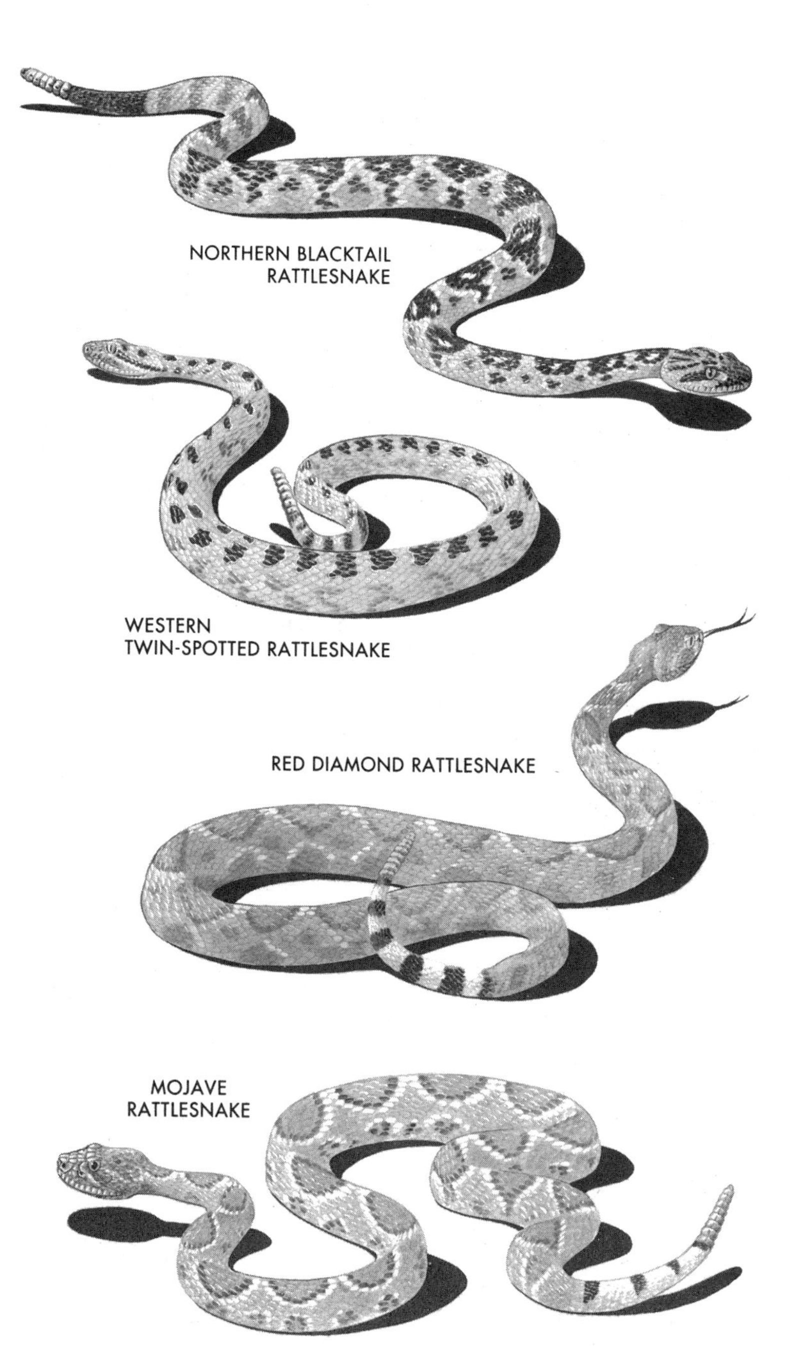
NORTHERN BLACKTAIL
RATTLESNAKE
WESTERN
TWIN-SPOTTED RATTLESNAKE
RED DIAMOND RATTLESNAKE
MOJAVE
RATTLESNAKE

RATTLESNAKES

TIGER RATTLESNAKE *Crotalus tigris* **Pl. 5**

From 18–36 in., rarely larger. Broad head, but smaller in proportion to body than other rattlesnakes in the region; rattle tends to be durable and hence long. Body color variable; blue, pink, gray, paler on sides. Crossbands irregular, but *more distinct* than in other western species.

Where found: Arid regions, canyons, desert mountain foothills, from lower slopes to timbered areas. S.-cen. Ariz. into Mexico.

Comments: Both diurnal and nocturnal. Often hunts after rainstorms. Unpredictable disposition, but venom not considered lethal in most cases.

Antivenin: #1

PRAIRIE RATTLESNAKE *Crotalus viridis viridis* **Pl. 6**

From 15 in. to well over 5 ft.; usually 3–4 ft. Highly variable; ground color ranges from cream and yellow through pink, green, and brown to almost black. Tail ringed. Dark blotches on back, usually outlined in lighter color. Light stripe from behind eye to corner of mouth. Broad head and rattle.

Where found: May be encountered in almost any habitat within its enormous range: prairies, cultivated land, brush, sandy areas, woodland, forest, rocky slopes, along streams, caves; avoids true desert. Alta., Sask., Idaho, N.D., S.D., Wyo., Neb., Kans., Colo., Okla., Texas, N.M., extreme e. Ariz., extreme se. Utah, to n. Mexico.

Comments: The Prairie Rattlesnake is one of the largest subspecies of *Crotalus viridis*, the Western Rattlesnake. Often found in dense populations. Active at night on or near paved roads. At higher altitudes, seeks deep caves. Seriously venomous.

Antivenin: #1

GRAND CANYON RATTLESNAKE *Crotalus viridis abyssus* **Pl. 6**

30–36 in.; rarely larger. Tan to salmon ground color; obscure dorsal body splotches with dark borders and pale centers. Broad head, rattle. Base of rattle is black.

Where found: Rarely encountered. Range restricted to Grand Canyon, Ariz.

Comments: Relatively little is known about this Western Rattlesnake subspecies , but it must be considered dangerously venomous.

Antivenin: #1

ARIZONA BLACK RATTLESNAKE *Crotalus viridis cerberus* **Pl. 6**

To 4 ft. Charcoal black ground color; some may be slightly paler. Markings on back are close to ground color, may be outlined with yellow. Broad head, rattle.

Where found: Higher elevations of the Central Arizona Plateau from pine zones to adjacent canyon floors. Ariz., extreme w. edge of N.M.

Antivenin: #1

TIGER RATTLESNAKE
PRAIRIE
RATTLESNAKE
GRAND CANYON
RATTLESNAKE
ARIZONA BLACK
RATTLESNAKE

MIDGET FADED RATTLESNAKE *Crotalus viridis concolor* **Pl. 6**

Almost always smaller than 24 in. Cream to yellow ground color, faint to absent blotches. Broad head, rattle.

Where found: Uses almost any suitable hunting habitat within its range. Grasslands, cultivated areas, remote brushy areas, rocky outcroppings and slopes. Limited to e. Utah, w. Colo., and sw. Wyo., possibly to the n. edge of Ariz.

Comments: One of the smallest subspecies of *Crotalus viridis*. Generally considered not lethal. Can be cranky and quick to defend itself.

Antivenin: #1

SOUTHERN PACIFIC RATTLESNAKE *Crotalus viridis helleri* **Pl. 7**

A robust, powerful snake, to 4½ ft. Ground color gray, with various shadings, often strongly marked. Broad head, rattle. Last tail ring is wider than others, but this feature is often indistinct. Immature specimens with yellow toward end of tail.

Where found: Sw. and coastal s. Calif., Santa Catalina Island and other offshore islands, south into Baja. Any suitable hunting ground where rodents are found. Heavy growth, arid areas, rocky slopes and outcroppings, cultivated areas.

Comments: A large, agile, dangerous snake. Its size alone makes it dangerous for the amount of venom it may deliver.

Antivenin: #1

GREAT BASIN RATTLESNAKE *Crotalus viridis lutosus* **Pl. 7**

To 5 ft. A large, heavy snake. Drab ground color, with olive to brown splotches. In some parts of its range, splotches are much wider than long, often 3:1 ratio. Pale side markings alternate with dorsal markings. Broad head, rattle.

Where found: Adaptable, like most Western Rattlesnake forms. Habitats vary from brush to timber to desert areas. Fairly widespread; s. Idaho, w. Utah, most of Nev., ne. corner of Calif., se. Ore., into nw. Ariz.

Comments: Very abundant in some areas, and may force out other rattlesnake forms. Large, can be aggressive.

Antivenin: #1

HOPI RATTLESNAKE *Crotalus viridis nuntius* **Pl. 7**

Rarely to 30 in., usually considerably smaller. Pinkish tan to salmon or brown ground color. Dark brown dorsal blotches become rings near tail. Broad head, rattle.

Where found: Habitats are variable, but tends toward drier areas within its range. S.-cen. Utah, possibly sw. Colo., nw. N.M., ne. Ariz.

Comments: Will usually try to retreat if disturbed, but can strike quickly and repeatedly if cornered or surprised.

Antivenin: #1

MIDGET FADED
RATTLESNAKE

SOUTHERN PACIFIC RATTLESNAKE
GREAT BASIN RATTLESNAKE
HOPI RATTLESNAKE

RATTLESNAKES

NORTHERN PACIFIC RATTLESNAKE *Crotalus viridis oreganus* **Pl. 7**

To $4\frac{1}{2}$ ft. Strong dorsal blotches against ground color ranging from gray to yellow. Dark rings on tail are consistent in width. Immature individuals with yellow tails pronounced in some specimens. Broad head, rattle.

Where found: Variable habitats to almost 11,000 ft. Rocky outcroppings, cultivated areas, brush. May turn up almost anywhere within its range. Roughly the n orthern two thirds of Calif., Ore., Wash., into n. Idaho, and well into B.C. Coastal only in Calif. and extreme sw. Ore.

Comments: A large and potentially dangerous rattlesnake. The largest specimens, though unusual, can deliver a large amount of venom.

Antivenin: #1

ARIZONA RIDGENOSE RATTLESNAKE *Crotalus willardi willardi* **Pl. 5**

Small; usually considerably under 2 ft. Brown or grayish ground color, very large blotches, usually 8 times longer than intervening ground color. Vertical white line on snout. Ridge runs around tip of snout. Broad head, rattle.

Where found: Usually 5,000–9,000 ft. Prefers forested areas: Douglas-fir, oak, pine, and aspen. May be found near streams. Range is extremely limited: Cochise, Santa Cruz and Pima counties in extreme se. Ariz., thence into Mexico.

Comments: Uncommon and little known.

Antivenin: #1

NEW MEXICO RIDGENOSE RATTLESNAKE **Pl. 5**
Crotalus willardi obscurus

Large at 2 ft. Broad head, rattle, prominent ridge around tip of snout. *No* vertical white line on snout. Variable ground color, blotches broader than intervening areas of ground color.

Where found: Usually at 6,500–7,000 ft. altitude. Apparently prefers forested areas to open, rocky slopes. Found only in the Animas Mountains, from Mexico north into the extreme sw. corner of N.M.

Comments: This *threatened subspecies* is very restricted in range, unlikely to be encountered.

Antivenin: #1

EASTERN MASSASAUGA *Sistrurus catenatus catenatus* **Pl. 8**

$1\frac{1}{2}$–$2\frac{1}{2}$ ft.; rarely larger. Ground color gray to brownish gray; dark blotches on back give a *spotted effect.* Larger scales on top of head will distinguish this snake from Timber Rattlesnake where their ranges overlap. Broad head, rattle.

Where found: May show up almost anywhere in its range but seems to prefer wet areas: swamps and marshes in eastern part of range, wet grasslands further west. Found where mice gather. Widespread; w. N.Y., w. Pa., Ohio, Ind., Ill., Mo., Iowa, extreme se. Minn., s. Wisc., Mich., Ont. to north of Lake Huron including Manitoulin Island.

Comments: Not lethal. Milder in disposition than Timber Rattlesnake, slow to sound rattle in some instances.

Antivenin: #1

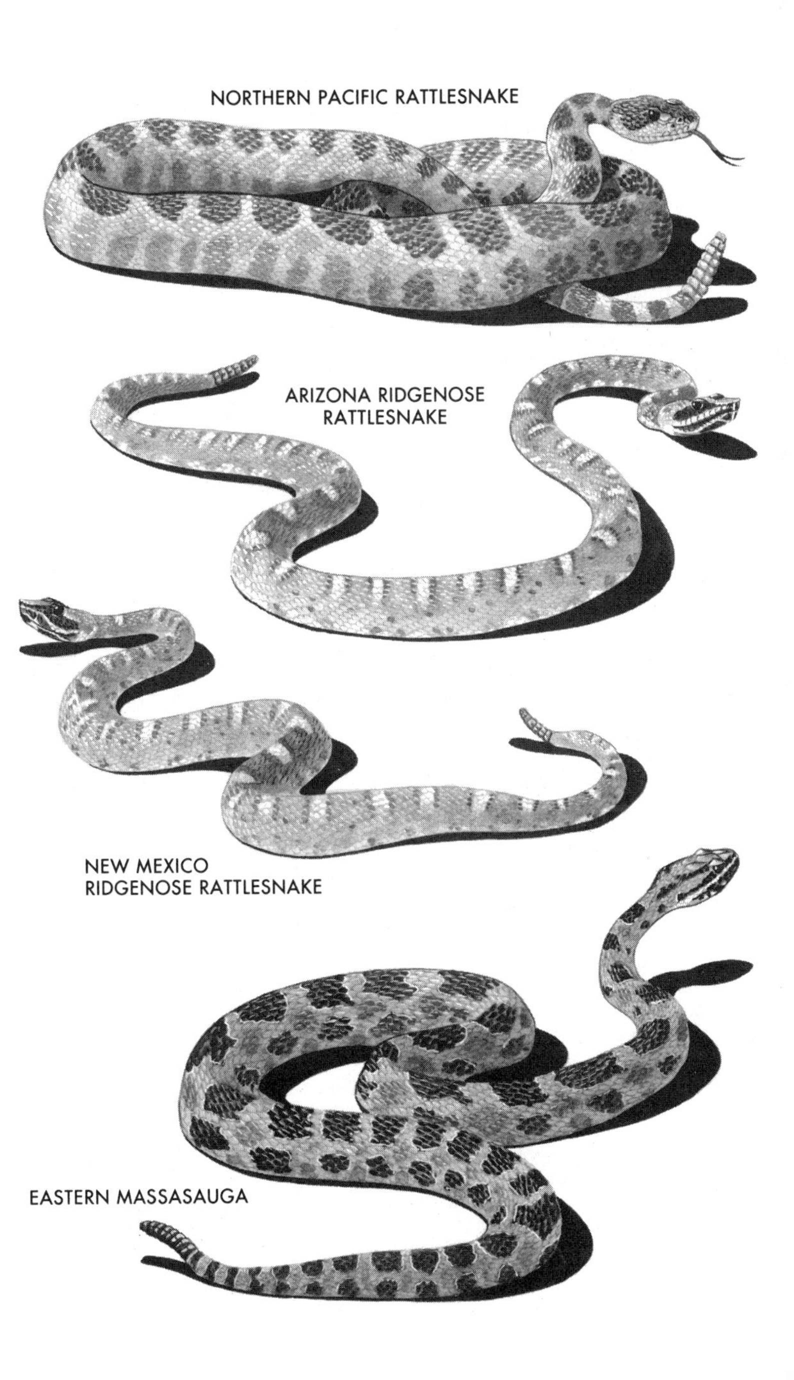
NORTHERN PACIFIC RATTLESNAKE
ARIZONA RIDGENOSE RATTLESNAKE
NEW MEXICO RIDGENOSE RATTLESNAKE
EASTERN MASSASAUGA

MASSASAUGAS AND PIGMY RATTLESNAKES

WESTERN MASSASAUGA *Sistrurus catenatus tergeminus* **Pl. 8**

18–26 in., occasionally a few inches longer. Broad scales on top of head, tan to gray ground color, blotches dark brown and usually pronounced, although overall effect is paler version of Eastern Massasauga. Broad head, rattle.

Where found: Grasslands, cultivated areas, moist areas where available, and rocky outcroppings. Sw. Iowa, nw. Mo., se. Neb., Kans., e. Colo., Okla., Texas.

Comments: Bite is not lethal to an adult. A relatively mild-mannered snake, but it cannot be handled safely.

Antivenin: #1

DESERT MASSASAUGA *Sistrurus catenatus edwardsii* **Pl. 8**

Small; to 21 in. A very pretty snake with light brown ground color, marked with about 30 dark brown dorsal blotches and smaller side spots. Nine large scales on top of head distinguish it from other rattlers in its range. Face marks extend onto neck. Broad head, rattle.

Where found: Seeks moist areas, marshes, water holes. Se. Colo., sw. Kans., w. Okla., w. half of Texas, N.M. and extreme se. Ariz.

Comments: A small, retiring rattlesnake that apparently feeds largely on amphibians. Distinctly venomous, but sublethal.

Antivenin: #1

CAROLINA PIGMY RATTLESNAKE *Sistrurus miliarius miliarius* **Pl. 9**

15–20 in. *Tiny, buzzy rattle,* brown to light gray ground color. Also a distinctly reddish phase in N.C. Dorsal marks distinct. Large scales or plates on top of head. *Tail slender,* head broad.

Where found: Wooded areas including pine flats, coastal plain, also pine and oak woods further west. N.C., S.C., n. Ga., cen. Ala., extreme e. Miss.

Comments: Disposition is unreliable, but venom is sublethal.

Antivenin: #1

DUSKY PIGMY RATTLESNAKE *Sistrurus miliarius barbouri* **Pl. 9**

15–31 in.; usually under 2 ft. Broad head, small rattle; brownish gray ground color with dark dorsal spots and smaller side rows. May be a distinct brown stripe down back. Large plates on head.

Where found: Flat woodlands; common near marshes and ponds. Extreme se. tip of S.C., s. Ga., Fla., s. Ala., se. Miss.

Comments: Very common in some areas. Venom is not lethal, but this snake often has an aggressive disposition and should not be handled. Rattle sound is very thin and easily missed or mistaken for insects.

Antivenin: #1

WESTERN PIGMY RATTLESNAKE *Sistrurus miliarius streckeri* **Pl. 9**

15–20 in.; occasionally a few inches longer. Broad head, *very small rattle, slender tail,* broad scales on top of head. Light gray to grayish brown ground color. May have reddish brown dorsal stripe, and irregular dark spots may be narrow or form crossbars.

Where found: Usually in wet areas; swamps, waterways. Extreme sw. Ky., w. Tenn., nw. Ala., Miss., La., eastern Texas, eastern Okla., Ark., s. Mo.

Comments: Frequently irritable and quick to strike. Venom is powerful but not lethal. Sound of rattle is likely to be missed.

Antivenin: #1

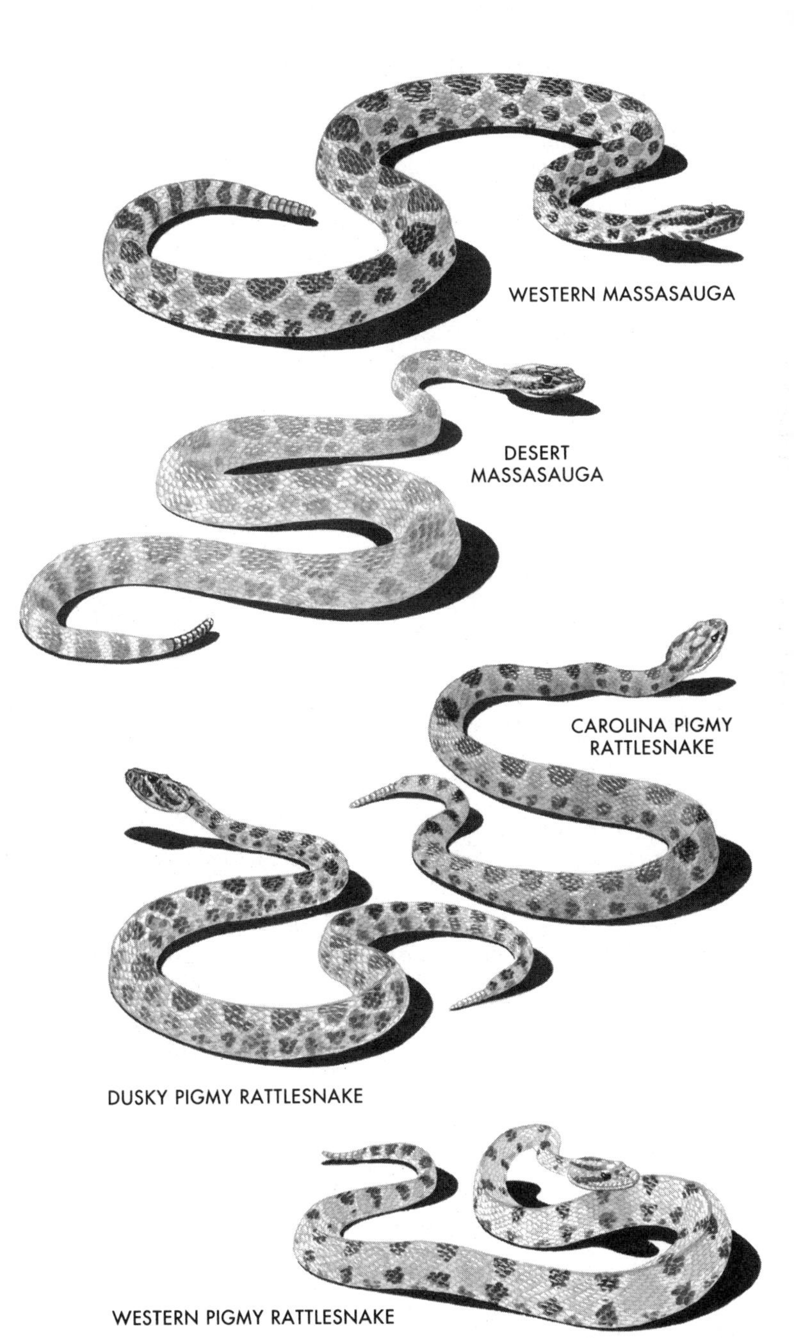
WESTERN MASSASAUGA
DESERT MASSASAUGA
CAROLINA PIGMY RATTLESNAKE
DUSKY PIGMY RATTLESNAKE
WESTERN PIGMY RATTLESNAKE

COTTONMOUTHS

EASTERN COTTONMOUTH *Agkistrodon piscivorus piscivorus* **Pl. 10**
Both the Eastern Cottonmouth and the subspecies Florida Cottonmouth *(A. p. conanti)* 2½–4 ft., rarely larger. Also called "water moccasin." Colors and patterns variable. Ground color from dark olive and brown to black. Crossbands sometimes present, but older specimens may be uniformly very dark. Crossbands may have borders, which may be indistinct. Center of crossbands may be lighter, especially if the ground color is light. Florida Cottonmouth has more distinct facial markings, a conspicuous dark brown cheek stripe bordered by light lines. Broad head, chunky body, tapering abruptly to a short, sharp tail. When threatened, this snake may coil, hold its head far back, and reveal the *white interior* of its mouth that gives it its common name. Has no rattle but will move its tail nervously when aroused. Not likely to retreat as other water snakes will.
Where found: Found in and near swamps, lakes, streams. Also in rice fields, drainage ditches, and any other place where water accumulates. Will move away from water in search of food, which ranges from fish, reptiles, and amphibians to birds and mammals. Southern lowlands; se. Va., e. N.C., S.C., Ga., and e. Ala. Florida Cottonmouth is found in s. Ala. and Ga. and throughout Fla. Mistaken reports of the Eastern Cottonmouth appearing north of Va. and in New England are usually traceable to the pugnacious but nonvenomous Water Snakes (*Nerodia* species, also often mistakenly called "water moccasins").
Comments: A cranky, unpredictable snake that will strike suddenly and repeatedly. Its venom is highly destructive, causing so much tissue damage that amputation may be necessary. Avoid all unidentified water snakes within its range.
Antivenin: #1

WESTERN COTTONMOUTH *Agkistrodon piscivorus leucostoma* **Pl. 10**
A heavy snake, 2½–3½ ft. Broad, flat head; chunky body abruptly tapering to short tail. White mouth lining is displyed as a warning. Color and pattern vary, as with Eastern Cottonmouth. Young are more strongly banded than most adults. Belly may be very dark and not as distinct from ground color as in the eastern forms. May have lighter lower jaw. No rattle. Nervous tail-twitching and unwillingness to give ground are characteristic.
Where found: Habitats are much the same as the Eastern Cottonmouth's, but this species tends to move more into upland areas. Much of Ala. except se. corner, all of Miss., La., to cen. Texas, eastern Okla., all of Ark., w. Tenn. and Ky., s. Mo., and extreme s. Ill. Rarely and questionably ever further north.
Comments: A nervous, cranky snake with highly destructive venom. Likely to be encountered by fishermen; may be found on low branches overhanging water and on partially submerged logs. Very common in some areas, particularly in La.
Antivenin: #1

EASTERN
COTTONMOUTH

WESTERN
COTTONMOUTH

SOUTHERN COPPERHEAD *Agkistrodon contortrix contortrix* **Pl. 11**

Normally 2 to 3 ft.; record over 4 ft. Slender neck, broad head. Palest of the Copperheads, often with a pinkish ground color. Brown *hourglass markings* are narrow and may break into separate halves on the back. Scales are very slightly keeled.

Where found: Frequently found in wet, lowland areas, especially along cypress-lined streams. May also ascend into hilly regions and may be found in the company of rattlesnakes in high, rocky ground. Also found in hammocks, floodplain forests, fields. E. N.C. and S.C., s. Ga., extreme n. Fla., Ala., Miss., e. Texas and Okla., La., Ark., s. Mo., extreme w. Tenn., sw. Ill. Intergrades with Northern, Osage, and Broad-banded subspecies.

Comments: No rattle, but vibrates tail rapidly if disturbed, producing in dry ground litter a sound a little like a rattlesnake's rattle. Usually rather lethargic, but will strike when aroused.

Antivenin: #1

NORTHERN COPPERHEAD *Agkistrodon contortrix mokasen* **Pl. 11**

Usually 2–3 ft.; record over 4 ft. Slender neck, broad head. Coppery head and ground color, with dark chestnut crossbands often forming distinct *hourglass markings.* Osage Copperhead *(A. c. phaeogaster)* is similar but with more strongly contrasting dark crossbands, often edged with white.

Where found: Generally uplands; rocky, wooded hillsides, mountains, abandoned lumbering operations, stone walls, farmland. Mass., Conn., N.Y., N.J., Pa., Md., Va., N.C., S.C., Ga., Ala., Ohio, W. Va. Also reported in e.Texas. Osage Copperhead ranges northwest of other Copperheads, e. Mo. west to e. Kans. and south to ne. Okla.

Comments: Bites are frequently reported, but they are generally mild, not lethal.

Antivenin: #1

BROAD-BANDED COPPERHEAD **Pl. 11**

Agkistrodon contortrix laticinctus

A small Copperhead, usually 22–30 in. Slender neck, broad head. Distinct reddish brown to chestnut crossbands are more straight-sided than corresponding markings on other Copperheads. Tail tip yellowish green to greenish gray.

Where found: Generally but not always toward uplands. Near streams through flat, low country, in well-wooded or rocky areas. Cen. Texas, Okla., extreme s. Kans.

Comments: May be quick to bite, but venom is not lethal.

Antivenin: #1

TRANS-PECOS COPPERHEAD *Agkistrodon contortrix pictigaster* **Pl. 11**

The smallest Copperhead, 20–30 in. Slender neck, broad head. Underside is strongly patterned, and there is a pale area at the base of each dark crossband.

Where found: Desert oases, arid areas, river bottoms, canyons. Sw. Texas, with isolated w. Texas populations.

Comments: May be nervous. Venom generally not lethal.

Antivenin: #1

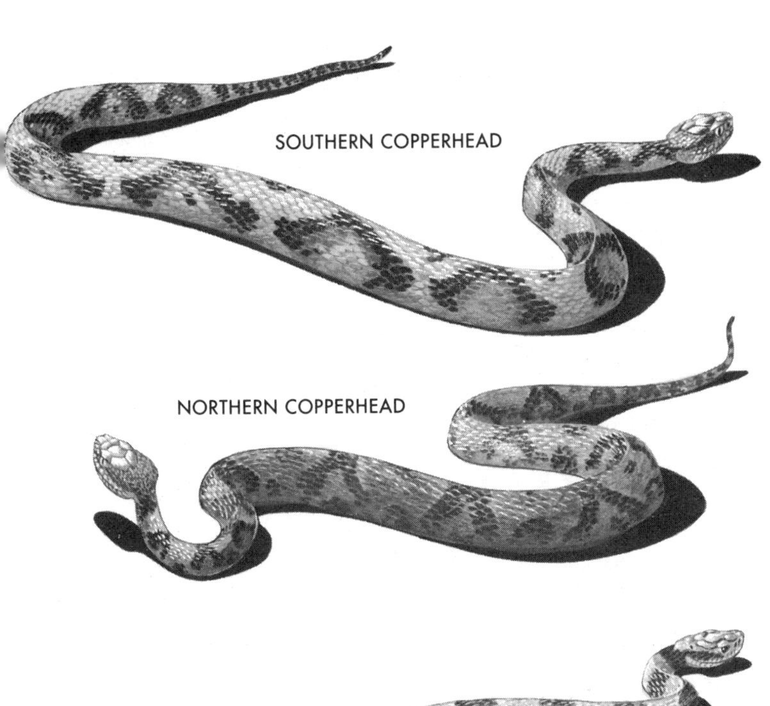
SOUTHERN COPPERHEAD
NORTHERN COPPERHEAD

BROAD-BANDED
COPPERHEAD

TRANS-PECOS
COPPERHEAD

EASTERN CORAL SNAKE *Micrurus fulvius fulvius* **Pl. 12**

A small, slender snake with a small head; usually 20–30 in., reported record 48 in. *Red, yellow, and black rings* with *red and yellow rings touching;* red rings spotted with black. Some specimens from south Florida have no black markings in the red ring.

Where found: Varies. Usually active in morning, hunting smaller snakes, lizards, and amphibians. More often under leaves and debris than on top of it. Pine woodlands, sandy open areas to pond and marshland border areas. Hardwood stands and hammocks in very wet areas. Se. N.C., e. S.C., s. half of Ga., all of Fla. (except s. tip), s. two thirds of Ala., widespread in Miss., and rare in extreme e. La.

Comments: Seriously venomous, but bites are unlikely unless the snake is handled.

Antivenin: #2

TEXAS CORAL SNAKE *Micrurus tener* **Pl. 12**

Small, slender, tubular snake; 20–30 in., rarely to 45 in. With encircling rings of black, yellow, and red; red and yellow always touch. Black snout. Black markings within red ring prominent.

Where found: Lowlands along coast, to plateau country of cen. Texas. Habitats variable: rocky outcrops, rock slides, canyons, cedar brakes. May seek moist areas, stream banks, edges of marshland, riverine islands. W. La. except for coast, s. third of Ark., e. and cen. Texas to Mexico.

Comments: Seriously venomous, but bites are virtually unknown unless snake is handled.

Antivenin: #2

ARIZONA CORAL SNAKE **Pl. 12**
Micruroides euryxanthus euryxanthus

Small: under 20 in., rarely more than 15 in. Slender body, small head. Black, red, and yellow rings encircling body. Red always touches yellow, black never touches red. Head black, followed by narrow yellow ring.

Where found: Habitat varies; in rotting logs, under debris, in burrows of other animals, near vegetation. N.M., Ariz.

Comments: Seriously venomous, but bites are unknown unless snake is being handled.

Antivenin: None.

SCARLET KINGSNAKE *Lampropeltis triangulum elapsoides* **Pl. 12**

This snake is harmless; mimics coloration of Coral Snakes. Note that red and yellow rings *do not touch.*

Where found: Extremely widespread; many subspecies.

EASTERN CORAL SNAKE
South Florida form
TEXAS CORAL SNAKE
ARIZONA CORAL SNAKE
SCARLET KINGSNAKE

REAR-FANGED SNAKES

TEXAS LYRE SNAKE *Trimorphodon biscutatus vilkinsonii* **Pl. 13**
A shy, slender snake with a blotched appearance, usually 18–30 in.; to 40 in. reported. Lyre-shaped marking on head that is typical of the group is faint or absent in this subspecies. Ground color brownish gray to gray, blotches relatively faint, widely spaced, and brownish. Vertical pupil.
Where found: Rocky outcroppings and prominences where small lizards may be found. Secretive; usually near wooded areas or on rocky slides where plenty of cover is available. Extreme s. N.M. south into Mexico, extreme w. Texas along Mexican border; north only as far as N.M.
Comments: Fangs in rear of upper jaw. Unquestionably venomous, but not dangerous to humans unless carelessly handled.
Antivenin: None.

SONORAN LYRE SNAKE *Trimorphodon biscutatus lambda* **Pl. 13**
Usually 24–40 in. Slender body. Pronounced *lyre-shaped mark* on top of head, grayish brown to gray ground color, blotches brown to brown-gray. Vertical pupil.
Where found: Habitats variable: desert grasslands, evergreen stands, rocky outcroppings and rockslides on slopes, often in rocky canyons, Ponderosa Pine forests. Extreme se. corner of Calif., s. Nev., extreme sw. corner of N.M., south into Mexico.
Comments: Like other rear-fanged snakes, not dangerous to humans unless handled, and even then bites only rarely.
Antivenin: None.

CALIFORNIA LYRE SNAKE **Pl. 13**
Trimorphodon biscutatus vandenburghi
From 24 in. to 43 in. A slender snake. *Lyre mark pronounced,* ground color pale to brown or both, approximately hexagonal blotches brownish and pronounced.
Where found: To 3,000 ft. A rock dweller often found in massive boulder piles and on rocky slopes. Often near bushy or forested areas where rock slides have occurred. Southern Calif. only. Along coast in extreme sw. corner of state down to Baja; small number in Inyo Co. on Nev. border, Funeral Mountains.
Comments: Venomous, but harmless unless mishandled.
Antivenin: None.

NORTHERN CAT-EYED SNAKE
Leptodeira septentrionalis septentrionalis
18–24 in. Slender; light cream to tan with *strong crossbands* of dark brown to almost black. Vertical pupil.
Where found: Prefers areas near water, streams, marshes; may climb low trees and bushes. Extreme s. tip of Texas, to coast.
Comments: Distinctly venomous but rear-fanged and not dangerous unless handled.
Antivenin: None.

TEXAS LYRE SNAKE
SONORAN LYRE SNAKE
CALIFORNIA LYRE SNAKE
NORTHERN
CAT-EYED SNAKE

SPIDERS

BLACK WIDOW SPIDER *Latrodectus mactans* **Pl. 14**

Easily recognized by the large, bulbous, shiny black abdomen, with a *bright red hourglass* beneath. Females generally twice as large as males; female body to 3/8 in.

Where found: Sandy soil, beneath leaf litter, in debris such as old tin cans, beneath cardboard, etc., in dark places. Throughout most the U.S.; Mass. to Fla., Texas, to Kans., west to Calif; most common in the South.

Comments: Female often eats the smaller male following mating, hence the name "widow." Venom, drop for drop, is one of the most toxic to humans. Bite of female more serious than bite of male because of her much larger size. Female is more likely to attempt escape than to bite, except when defending her egg sac; bites reported for males questionable. Before the advent of indoor plumbing, many bites were reported from Black Widows that lived beneath the rim of outhouse seats. Seriously venomous. Antivenin available.

BROWN WIDOW SPIDER *Latrodectus geometricus*

Female body to 3/8 in. Brownish to gray, sometimes black, with dark brown spots surrounded by lighter brown-yellow markings. Hourglass marking on underside of abdomen is *orange rather than red.*

Where found: A cosmopolitan species found in and around buildings. Introduced to Fla.

Comments: Less likely to bite than its black cousin.

RED-LEGGED WIDOW SPIDER *Latrodectus bishopi*

Female body to 3/8 in. Body shape similar to above species. Blackish, with orange spots surrounded by white; *legs distinctly reddish.*

Where found: Sandy soil, ground litter and nearby buildings; in palmetto and pine scrub of cen. and s. Fla.

Comments: Bites in humans are rare and milder than those of the Black Widow.

BROWN RECLUSE SPIDER *Loxosceles reclusa* **Pl. 14**

Also called "fiddleback." Females to 3/8 in. Head orange-yellow with a *dark brown fiddle-shaped marking,* abdomen dark brown.

Where found: Common in sheltered natural habitats, brush and loose debris; indoors frequently in dark areas of barns and houses: closets, basements, attics, beneath furniture, etc. Can be found in any of the 48 contiguous states; common in South.

Comments: Bites when disturbed, a common event given the spider's propensity to hide in stored clothing, in bookshelves, etc. Symptoms often develop 2 to 8 hours later. Wound usually localized; fatalities rare. Necrosis may develop at wound site, producing a large, deep area of dead cell tissue. A crusty, red wound may develop that can take months to heal and often leaves a permanent scar.

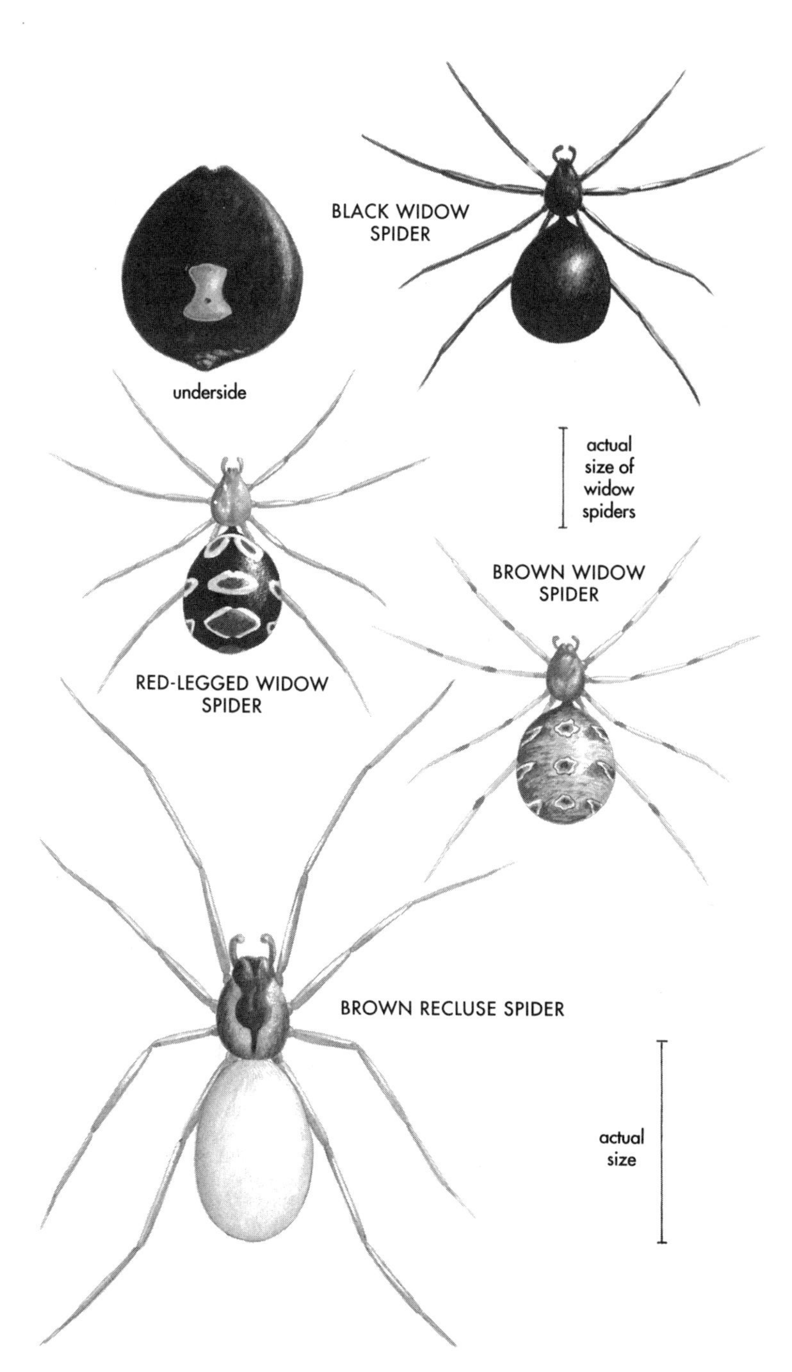
BLACK WIDOW SPIDER
underside
actual size of widow spiders
BROWN WIDOW SPIDER
RED-LEGGED WIDOW SPIDER
BROWN RECLUSE SPIDER
actual size

COSTA RICAN TARANTULA *Aphoropelma chalcodes*

Also known as Desert Tarantula. Heavy, large, *hairy* spider; males to 2½ in. long, females to 2¾ in., with leg span to 4 in. Thorax grayish to dark brown, with brownish black abdomen. Beneath tip of each leg, iridescent hairs form pads. May live for 20 years or more.

Where found: Desert soils, N. M., Ariz. and s. Calif.

Comments: Nocturnal; tarantulas are seen near sunset or sunrise and hide by day in holes or under rocks. They usually avoid humans except when confronted. Venom is mild; a tarantula bite is generally no more serious than a bee sting.

CENTIPEDE **Pl. 14**

Phylum Athropoda, Class Chilopoda

Centipedes have flat bodies and many legs, with *one pair of legs for each body segment* (15 or more in adults). Centipedes can reach nearly a foot in length in the Southwest, while in the Southeast a large individual may be up to 4 in. long. Coloration varies from bright red or orange to olive brown. Centipedes have a pair of claws beneath the head that can inject venom. Hundreds of species occur in the United States.

Where found: Throughout North America, large species mostly in South and Southwest.

Comments: Centipedes are carnivorous, feeding on smaller invertebrates; the herbivorous, *nonvenomous* millipedes generally have *two pairs of legs per body segment.* Any centipede large enough to grip skin with the venomous claws can inject venom into humans. Bites are usually the result of handling; they will generally try to escape rather than bite. Bites not lethal unless victim is highly sensitive to foreign protein.

SCULPTURED SCORPION **Related species Pl. 14**

Centruroides exilcauda (C. sculpturatus)

Scorpions look like tiny lobsters, with a *long, up-curved tail* tipped with a poisonous stinger and a 12-segmented abdomen (five of these segments are the "tail"). Members of several genera occur in North America. The deadly Sculptured Scorpion, to 2 in., is dark brown to tan, often with a yellowish stripe on the side of cephalothorax, *toothed beneath venom bulb.*

Where found: Some species live on or in the ground, others in vegetation, others beneath bark of dead logs or in human dwellings. Of about 1,500 species worldwide (only 25 with lethal venom), more than 70 species occur in the U.S., some in the South but especially in the Southwest.

Comments: Centruroides scorpions (*Centruroides* species of the Buthidae Family, the largest scorpion family and the one to which all species lethal to humans belong) occur in the Southwest. The Sculptured Scorpion of Ariz. is the most poisonous in the U.S.; its sting can be fatal. Venom is primarily neurotoxic. The sting of most U.S. species produces symptoms much like a wasp sting: painful swelling and discoloration. Most do not require medical attention.

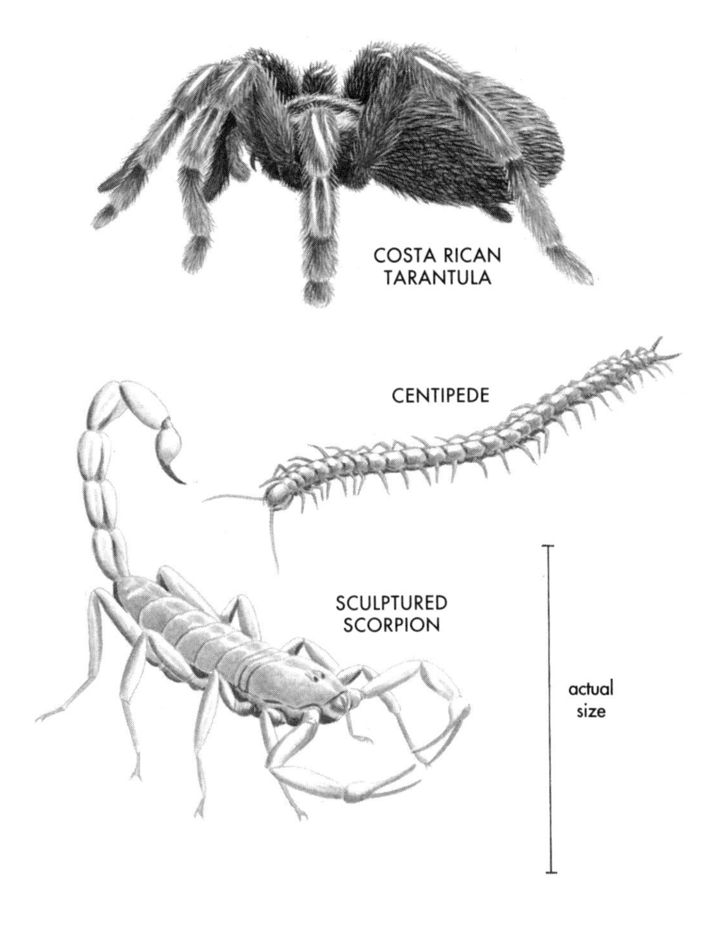
COSTA RICAN
TARANTULA
CENTIPEDE
SCULPTURED
SCORPION
actual
size

PUSS MOTH CATERPILLAR *Megalopyge opercularis*

Also called Tree Asp. Densely hairy caterpillar with soft brown hair.
Where found: Among trees, shrubs, orchards, and other vegetation from Md. to Fla., west to Mo. and Tex. Common in the South. Seen year round.
Comments: Beneath the hairs are numerous poisonous spines, which can cause severe skin irritation and stinging. Each hair has its own supply of venom, sometimes contained in a single cell. Severity of reaction, ranging from a slight rash to shock, depends upon the individual's sensitivity. Sensitivity may increase with repeated exposure. Common symptoms include fainting, nausea, and vomiting, and in some cases severe shooting pains.

IO MOTH CATERPILLAR *Automeris io* **Pl. 15**

$2^1/_2$–3 in. A green caterpillar with tufts of bristly, branched, hairlike spines, with *reddish pink stripes* edged in white stripe on sides.
Where found: Meadows, fields, open woods; commonly feeds on trees and shrubs; Me., s. Que. to Fla., west to Texas and Man.
Comments: Spines embedded in hair tufts can inflict painful stings if caterpillar is touched or handled.

SADDLEBACK CATERPILLAR *Sibine stimulea* **Pl. 15**

Brilliant green caterpillar to about 1 in. long. Dark brown at both ends, with a distinct brown *saddlelike mark* on its back, outlined in white and a thin black line; many tufts of bristly, stinging hairs. This moth larva has small thoracic legs and no prolegs, and moves with a creeping motion.
Where found: Gardens, edge of woods, orchards, and cultivated ground; Mass. to Fla., La., to Ill.
Comments: The hairs cause skin irritation and relatively mild stinging if caterpillar is touched or handled.

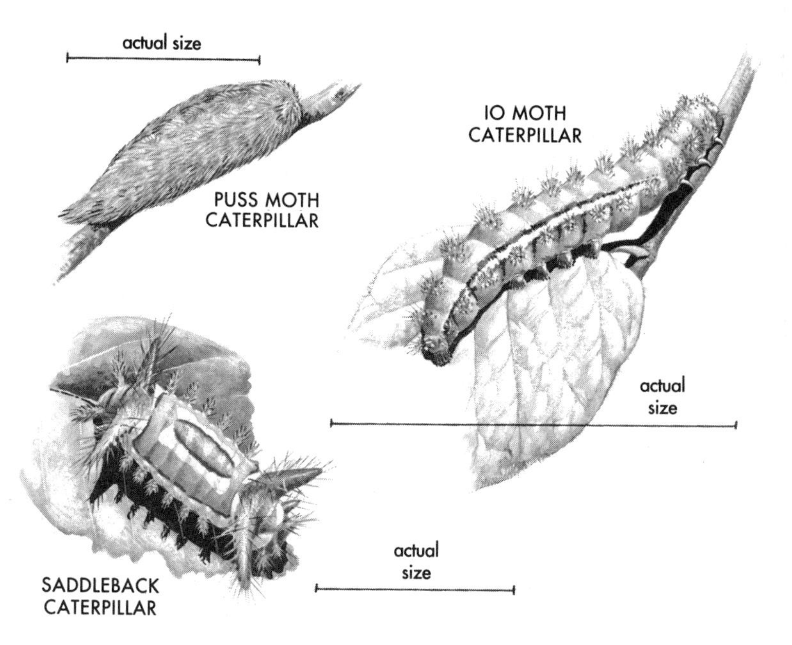
actual size
PUSS MOTH
CATERPILLAR
IO MOTH
CATERPILLAR
actual
size
SADDLEBACK
CATERPILLAR
actual
size

BUMBLEBEE *Bombas* spp. **Pl. 16**

Bumblebees are common and familiar, easily recognized by their large size, about twice that of the Honey Bee. Their bodies are stout, hairy, and marked with contrasting *black and yellow stripes* around the abdomen.

Where found: Various species found in woods, open fields, gardens, etc., throughout North America.

Comments: Unless disturbed, the bumblebee goes about its own business. Most often seen working flowers; will sting if accidentally disturbed or handled. Bumblebee nests are built in the ground, often in abandoned rodent dens. The bees will defend their nest if it is disturbed.

HONEY BEE *Apis mellifera* **Pl. 16**

Of all venomous species of North America, the Honey Bee is perhaps the most welcome. The familiar Honey Bee seen collecting nectar and pollen is the female worker. She is variably brown to yellowish orange to black in coloration, marked with orange-yellow rings on the abdomen. Female workers 3/8–5/8 in.; male drones 5/8 in.; queen to 3/4 in.

Where found: Native to Europe and introduced to North America with early settlers. Now found throughout the continent.

Comments: The Honey Bee is unique among insects in North America in that it has a barbed stinger that stays in the skin of the victim. When the bee pulls away after stinging, vital body tissue is torn out, and the bee dies shortly thereafter. Wasps and yellowjackets, by contrast, have smooth stingers, which they can use repeatedly at will. Honey Bee workers will vigorously defend their hives, dying to protect the colony and its precious food stores. Stinging is also a defensive response to being disturbed by handling, being stepped on, etc.

Honey Bees are so common and widespread that nearly everyone is stung at one time or another. Some individuals are severely allergic to the stings of bees, wasps, and hornets, sometimes suffering anaphylactic shock and even death. Sensitive individuals should carry an insect sting treatment kit. The kit includes syringes containing epinephrine and antihistamine, a tourniquet, alcohol swabs, and directions. The insect's stinger is not absorbed by the body and will work its way deep into the skin. The stinger should be removed from skin by scraping with a knife or fingernail. Avoid grasping the stinger with the fingers or tweezers; this will only inject more venom, since the entire stinging apparatus, including the venom sac, is left in the skin.

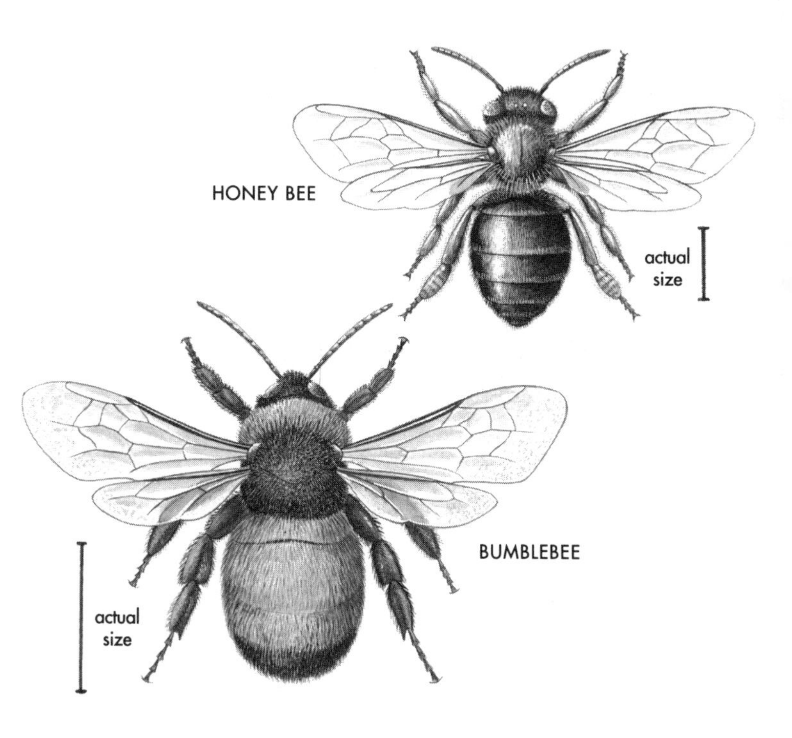
HONEY BEE
actual
size
BUMBLEBEE
actual
size

INSECTS: WASPS

Over 4,000 wasp species are found in North America. Only the females have stingers. Rather than simply jabbing, the stinger has two barbed shafts (lancets) that alternately work deeper into the victim; then the poison duct between the lancets opens, injecting the venom. Wasp stings can cause serious reactions in susceptible individuals, including respiratory arrest and anaphylactic shock. Many human fatalities have resulted. The extremely complex venom consists of numerous chemical compounds.

BALD-FACED HORNET *Vespula maculata* **Pl. 16**

Wasp with *black and white patterns* on face, thorax, abdomen, and first segment of antennae, 1/2–5/8 in. Wings smoky colored; head broad.

Where found: Fields, lawns, wood edges, throughout North America.

Comments: Females build small, gray, pendent nests out of chewed wood. Females are quick to defend the nest if disturbed and will sting repeatedly.

PAPER WASP *Polistes* spp. **Pl. 16**

Brownish or bronze-colored, long-legged wasp, to 1 1/4 in. long; middle tibia with 2 spurs; first abdominal segment is conical, not threadlike as in other wasps.

Where found: In vegetation near water, in barns, outbuildings, and houses, wooded areas. Common throughout.

Comments: Builds nests out of paper made from chewed wood mixed with saliva. Less aggressive in defending nests than hornets. Inflicts a painful sting.

SPIDER WASP *Anoplius* spp.

Long-legged, large, dark-colored wasps, 1/2–2 in., with middle leg segment (femur) to end of abdomen. Wings are folded laterally at rest; many species have dark wings.

Where found: Fields, grasslands, etc.; larger species in w. U.S.

Comments: Females run across ground, wings flicking, in search of spiders to prey upon. Inflicts a painful sting.

YELLOWJACKET *Vespula* spp. **Pl. 16**

1/2–5/8 in. *Abdomen banded with black and yellow* (or white); wings folded lengthwise along body when at rest. Middle leg (tibia) with 2-spurred tip. Body wider than head.

Where found: Fields, wood edges; builds nests in ground and in stumps and old logs. Throughout North America.

Comments: Female yellowjackets inflict a painful sting and will sting repeatedly with the slightest disturbance or provocation. These pesky wasps steal morsels of food from barbecues and picnics and are ubiquitous in parks, cities, and recreational areas. Serious envenomation can result from unwittingly stepping on a nest while walking through a field or forest.

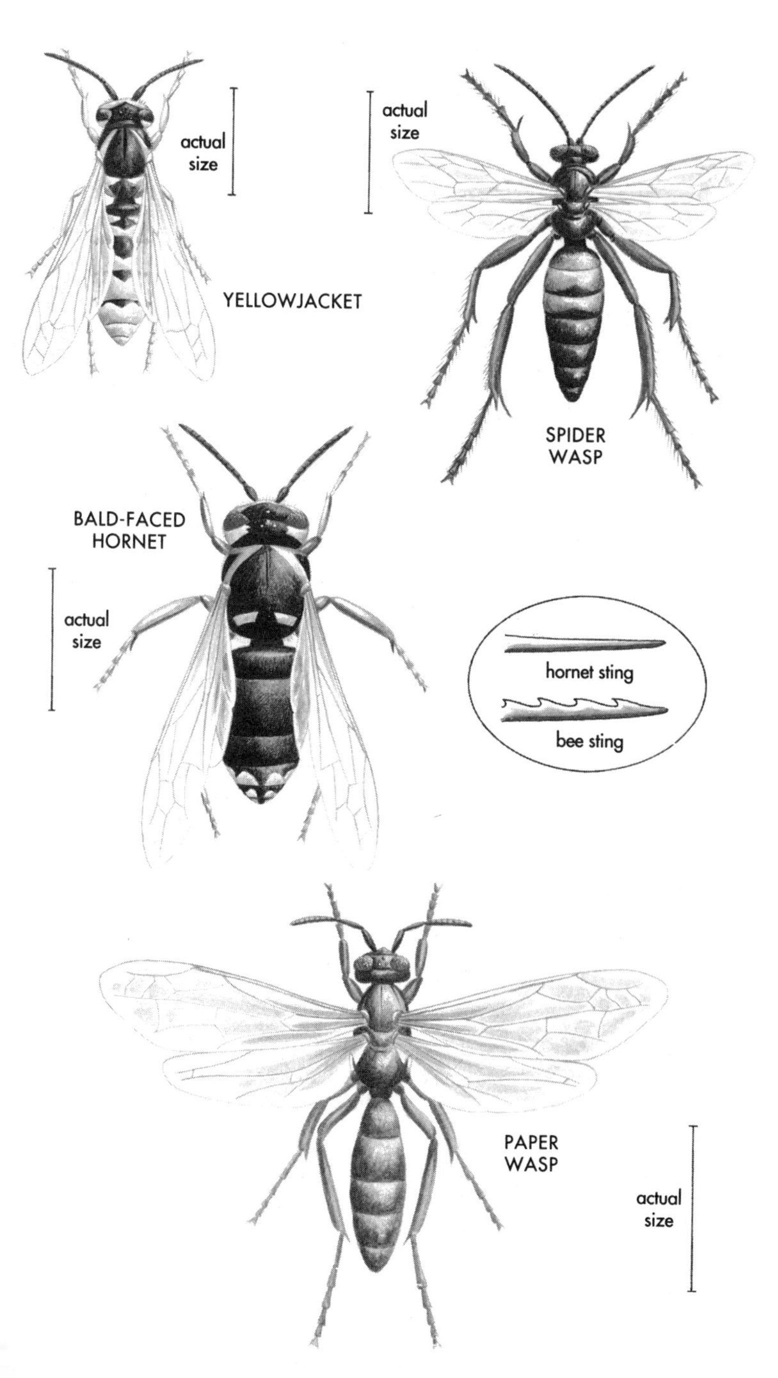
actual size
YELLOWJACKET
actual size
SPIDER WASP
BALD-FACED HORNET
actual size
hornet sting
bee sting
PAPER WASP
actual size

INSECTS: ANTS AND ANTLIKE WASPS

FIRE ANT *Solenopsis geminata*

Dirty yellow to red or black ants; workers of different castes are different sizes, ranging from $1/16$–$1/4$ in. in length, with two segmented humps between thorax and abdomen. Head large, with incurved jaws. The imported South American *S. invicta* is becoming increasingly common in the southeast.

Where found: Fields, woods, croplands. Fla., across the South, north to B.C.

Comments: Female Fire Ants can both bite and sting. If one steps on a mound, the ants will cling and are difficult to brush off. A single Fire Ant sting can cause shortness of breath, dizziness, or anaphylactic shock. Severity of reaction depends upon individual sensitivity. May cause local pain, swelling, itching, or a pus-filled wound that persists for several days or weeks.

HARVESTER ANT *Pogonomyrmex* spp.

Harvester ants range from pale yellow to reddish brown; length from $1/4$–$1/2$ in. long, depending upon worker caste; with 2-segmented humps between waist and abdomen (less prominent than in Fire Ants).

Where found: Various species distributed in South and Southwest.

Comments: Harvesters can swarm over cropland and cut down plants, clearing a relatively large area. Workers are active during the day. Can inflict both a painful bite and sting.

VELVET ANT *Dasymutilla* spp. **Pl. 15**

Actually wasps with an antlike appearance. Males are winged, females are *wingless.* Densely hairy; mostly *bright colored* with red, yellow, or orange markings; only slightly constricted between thorax and abdomen. *D. gloriosa* is $1/2$–$5/8$ in. long, black with long, loose, white hair. *D. magnifica* female to $7/8$ in long; female brownish orange to yellow-orange. *D. occidentalis*, to 1 in. long, with brilliant orange-red hair on head, thorax, and abdomen.

Where found: *D. gloriosa* and *D. magnifica*: arid lands; Texas, Utah, Nev. to Calif. *D. occidentalis*: fields, forest edges; N.Y. to Fla., to Texas.

Comments: Female velvet ants can inflict a sting that is among the most painful of the North American wasps; also called "cow killer." Winged males appear intimidating but are harmless. The black stinger of *D. occidentalis* is almost half the length of the body. Causes acute pain, usually of short duration, though swelling and pain can last for several hours. Some individuals may have allergic reactions to the sting.

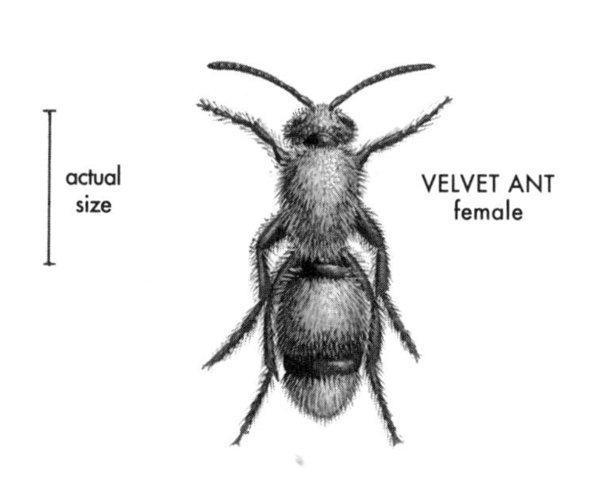

actual size

actual size

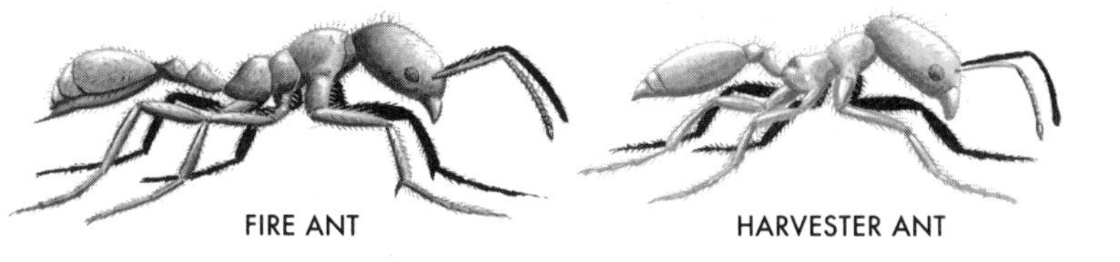

NONVENOMOUS INSECTS: BITING INSECTS AND BLISTER BEETLES

WHEEL BUG *Arilus cristatus*

Grayish brown to black, to about 1³/₈ in. Thorax with rounded plate *toothed* along the edge, hence the common name.

Where found: Fields, pastures; s. Canada to the South, east of the Rockies.

Comments: While not venomous, the Wheel Bug is a predator that can inflict a painful bite if disturbed or handled.

BLOOD-SUCKING CONE NOSE *Triatoma sanguisuga*

Dark brown or black, to ³/₄ in., with 6 *orange dots* atop and beneath the abdomen. Head with a *conelike, blunt-tapered protrusion* in front of eyes.

Where found: Nests of small animals, feeding on mammal blood. Fla., west to Texas, north to Ont.

Comments: Nonvenomous, but its painful bite may cause severe allergic reactions in susceptible individuals. In South and Central America, other species of the genus transmit parasites, resulting in Chagas' disease, sometimes causing permanent physical and mental defects and, rarely, fatalities. The parasitic infection (South American trypanosomiasis) is caused by the protozoa *Trypanosoma cruzi*. Other members of the *Trypanosoma* protozoa genus produce African sleeping sickness, spread by the bite of the tsetse fly.

BLISTER BEETLE *Meloe* spp.

Distinctively shaped, soft-bodied, relatively large beetles, ¹/₂–1¹/₂ in. Males smaller than females. Usually leathery, black to bluish. The front wings are short (²/₃ length of abdomen) and overlapping; head is broad; antennae beadlike, sometimes modified into relatively large segments in female. Twenty species in North America.

Where found: E. U.S., often on crops.

Comments: When threatened, the insect plays dead, falling on its side. An irritating oily substance exuded from the leg joints can cause blistering on human skin.

BLISTER BEETLE **Pl. 15**

Meloe, Epicuata, Zonitis, Lytta, Eupompha spp., etc.

Plant-eating, medium to large beetles. Typically black, brown, or gray; some species with clay yellow and black markings. In all, 26 genera with over 330 species identified in North America; *Epicuata* is the largest genus, with about 100 species. Distinctively shaped, soft-bodied, usually leathery, with long slender body, the pronotum (platelike coverings of the prothorax) are usually narrower than the front wing; front wings rolled; the head is somewhat squarish, broad; usually broader than the pronotum; antennae threadlike.

Where found: Throughout U.S., often on crops.

Comments: Blister beetles exude reddish canthardin from leg joints when disturbed as a defensive device against predators. Canthardin is extracted from certain species of beetles and used medicinally. Can cause blistering of the skin.

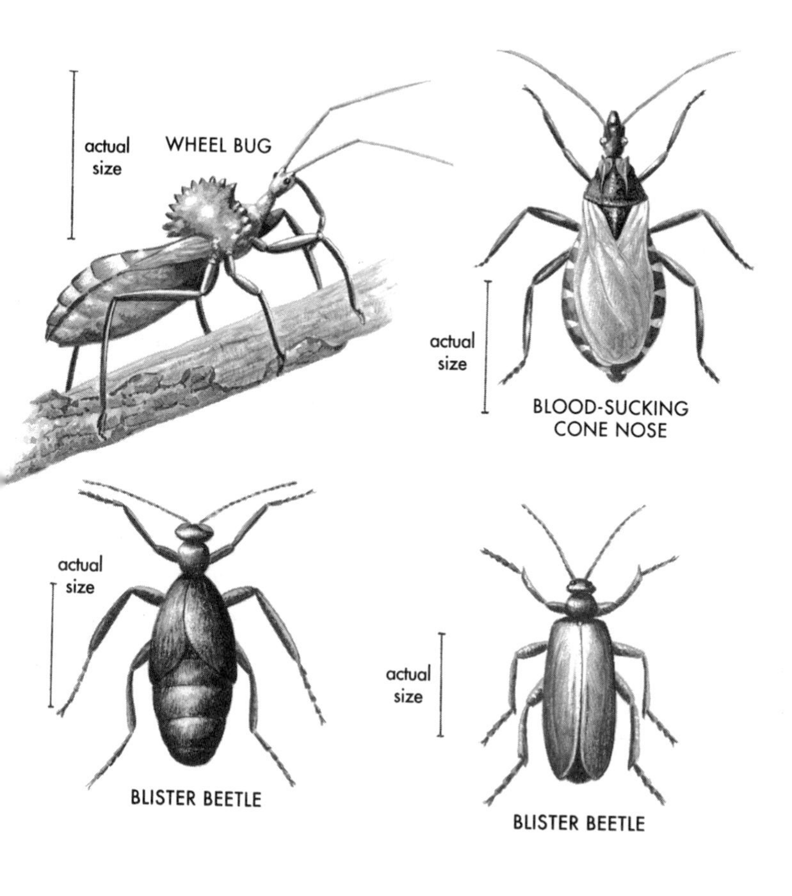
actual
size
WHEEL BUG
actual
size
BLOOD-SUCKING
CONE NOSE
actual
size
BLISTER BEETLE
actual
size
BLISTER BEETLE

NONVENOMOUS BITING INSECTS: FLIES

BLACK FLY *Simulium* spp.

Stocky, somewhat humpbacked, black to brownish, small flies, generally less than $^{3}/_{16}$ in. long. Segments of front legs elongated. Wings with distinctive broad base, veins near front edge heavy. Antennae *short.*

Where found: Woods, fields, near water; Canada to mountains of Ga., west to Calif; Mexico.

Comments: Adults emerge in late spring and early summer. Females are vicious biters and are extremely pesky. Nonvenomous.

BITING MIDGE

Family Ceratopogonidae

Also called "punkies" and "no-see-ums." Tiny flies, generally $^{2}/_{16}$ in. or less, *distinctly humpbacked* in side view.

Where found: Woods and fields, over much of our range.

Comments: Some species bite humans and are extremely annoying; small enough to pass through window screens.

DEER FLY *Chrysops* spp.

Large flies, $^{3}/_{8}$–$^{5}/_{8}$ in. long; broad and flattened. Yellow-green markings; *eyes patterned in gold or green.* Wings distinctively patterned.

Where found: Forests, meadows. Throughout.

Comments: Females are blood-sucking and are annoyingly persistent in attempting to bite. Males feed on vegetable juices.

HORSE FLY *Tabanus* spp.

Large, stout, broad flies, $^{3}/_{4}$–$1^{1}/_{8}$ in. long, with bulging eyes. The American Horse Fly *(T. americanus)* has a tannish head with green eyes and a blackish red-brown abdomen. Wings smoky near base. Black Horse Fly *(T. atratus)* is jet black with a bluish-tinted abdomen.

Where found: American Horse Fly, near water from Nfld. to Fl., west to Texas, Northwest Territories. Black Horse Fly is found in meadows, fields, marshy ground, Que. to Fla., N.M. to B.C.

Comments: Bite of the female American Horsefly can bleed for several minutes, as her saliva contains an anticoagulant. The Black Horsefly bites on the neck, head, or back, and sucks blood. Mouths have slicing bladelike parts. Nonvenomous.

SAND FLY

Subfamily Phelbotominae

Small, *very hairy flies,* $^{3}/_{16}$ in. or less in length; wings held together above body when at rest.

Where found: Near water. South.

Comments: Sand flies are persistent biters, and in tropical areas they carry numerous parasitic diseases.

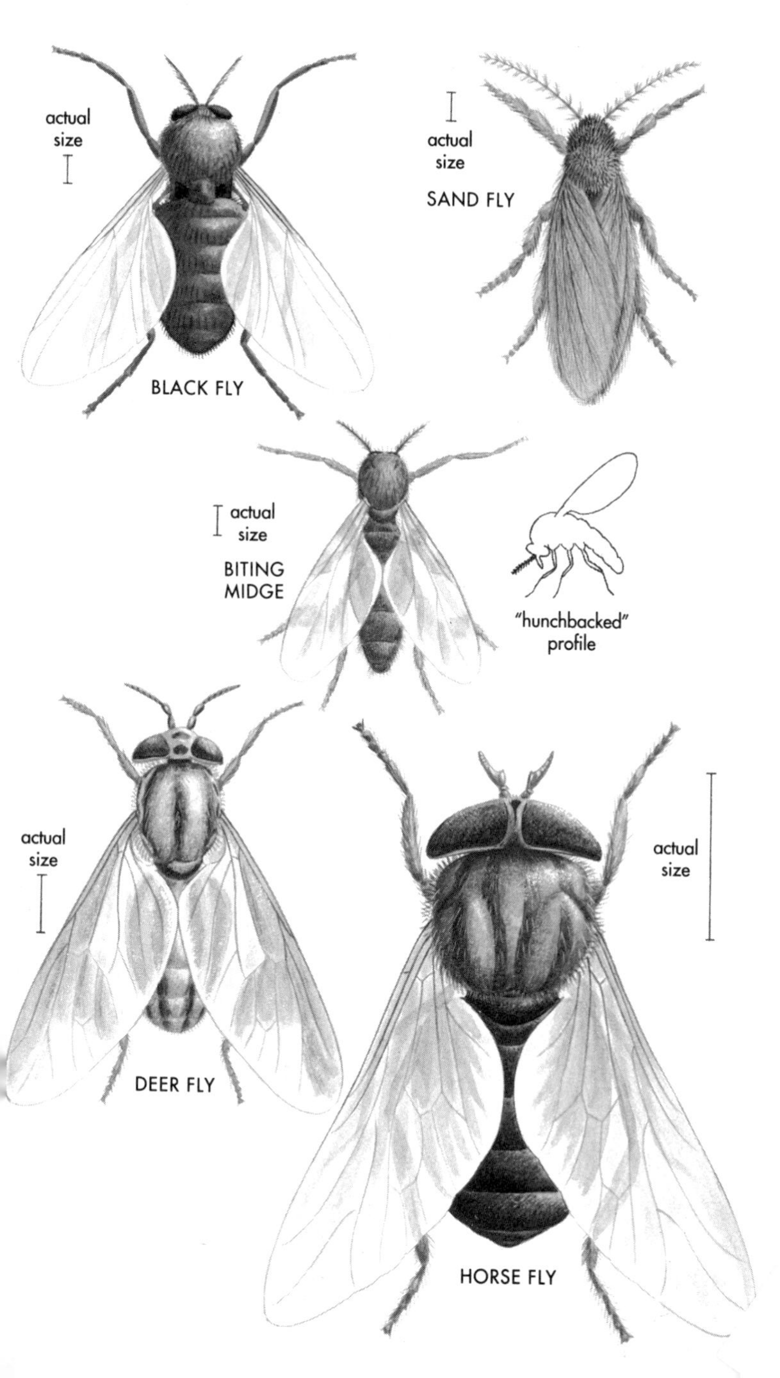
actual
size
BLACK FLY
actual
size
SAND FLY
actual
size
BITING
MIDGE
"hunchbacked"
profile
actual
size
DEER FLY
actual
size
HORSE FLY

SHREW AND GILA MONSTERS

SHORTTAIL SHREW *Blarina brevicauda* p. 10

This shrew, like an unknown number of insect-eating mammals, has poisonous saliva. The tiny animal's 32 black-tipped teeth create a wound into which a neurotoxic venom is injected. The shrew is secretive and bites humans only if it is being handled. Its venom is too mild to pose a danger to an animal the size of a human being.

RETICULATE GILA MONSTER
Heloderma suspectum suspectum p. 12

A very seriously venomous lizard, shy and retiring. It should not be disturbed, not only because of its bite but also because this rare reptile is protected by law in the states in which it occurs. Although it is not known to bite unless disturbed, if molested it can lunge forward or strike with a quick sideways head snap. Then it holds on with a tenacious bulldog-like grip, "chewing in" its neurotoxic venom (which has been compared with that of cobras). Gila Monsters have 8–10 grooved teeth in the lower jaw; venom is secreted from glands at the base of the teeth and injected via grooves in the teeth. Fatalities are rare. The Reticulate Gila Monster is primarily found in rocky, semiarid regions, in foothills and washes and near streams. It may intrude in cultivated areas.

BANDED GILA MONSTER *Heloderma suspectum cinctum* p. 12

Similar in habits and appearance to Reticulate Gila Monster, except tail is strongly banded. Often found on rocky ridges and in desert areas. Attacks by unmolested free specimens are unknown; bites are uncommon. Gila Monsters are increasingly rare and should not be disturbed.

SHORTTAIL SHREW

RETICULATE GILA MONSTER

BANDED GILA MONSTER

RATTLESNAKES: DIAMONDBACKS AND ROCK RATTLESNAKES

EASTERN DIAMONDBACK RATTLESNAKE **p. 14**
Crotalus adamanteus

The longest and bulkiest North American rattlesnake; typically to 6 feet, rarely to 8 feet. The only species in the East with a diamondlike pattern, ringed tail, and light facial stripes. *Extremely dangerous, seriously venomous,* and highly unpredictable. Does not always give a warning rattle. Venom destroys red blood cells. Mortality rate from bites is as high as 40 percent, and many other victims are crippled. This snake is a strong swimmer and is often found in coastal lowlands, palmetto growth, and pinelands, in or near both fresh and salt water.

WESTERN DIAMONDBACK RATTLESNAKE *Crotalus atrox* **p. 14**

Second in size only to the Eastern Diamondback in North America; average 2½–6 feet. Diamond pattern is less clearly defined than in Eastern Diamondback. This snake is extremely dangerous, aggressive, and highly irritable. Has a loud, persistent rattle. Bite is potentially lethal; causes hemorrhaging, and venom also contains neurotoxic components and tissue-digesting enzymes. The Western Diamondback frequents sites of human habitation, and bites are not uncommon. Undoubtedly responsible for more human deaths in the United States than any other snake. Found in any relatively undisturbed habitat within its range; desert, rocky areas, canyons, cliffs, pastureland, and river bottoms.

MOTTLED ROCK RATTLESNAKE *Crotalus lepidus lepidus* **p. 20**

A small rattlesnake, usually under 2 feet. Ground color is variable, matching its surroundings. Although its venom is powerful, the Mottled Rock Rattlensnake is rarely aggressive, and bites are rare. Most often found in mountainous habitats.

BANDED ROCK RATTLESNAKE *Crotalus lepidus klauberi* **p. 20**

Small, typically 1–2 feet. Ground color is variable, with widely spaced crossbands. This rattler is not generally aggressive, and its venom is not lethal.

EASTERN DIAMONDBACK RATTLESNAKE

WESTERN DIAMONDBACK RATTLESNAKE

MOTTLED ROCK RATTLESNAKE

BANDED ROCK RATTLESNAKE

RATTLESNAKES: SIDEWINDERS AND TIMBER RATTLESNAKE

MOJAVE DESERT SIDEWINDER *Crotalus cerastes cerastes* **p. 16**

The sideways locomotion of Sidewinder rattlesnakes, which reduces the contact between the snake's body and hot sand, leaves distinctive J-shaped marks in loose sand. Sidewinders are also recognizable by the "horns" over their eyes, which may reduce sun glare. Sidewinders are nocturnal, hiding from heat during the day. They are often nervous, and may move unpredictably. Bites are not generally lethal. This subspecies is found mainly in southeastern California and southern Nevada.

SONORAN SIDEWINDER **p. 16**
Crotalus cerastes cercobombus

Very similar to Mojave Desert Sidewinder, except first rattle segment is black in Sonoran Desert adults rather than brown as in Mojave Desert Sidewinder. Found in southwestern Arizona, south into Mexico.

COLORADO DESERT SIDEWINDER **p. 16**
Crotalus cerastes laterorepens

This Sidewinder has the first rattle segment dark to black in adults. Range south of Mojave Desert Sidewinder, west of Sonoran Desert Sidewinder.

TIMBER RATTLESNAKE *Crotalus horridus* **p. 18**

Also called "canebrake rattler." A variable snake with several color phases. Tail is usually much darker than the rest of the snake, and often velvety looking. Head is broad, without distinctive markings. The Timber Rattlesnake's disposition is also variable—do not count on it to rattle or retreat—and it should always be considered dangerous. Venom destroys red blood cells, causing pain, swelling, breathing difficulty, blood coagulation, and shock. Fatalities have been reported but are relatively rare. The only rattlesnake in much of its enormous though discontinuous range in the eastern United States; probably has the widest range of any North American rattlesnake. Usually in rocky or timbered terrain, also river bottoms.

MOJAVE DESERT SIDEWINDER

SONORAN SIDEWINDER

COLORADO DESERT SIDEWINDER

TIMBER RATTLESNAKE

RATTLESNAKES: SPECKLED, BLACKTAIL, AND TWIN-SPOTTED

SOUTHWESTERN SPECKLED RATTLESNAKE **p. 20**
Crotalus mitchellii pyrrhus

The ground color of this handsome snake is highly variable, with darker, speckled crossbands. Speckled Rattlesnakes are nervous in disposition, and will hold their ground and strike readily if cornered.

PANAMINT RATTLESNAKE *Crotalus mitchellii stephensi* **p. 20**

Similar to Southwestern Speckled Rattlesnake, frequently with a salt-and-pepper appearance. Ranges north of Southwestern Speckled Rattlesnake, in southeastern California and southwestern Nevada.

NORTHERN BLACKTAIL RATTLESNAKE **p. 22**
Crotalus molossus molossus

A medium-sized rattler, average 2–4 feet. Tail is usually very dark to black. Hunts both day and night. Not a particularly aggressive snake but agile and unpredictable. Mainly found in mountainous habitats.

WESTERN TWIN-SPOTTED RATTLESNAKE **p. 22**
Crotalus pricei pricei

A small rattlesnake, usually less than 2 feet. Recognizable by the two rows of brown spots on its back. Lives in the mountains of southeast Arizona and into Mexico; hunts by day at higher altitudes. Rattle makes a weak sound. This rattlesnake is generally mild-mannered, and its venom is not considered lethal.

SOUTHWESTERN SPECKLED RATTLESNAKE

PANAMINT RATTLESNAKE

NORTHERN BLACKTAIL RATTLESNAKE

WESTERN TWIN-SPOTTED RATTLESNAKE

PLATE 5

RATTLESNAKES: RED DIAMOND, MOJAVE, TIGER, AND RIDGENOSE

RED DIAMOND RATTLESNAKE *Crotalus ruber* **p. 22**

A reddish, heavy-bodied rattler, to $5^1/_2$ feet. Similar to Western Diamondback Rattlesnake (see p. 14), with a conspicuous black and white ringed tail. Generally mild-mannered, but its venom is powerful. Usually found at low altitudes, in desert, brush, and coastal habitats, but the Red Diamond Rattlesnake may also be found in cultivated areas.

MOJAVE RATTLESNAKE *Crotalus scutulatus* **p. 22**

A medium-sized, excitable rattlesnake, 2–4 feet, with a well-defined diamondback pattern. Considered our most lethal rattler: its virulent venom is highly neurotoxic, affecting the heart and skeletal muscles by blocking nerve-to-muscle impulses. This is one of our few truly lethal snakes, with venom 10 times more toxic than any other North American crotaloid (the pit vipers, including Copperheads, Cottonmouths, and rattlesnakes). The Mojave Rattlesnake is often but not exclusively found in high desert.

TIGER RATTLESNAKE *Crotalus tigris* **p. 24**

Back is marked with distinct crossbands ("tiger stripes"). Its head is smaller and its rattle larger than other rattlesnakes. Hunts both day and night, often after rainstorms. Its venom is probably not lethal.

ARIZONA RIDGENOSE RATTLESNAKE **p. 28**
Crotalus willardi willardi

A small snake, usually under 2 feet, with a ridge running around the tip of its snout. Found only in extreme southern Arizona, usually in mountainous habitats.

NEW MEXICO RIDGENOSE RATTLESNAKE **p. 28**
Crotalus willardi obscurus

A *threatened subspecies* of the Ridgenose Rattlesnake found only in the Animas Mountains of New Mexico. Should not be disturbed in the unlikely event of an encounter.

RED DIAMOND RATTLESNAKE

MOJAVE RATTLESNAKE

TIGER RATTLESNAKE

ARIZONA RIDGENOSE RATTLESNAKE

NEW MEXICO RIDGENOSE RATTLESNAKE

WESTERN RATTLESNAKES

PRAIRIE RATTLESNAKE *Crotalus viridis viridis* **p. 24**
The most wide-ranging member of the large Western Rattlesnake group. The ground color of this large snake is variable, usually with more or less square dark blotches with light borders. An extremely irritable rattlesnake, quick to strike. Its bite is generally sublethal, but the complex venom can destroy tissue and produce cardiovascular and neurological symptoms. Uses many habitats: prairies and cultivated areas, woodlands, rocky slopes, along streams, near caves. May be encountered anywhere within its enormous range from southern Canada to northern Mexico.

GRAND CANYON RATTLESNAKE *Crotalus viridis abyssus* **p. 24**
A subspecies of the Western Rattlesnake found only in Grand Canyon, Arizona. Dangerously venomous. Ground color is tan to salmon; obscure blotches have dark borders and pale centers.

ARIZONA BLACK RATTLESNAKE *Crotalus viridis cerberus* **p. 24**
This Western Rattlesnake subspecies usually has a charcoal black ground color. Its markings are close to the ground color and may be outlined with yellow. Its range is in the mountains of Arizona and extrme western New Mexico.

MIDGET FADED RATTLESNAKE *Crotalus viridis concolor* **p. 26**
One of the smallest subspecies of the Western Rattlesnake, almost always smaller than 2 feet. Ground color is cream to yellow; blotches are faint or absent. Although it can be cranky and quick to defend itself, its bite is generally not lethal.

PRAIRIE RATTLESNAKE

GRAND CANYON RATTLESNAKE

ARIZONA BLACK RATTLESNAKE

MIDGET FADED RATTLESNAKE

WESTERN RATTLESNAKES

SOUTHERN PACIFIC RATTLESNAKE *Crotalus viridis helleri* **p. 26**

A robust member of the Western Rattlesnake group, to 4½ feet. Ground color is gray, often with strong markings. Young individuals have a yellow tail. A large, agile, dangerous snake. Range is in southwestern and coastal southern California, including offshore islands, south into Baja.

NORTHERN PACIFIC RATTLESNAKE **p. 28**
Crotalus viridis oreganus

Like the Southern Pacific Rattlesnake, but dark rings on tail are well defined and consistent in width. Young individuals with a yellow tail. This is a large snake, to 4½ feet, with the potential to deliver a large amount of venom. Range includes the northern two thirds of California, Oregon, Washington, northern Idaho, and well into British Columbia.

GREAT BASIN RATTLESNAKE *Crotalus viridis hutosus* **p. 26**

A large, heavy snake, to 5 feet. Ground color is drab, blotches are olive to brown. A dangerous snake if only for its size, and it can be aggressive. Like most Western Rattlesnake forms, this one is adaptable, and uses habitats ranging from timber to desert. Can be very abundant in some areas.

HOPI RATTLESNAKE *Crotalus viridis nuntius* **p. 26**

A small form of Western Rattlesnake, usually less than 30 inches. Pinkish to brown ground color, with dark brown blotches on back becoming rings near the tail. The Hopi Rattlesnake usually tries to retreat if disturbed, but if surprised or cornered it may strike quickly and repeatedly. It is usually found in drier habitats.

SOUTHERN PACIFIC RATTLESNAKE

NORTHERN PACIFIC RATTLESNAKE

GREAT BASIN RATTLESNAKE

HOPI RATTLESNAKE

RATTLESNAKES: MASSASAUGAS

EASTERN MASSASAUGA *Sistrurus catenatus catenatus* **p. 28**
Massasaugas (and Pigmy Rattlesnakes) are distinguishable from other rattlers by the large plates on top of their heads, which have small scales. The Eastern Massasauga is small, rarely longer than 2½ feet, and has a spotted appearance. Ground color is usually gray, with dark blotches. Most individuals are mild-tempered and slow to rattle. Venom is not lethal. Prefers wet habitats such as swamps and wet prairies. Often found in areas where mice gather, such as grain fields.

WESTERN MASSASAUGA *Sistrurus catenatus tergeminus* **p. 30**
Similar to Eastern Massasauga but paler, with tan to gray ground color and brown blotches. Found in plains and prairies; prefers moist areas where they are available. The Western Massasauga is relatively mild-mannered, and its bite is not lethal to an adult; still, it should not be handled.

DESERT MASSASAUGA *Sistrurus catenatus edwardsii* **p. 30**
A small, slender, paler subspecies of Western Massasauga, to about 21 inches. This pretty snake is marked with about 30 dark brown dorsal blotches and smaller spots on its sides. The facial markings extend onto the neck. Seeks moist areas in desert grasslands, and apparently feeds largely on amphibians. The venom of this shy snake is not lethal.

EASTERN MASSASAUGA

WESTERN MASSASAUGA

DESERT MASSASAUGA

RATTLESNAKES: PIGMY RATTLESNAKES

CAROLINA PIGMY RATTLESNAKE p. 30
Sistrurus miliarius miliarius

Like Massasaugas, Pigmy Rattlesnakes have large plates on top of their heads rather than small scales, as do other rattlesnakes. Pigmies also have a distinctly slender tail. The rattle of a Pigmy Rattlesnake is thin and weak, easily missed or mistaken for the buzzing of insects. This subspecies is brown to light gray with distinct markings. Its disposition varies widely, from placid to pugnacius. Venom is sublethal.

DUSKY PIGMY RATTLESNAKE *Sistrurus miliarius barbouri* p. 30

May grow a little larger than other Pigmies, but still usually less than 2 feet long. Brownish gray ground color has a "dusky" look, and there may be a distinct brown stripe down the back. This snake can be quite common near marshes and ponds in the South. Although its venom is not lethal, the Dusky Pigmy often has an aggressive disposition. Its rattle is thin and weak.

WESTERN PIGMY RATTLESNAKE p. 30
Sistrurus miliarius streckeri

This little rattler is usually less than 20 inches long. Its ground color is pale, often light gray, and usually with a reddish stripe down the back. Irregular dark spots may form crossbars. This Pigmy is often irritable and quick to strike. Its venom is powerful but not lethal. Usually found in wet habitats, such as swamps. The thin sound of its rattle is easy to miss.

CAROLINA PIGMY RATTLESNAKE

DUSKY PIGMY RATTLESNAKE

WESTERN PIGMY RATTLESNAKE

COTTONMOUTHS

EASTERN COTTONMOUTH p. 32

Agkistrodon piscivorus piscivorus

A fat, dark snake, to about 4 feet, with or without crossbands. If threatened, a Cottonmouth may coil, hold its head far back, and open its mouth wide to reveal the white interior that gives it its name. Has no rattle, but will vibrate its tail nervously, producing in dry ground litter a sound like a rattlesnake's rattle. This very dangerous snake is short-tempered, cranky, unlikely to retreat, and often aggressive. Its potent venom can cause extensive tissue damage and destroy red blood cells. It also can result in tetanus or gangrene; as many as 50 percent of Cottonmouth bites necessitate amputation of limbs or digits because of gangrene. Also called "water moccasin"; found in and near swamps, rivers, lakes and streams; in rice fields, drainage ditches, and other places where water accumulates.

FLORIDA COTTONMOUTH p. 32

Agkistrodon piscivorus conanti

See under Eastern Cottonmouth. This subspecies is as short-tempered as the Eastern Cottonmouth and similar in appearance, but with a conspicuous dark brown cheek stripe bordered by light lines. Found throughout Florida and in southern Alabama and Georgia.

WESTERN COTTONMOUTH p. 32

Agkistrodon piscivorus leucostoma

A heavy snake, a little smaller than Eastern Cottonmouth. Its chunky body tapers abruptly to a short tail. Like Eastern Cottonmouth, its color and pattern are variable. Nervous tail-twitching and unwillingness to give ground are characteristic. This is a nervous, cranky snake with highly destructive venom. Usually found near water, in habitats much the same as Eastern Cottonmouth's but a little more upland.

EASTERN COTTONMOUTH

FLORIDA COTTONMOUTH

WESTERN COTTONMOUTH

COPPERHEADS

SOUTHERN COPPERHEAD p. 34

Agkistrodon contortrix contortrix

Copperheads are probably responsible for more venomous snakebites than any other snake in the United States. Yet although they will strike when aroused or stepped on, they are usually rather lethargic. Copperhead venom causes hemorrhage, pain, swelling, breathing difficulty, vomiting, gangrene, headache, and unconsciousness. Venom is considered sublethal, though very rare human deaths have been reported. The Southern is the palest of the Copperheads, often with pinkish to light gray background and dark hourglass markings that may break into separate halves on the back. No rattle, but vibrates tail rapidly if disturbed. Often found in association with rattlesnakes. Its habitat is variable. Prefers lowland areas near swamps and streams, but may ascend into rocky hills and outcroppings.

NORTHERN COPPERHEAD p. 34

Agkistrodon contortrix mokasen

Usually 2–3 feet in length, with coppery head and ground color and dark chestnut crossbands often forming distinct hourglass markings. This snake is seriously venomous, but its venom is considered sublethal. Normally lethargic, but will strike if aroused. Habitat generally in upland areas.

OSAGE COPPERHEAD p. 34

Agkistrodon contortrix phaeogaster

See under Northern Copperhead. Similar to Northern Copperhead but with more strongly contrasting dark crossbands, often edged with white. Range is northwest of other Copperheads.

BROAD-BANDED COPPERHEAD p. 34

Agkistrodon contortrix laticinctus

A small Copperhead, usually about 2 feet long. Crossbands are more straight-sided than the hourglass markings of other Copperheads.

TRANS-PECOS COPPERHEAD p. 34

Agkistrodon contortrix pictigaster

The smallest Copperhead. Resembles the Broad-banded Copperhead, but the underside is strongly patterned. Range is west of other Copperheads.

SOUTHERN COPPERHEAD

NORTHERN COPPERHEAD

OSAGE COPPERHEAD

BROAD-BANDED COPPERHEAD

TRANS-PECOS COPPERHEAD

CORAL SNAKES AND KINGSNAKE

EASTERN CORAL SNAKE *Micrurus fulvius fulvius* **p. 36**

This slender candy-cane snake is marked with rings of red, yellow, and black encircling the body, red touching yellow. The red rings are flecked with black. The venom of Coral Snakes is very powerful, but they are unlikely to bite humans; they cannot strike and stab like a rattlesnake because they have small, fixed fangs. But this snake will bite if handled or inadvertently poked with fingers or toes. Its mouth is so small that bites elsewhere on the body are unlikely. Found under leaves and debris in pine woodlands, hardwood stands and hammocks.

TEXAS CORAL SNAKE *Micrurus fulvius tener* **p. 36**

Small, slender, tubular snake to 30 inches. Rings of black, yellow, and red; red and yellow always touch. The venom of the Texas Coral Snake is powerful and essentially neurotoxic, so the snake is dangerous to handle. But it is retiring and unlikely to be encountered; accidents are almost impossible unless the snake is inadvertently poked with an exposed finger or toe. Uses variable habitats from rock slides to steam banks.

ARIZONA CORAL SNAKE **p. 36**
Micruroides euryxanthus euryxanthus

Shy and retiring. It is almost impossible for the snake to bite unless it is handled; it must find an area of thin skin to get a purchase, and then must chew to inject venom. Still, it is a seriously venomous snake to be left in peace. Found in protected habitats; rotting logs (presumably in search of insect larvae), under debris. Frequently nocturnal. Not found in desert sand areas without vegetation. Extreme southwestern corner of New Mexico, scattered areas in southeastern Arizona and north and west to Mohave and Yavapai counties.

SCARLET KINGSNAKE *Lampropeltis triangulum* **p. 36**

Harmless. This is one of the Milk Snakes, shown here to demonstrate mimicry of the Coral Snake group by harmless snakes. Looks remarkably like the Eastern Coral Snake, but the snout is red rather than black, and the black and red rings touch. Yellow and red rings are contiguous only in Coral Snakes. Milk Snakes are extremely widespread, especially in the eastern two thirds of the U.S. There are numerous subspecies.

EASTERN CORAL SNAKE

TEXAS CORAL SNAKE

ARIZONA CORAL SNAKE

SCARLET KINGSNAKE

REAR-FANGED SNAKES

TEXAS LYRE SNAKE *Trimorphodon biscutatus vilkinsonii* **p. 38**

Although rear-fanged snakes are distinctly venomous, their fangs are in the rear of the jaw, so it is difficult for them to bite even exposed parts of human anatomy. The Texas Lyre Snake is shy and usually nocturnal, hiding in rocky places by day and hunting lizards and other small animals by night. The name comes from a lyre-shaped marking on the head that is characteristic of other Lyre Snakes, but the marking is faint or absent in this subspecies. It has a vertical pupil, like a cat.

SONORAN LYRE SNAKE *Trimorphodon biscutatus lambda* **p. 38**

The characteristic lyre- or V-shaped mark is pronounced in this subspecies. Like other Lyre Snakes, this one is most often found in rocky habitats that offer plenty of cover and abundant small rock-dwelling lizards to hunt. Usually nocturnal, hiding by day. This range of this subspecies is chiefly in Arizona and south into Mexico.

CALIFORNIA LYRE SNAKE **p. 38**
Trimorphodon biscutatus vandenburghi

This subspecies also has a distinct lyre-shaped marking on top of its head and a vertical pupil. Its brownish blotches are roughly hexagonal in shape. It is a rock dweller, found to 3,000 feet in massive boulder piles and on rocky slopes. It is found in Southern California, largely near the coast, and south into Baja.

TEXAS LYRE SNAKE

SONORAN LYRE SNAKE

CALIFORNIA LYRE SNAKE

PLATE 14

SPIDERS, CENTIPEDES, AND SCORPIONS

BLACK WIDOW SPIDER *Latrodectus mactans* **p. 40**

Easily recognized by the large, bulbous, shiny black abdomen with a bright red hourglass beneath. The female is twice as large as the male, and her bite is more serious because of the larger amount of venom she delivers. Black Widow venom, drop for drop, is one of the most toxic to humans known. Lives in dark places; beneath leaf litter and debris such as old tin cans, cardboard, etc. Most common in the South. Antivenin is available.

YELLOW SAC SPIDER *Cheiracanthium mildei*

Recently introduced from Europe, this spider lives almost exclusively indoors in North America and has become very common in some parts of its range. It is most abundant in fall. Its bite produces sharp pain and swelling at the site, followed by localized necrosis. The wound may take several weeks to heal.

BROWN RECLUSE SPIDER *Loxosceles reclusa* **p. 40**

Also called "fiddleback." Note the dark brown fiddle or violin marking on head; abdomen dark brown. Found in sheltered natural habitats such as beneath fallen trees, brush, and loose debris; indoors, frequently hides in dark areas of barns and houses, especially closets, stored clothing, basements, attics, book shelves, beneath furniture, etc. Common in South, now possibly found in all states, as humans have moved them in furniture and clothing. Bites when disturbed. Fatalities are rare. Necrosis may develop at wound site, producing a large, deep area of dead cell tissue. The crust-covered wound may take several months to heal and often leaves a permanent scar.

CENTIPEDE **p. 42**

Centipedes have flat bodies and numerous legs, one pair on each body segment. The largest American species are in the southwestern U.S., where they can reach a foot in length. They inject venom through a pair of claws beneath the head. Any individual large enough to grip skin with the claws can bite; the severity of the bite depends partly on the size of the centipede and therefore on the amount of venom delivered. Bites are not lethal unless the victim is highly sensitive to foreign protein.

SCORPION *Hadrurus* ssp. **p. 42**

Scorpions are immediately recognized by the long, up-curved "tail" (actually part of the 12-segmented abdomen) tipped with a venomous stinger. More than 70 species occur in the U.S., some in the South, but especially the Southwest. Most U.S. species, like the one shown, produce a painful swelling around the sting and discoloration, much like a wasp sting, with recovery in a few minutes to days, usually not requiring medical attention. Scorpion venom is primarily neurotoxic.

BLACK WIDOW SPIDER

YELLOW SAC SPIDER

BROWN RECLUSE SPIDER

CENTIPEDE

SCORPION

PLATE 15

STINGING INSECTS AND BLISTER BEETLES

IO MOTH CATERPILLAR *Automeris io* **p. 44**

This large, green caterpillar has tufts of bristly, branched, hairlike spines, with reddish pink stripes edged in white on its sides. To 3 inches long. Found in meadows, fields, and open woods; feeds on many trees and shrubs such as maple, oak, beech, and roses. If disturbed, the caterpillar will roll up and drop to the ground, but sharp spines imbedded in hair tufts can inflict painful stings if touched or handled.

SADDLEBACK CATERPILLAR *Sibine stimulea* **p. 44**

Brilliant green caterpillar to about an inch in length, dark brown at both ends, with a distinct brown saddlelike mark on its back, outlined in white and a thin black line; has many tufts of bristly hairs. This moth larvae has small thoracic legs and no prolegs, moving with a creeping motion. Gardens, edge of woods, orchards, and cultivated ground. Has stinging hairs, causing skin irritation and relatively mild stinging if touched or handled. Its appearance is so unusual that children may be attracted to playing with it, leading to irritation. Matures into the small, brown Saddleback Caterpillar Moth.

VELVET ANT *Dasymutilla magnifica* **p. 50**

Actually a wingless female wasp, with a sting so painful that the Velvet Ant is also called "cow killer." The pain is usually of short duration, however, and the other effects, such as swelling and discoloration, generally disappear within a few hours. A very pretty insect, so densely hairy that it looks velvety. Some species are brilliant red in color; this one is more orange.

BLISTER BEETLE **p. 52**

More than 330 species of these plant-eating beetles are found in North America. Their shape is distinctive, with a long slender body and the platelike coverings of the prothorax usually narrower than the rolled front wings. When disturbed, Blister Beetles exude a reddish substance called canthardin from their leg joints. The substance, which can cause blistering of the skin, is extracted from some species for medicinal use.

IO MOTH CATERPILLAR

SADDLEBACK CATERPILLAR

VELVET ANT

BLISTER BEETLE

STINGING INSECTS

HONEY BEE *Apis mellifera* **p. 46**

Of all venomous species of North America, the Honey Bee is probably responsible for the most cases of envenomation, from the defensive stinging of life-sacrificing worker females. Native to Europe, the Honey Bee was introduced to North America with early settlers and is now found throughout the continent. The Honey Bee is unique among stinging insects in North America in that it has a barbed stinger that stays in the skin of the victim. When the bee pulls away, its vital body tissue is torn out, and the bee dies shortly thereafter. Honey Bees will vigorously defend their hives, stinging to protect the colony and its precious food stores. Stinging is also a defensive response to being disturbed, stepped on, etc.

BUMBLEBEE *Bombas* spp. **p. 46**

Bumblebees are common and familiar, easily recognized by their large size, about twice that of the Honey Bee. Their bodies are stout, hairy, and marked with contrasting black and yellow stripes around the abdomen. Various species are found in woods, open fields, and gardens throughout North America. The bumblebee minds its own business unless disturbed, but it can inflict a painful sting.

BALD-FACED HORNET *Vespula maculata* **p. 48**

This beelike wasp has black and white patterns on its face, thorax, abdomen, and first antenna segment, and has smoky-colored wings. Occurs in fields, lawns, and wood edges throughout North America. Females are quick to defend their small, gray, pendent nests, and will sting repeatedly. Wasps have caused many human fatalities, usually from multiple stings or in people highly sensitive to foreign protein.

PAPER WASP *Polistes* spp. **p. 48**

The long-legged Paper Wasp builds its nest out of paper that it makes itself out of chewed wood and saliva. Often found near water. Paper Wasps are less aggressive in defending their nests than hornets, but they can inflict a painful sting.

YELLOWJACKET *Vespula* spp. **p. 48**

The all too familiar Yellowjacket is banded with black and yellow (or white) around its abdomen. Various species are common throughout North America in fields and wood edges. They build their nests in the ground or in stumps or old logs. These pesky wasps are familiar to all who have eaten outside in the summer in areas where Yellowjackets are common. They steal morsels of food from barbecues and picnics and are ubiquitous in parks, cities, and recreational areas. Serious envenomation can result from unwittingly stepping on a nest while walking through a field or forest. Yellowjackets will sting repeatedly with the slightest disturbance or provocation.

HONEY BEE

BUMBLEBEE

BALD-FACED HORNET

PAPER WASP

YELLOWJACKET

POISONOUS PLANTS

POISONOUS PLANTS

There are three reasons why most people want to know how to identify poisonous plants: to avoid them; to identify a plant that might be involved in an accidental poisoning; and simply to learn more about the living things in our environment. Contact with poisonous plants can result in a wide range of uncomfortable or even deadly effects. There are plants that cause allergies, plants that cause dermatitis through direct or indirect contact, plants that cause internal poisoning if eaten, and plants that may cause injury as the result of spines, thorns, or stinging mechanisms.

Most of us would never knowingly ingest a known poisonous plant. The majority of plant-induced poisonings are accidental. But accidents do happen, especially when children are around. I learned this from experience.

I brought an eight-foot aluminum stepladder into the house to change a light bulb. Finished, I headed down the ladder. There at the bottom rung was my daughter, then 10 months old, who had crawled into the vicinity unseen and unheard. Clamped between her lips was some green leafy material. I scampered down to discover that she had gotten hold of some leaves of a trumpet vine *(Campsis radicans).* When I brought the ladder inside, I had raked it through the vine, and several leaves had become lodged in the ladder's brackets.

While trumpet vine is implicated in causing contact dermatitis, it is not generally involved in cases of internal poisoning. Still, I knew that the plant was not on anyone's list of salad ingredients, and certainly was not a dietary item with which my young child should experiment. She had not chewed the leaves or swallowed them by the time I wrested them from her lips. A potential accident, which may or may not have caused any kind of reaction, was avoided in this case.

But through this experience, I came to realize how easily an accidental plant poisoning could occur, not just from eating an unknown or misidentified plant in the wild but in my own home.

Poisonous plants, like weeds, wildflowers, and shade trees, are simply an unavoidable part of our surroundings. It is not possible to eliminate all the native, cultivated, or exotic plants that pose potential hazards from our environment. It is possible to learn about those potentially toxic plants, how to identify them and how to take precautions against accidents, just as one must learn to properly use, store, and recognize household chemicals. The major difference between other hazardous household substances, such as cleaning fluids or medicines, and hazardous plants is that the plants do not have bold warning labels or safety caps.

Although everyone should be aware of the presence and dangers of potentially poisonous plants, it is especially important for those who care for young children. Of the 12,000 reported cases of ingestion of potentially poisonous plants every year, over 60 percent involve small children. Children weigh much less than adults, so smaller amounts of poisonous substances will have a more profound effect on them. They are often curious about the edibility of leaves or berries, and children under five years of age may try to eat just about anything. Older children may experiment with plants out of curiosity or in play, such as infusing leaves for a "tea party."

Parents, teachers, day care center workers, and school employees should be aware of the potential dangers that the plant world may hold. Health care professionals who work in clinics, emergency rooms, and poison control centers must also know how to identify poisonous plants and how to care for those who come to them for treatment. Others who enjoy the outdoors—birders, hunters, campers, hikers, naturalists, and scouts—may also want to know which plants are edible and which to avoid.

The plant section of this book is a layman's guide to the identity and potential dangers of poisonous plants. It is a review and sampling of the available information on the more important plants known to poison both humans and livestock in North America. It is certainly not all-inclusive. Hundreds of additional species could have been added, especially those known to cause livestock poisoning. Just because a plant is not found in the pages of this book does not mean that it might not be potentially poisonous to someone, somewhere.

Is it Safe or Not?

Historical reports of poisoning are often suspect unless recent experimental or case history evidence has provided new insights into the potential or real toxicity of the plant. Many authors, myself included, feel obligated to provide as much information as the scope of the work will allow. In doing so, some errors or misconceptions may be perpetuated. Even with new evidence, it is difficult to definitively say whether or not a particular plant is poisonous.

A good case in point is information on the toxicity of Goldenseal *Hydrastis canadensis.* In a number of recent poison plant books, it has been noted that ingestion may irritate the mucous membranes or even produce ulcerous lesions. Tracing this information back to its original source leads one to a work first published in 1887, *American Medicinal Plants*, by Charles Millspaugh. Millspaugh's work, still widely available and often cited, lists apparently toxic "physiological actions" for dozens of native plants.

Millspaugh has this to say about Goldenseal: "When taken in large doses Hydrastis causes a train of symptoms due to a hyper-secretion of the mucous membranes. If persisted in, it causes severe ulceration of any surface it may touch; and a catarrhal inflammation of mucous surfaces, followed by extreme dryness and fission."

To understand this in context, one must take into account the subtitle of the first edition of Millspaugh's book: "*An illustrated and descriptive guide to the American plants used as homeopathic remedies . . .*" The key word here is *homeopathic.* Homeopathy is a system of medicine developed in the late 18th century, which continues to flourish in Europe and India. It is based on the theory that "like cures like," that is, a substance that will produce a certain set of symptoms in a well person will cure that same set of symptoms in a diseased person. The dosage forms used are so diluted that the actual chemical content of the remedy is undetectable, even using sophisticated modern analytical chemistry. Homeopathy's scientific basis, if one exists, has remained a mystery for 200 years.

In the context of homeopathy, all 180 plants treated by Millspaugh can be viewed as toxic. Millspaugh reports that peppermint, for example, causes "headache, with confusion; shooting pains in the region of the fifth-nerve terminals; throat dry and sensitive . . . ."

As a member of the buttercup family lacking definitive experimental data on its safety or toxicity, Goldenseal can be viewed with suspicion. Yet Goldenseal has been one of the best-selling herb products in health and natural food markets in the U.S. over the past 15 years. It has been used by thousands of people with no published reports of adverse reactions and would seem to be relatively safe. I have chosen not to include it in this book.

You can begin to see some of the problems met with by researchers, writers, and concerned lay people, not to mention the person in the most precarious position: the clinician who must decide quickly how to treat a case of possible poisoning based on the best available information in hand. This book is a guide to the identity and possible effects of typical poisonous plants of North America.

How to Use the Plant Section

Species and area covered

Over 250 species of wildflowers, weeds and exotic aliens, shrubs, trees, ferns, and mushrooms are covered in the plant section. The area covered includes the 48 contiguous states, with incomplete coverage of southern Florida and Texas. Readers in those areas may refer to specialized texts covering those subtropical areas. While a number of plants in the book will be familiar to readers in northern areas only as house plants or container plants, all selections, with a few exceptions, such as Poinsettia and Ginkgo, are known to occur without cultivation in the United States. If the plant is nonnative, its inclusion here is based on documented evidence of established populations, or its ability to persist without cultivation, by its citation in *A Synonymized Checklist of the Vascular Flora of the United States, Canada and Greenland* by John T. Kartesz and Rosemarie Kartesz. Scientific names of plants used here generally follow the primary names given by Kartesz and Kartesz.

Plant identification and species arrangement

As in other Peterson Field Guides, the plants in this book are arranged based on visual features, to help the reader quickly find the identity of a plant in hand. The herbaceous (nonwoody) plants are arranged first by flower color. They are further separated by numbers of petals, the plant's growth habit, leaf type or arrangement, and so on. The brief descriptive details of the text complement the illustrations. Pay particular attention to *italicized* details. Positive identification of any plant that might be ingested or involved in a case of poisoning is essential, especially for health care professionals who must decide on a course of treatment.

The flowers of any species vary tremendously in their coloration, and people see colors differently. What one person might interpret as "pink to red" may be viewed by another as "violet to blue." Be aware of these subtle differences. Multicolored flowers or flowers that come in more than one color—such as those of Foxglove, which typically has purple flowers but may also be white—will be found in both the "white" and "violet to blue" sections.

Flowering shrubs, trees, and vines (including both woody and herbaceous vines) are in separate sections following the wildflowers. Woody plants are generally excluded from the section on herbaceous plants (wildflowers). A plate on ferns or plants with fernlike leaves follows the vine section. Last is a section on poisonous mushrooms.

The mushroom section serves as a sampler of various types of poisonous mushrooms and is not intended to be all-inclusive. Identifying mushrooms is quite different from identifying flowering plants. Mushrooms are often best identified by spore prints or microscopic details. Since such detailed technical information is beyond the scope of this work, all mushroom species accounts contain page references to the excellent *Field Guide to Mushrooms of North America,* by McKnight and McKnight (see References). This book contains the technical details necessary for accurate identification of mushrooms. The mushroom nomenclature used here follows McKnight and McKnight.

Only experienced amateurs and experts should eat wild mushrooms. There are no foolproof methods for determining the edibility or toxicity of mushrooms whose identity is not definitively confirmed by an expert. Beginners wishing to enjoy wild edible mushrooms should refer to Lee Peterson's *A Field Guide to Edible Wild Plants*. Four mushrooms that can be safely identified and eaten by beginners are described on page 238 of that work and illustrated in the accompanying plate.

Given the difficulties of identifying mushrooms by visual features alone, the mushrooms are grouped according to types of toxic reactions they are known to produce. These include four general categories: deadly mushrooms; neurotoxic mushrooms, those that affect the central nervous system; mushrooms of miscellaneous toxicity, including those that are toxic when consumed with alcohol and those that are harmful to the immune system or kidneys; and various mushrooms known to cause gastroenteritis.

Species accounts

I have tried to keep technical terms to a minimum. Nevertheless, technical terms sometimes best describe a plant's details or better explain the poisonous nature of a plant. Terms unfamiliar to the general reader can be found in the Glossary.

The purpose of the species accounts is only to aid in the identification of typical poisonous plants. No information on treatments is given.

Plant names. Each species entry begins with a common name. The common names used in this book generally follow those used for the same plant in other Peterson Field Guides. If a plant does not appear in another Peterson Guide, then common names from *Hortus Third* (1976) are used. Unfortunately, the nomenclature for plants is not clearly defined, as it is for most animals, and many plants go by different names in different parts of the country.

The plant's scientific name follows its common name. The scientific name is composed of two parts. The first word is the name of the genus. The second word is the name of the species. Scientific names are followed by an abbreviation of the name(s) of one or more botanists who named the plant, or the "species author." If a plant is frequently referred to by a now obsolete scientific name, the old name is placed in brackets beneath or beside the preferred Latin name. As previously noted, the scientific names of flowering plants and ferns follow those of Kartesz and Kartesz (1980). In the rare instances in which a plant is absent from Kartesz and Kartesz, then *Hortus Third* is followed for scientific names.

Poisonous part. Following the common name is the part of the plant considered to be most toxic. In many cases the whole plant is poisonous. Even honey made from the flowers may be toxic.

Family names. The scientific or Latin name is followed by the common name for the plant family. Scientific names for plant families have not been included. Family names are not included in the mushroom section.

Description. The brief descriptions are based on primary visual characteristics. Each account includes the growth habit of the plant (annual, biennial, perennial, etc.) and height, followed by identifying features of the leaves. Next comes a description of the flowers, with flowering time. Information on seed-producing parts (such as fruits or pods) is included if they are important for identification or are a major poisonous part of the plant. Information on related species is included at the end of some species accounts.

Distribution. Under the heading **"Where found"** comes information on the plant's habitat and the boundaries of its distribution in Canada and the U.S. Each plant's range is given from northeast to southeast and from southwest to northwest. If the plant is alien (nonnative), its continent of origin is noted.

Poisonous aspects. Under the heading of **"Comments"** are brief notes on the poisonous nature of the plant, the symptoms it may produce, pertinent historical aspects, and often information on the components or chemical groups responsible for its toxicity. Except for the general discussion in this introduction, no information on treatments is given; if you need help, call the nearest poison control center.

Toxic Elements of Poisonous Plants

A vast array of chemical compounds is found in the plant world. A few major groups of chemicals are responsible for the toxicity of most poisonous plants. A toxic plant may have more than one of these compounds.

Alkaloids

Alkaloids are the most common toxic compounds found in plants. They are a large and varied group of complex nitrogen-containing compounds, usually alkaline, that react with acids to form soluble salts, many of which have positive (medicinal) or negative (toxic) physiological effects on humans. Well-known alkaloids include nicotine in tobacco and caffeine in coffee. Alkaloids often have a bitter taste and are insoluble in water.

Glycosides

Glycosides are a group of compounds that are converted by

hydrolysis to sugars and nonsugar residues. (Hydrolysis is the addition of the hydrogen and hydroxyl ions of water to a molecule, resulting in its breakdown into smaller molecules.) The nonsugar part of the molecule is responsible for toxicity in glycoside-containing plants. There are different types of glycosides. Cyanogenic glycosides, which yield cyanidelike compounds, are the toxic components of cherry leaves, peach pits, and apple seeds. Steroid glycosides include cardiac glycosides and saponin glycosides. Cardiac glycosides include the medicinally active—and toxic—components of Foxglove *(Digitalis purpurea),* as well as those found in Oleander *(Nerium oleander)* and dogbanes (*Apocynum* species). These compounds can increase heart function in medicinal dosages, but in toxic doses (very close to medicinal doses), they may cause cardiac arrest.

Saponin glycosides produce breakdown products that make suds when shaken in water. These glycosides, found in the nut of the Tung Tree *(Aleurites),* can cause corrosive irritation of the digestive tract. Though glycosides are not readily absorbed into the bloodstream, if they are absorbed they may destroy red bloods cells. Anthraquinone glycosides may be powerful cathartic (laxative) substances. These include the glycosides of buckthorn (*Rhamnus* species) and of senna leaves and pods (*Senna* species). These ingredients are often found in commercial laxative products.

Oxalates

Oxalates are a group of carbon-based acids, such as oxalic acid, that can have a corrosive effect on animal tissues. Members of the arum family (Araceae) are well known for containing calcium oxalate in the form of small, insoluble crystals. If chewed, these microscopic needles cause intense burning and irritation of the mouth. In the case of Dieffenbachia, the calcium oxalate crystals (raphides) are encased in highly specialized ejector cells, which cause a kind of microscopic mechanical injury as well as chemical irritation. (Dieffenbachia, one of the plants most commonly involved in incidents reported to poison control centers, is a virtual chemical factory. In addition to the oxalates, it contains other possible toxins including alkaloids, glycosides, and proteinlike substances.)

Proteins

Proteins known as phytotoxins or toxalbumins occur in a relatively small number of plants, but they are among the most powerful toxins. Included in this group is ricin, found in the seeds of the Castor plant *(Ricinus communis)*, and abrin, the toxin in the seeds of the Rosary Pea *(Abrus precatorius)*.

Amines

Amines are nitrogenous compounds. They are the toxic components found in the seeds of sweet peas and in American mistletoes.

Polypeptides

Polypeptides are the basic structural units of a protein. Toxic polypeptides include amatoxin, phallotoxin, and phalloidin, found in a number of deadly poisonous Deathcap mushrooms (*Amanita* species).

Resins and resinoids

These are substances that exude from the wounds of certain plants to protect them from insects and fungus. Resins may harden into glassy solids. Among the most important toxic resinoids is the oily substance known as urushiol, the poisonous component of Poison Ivy and related plants.

Mineral toxins

Mineral toxins usually result from the ability of certain weed species to accumulate toxins such as selenium or nitrates from the soil. Some poison vetches (*Astragalus* species) owe their toxicity to their ability to concentrate selenium.

Alcohols

Alcohols are common in plants, but only a few are responsible for plant toxicity, such as cicutoxin from Water-hemlocks (*Cicuta* species). A toxic alcohol in White Snakeroot *(Eupatorium rugosum)* is called tremetol. It can become concentrated in the milk of cows who feed on the plant and is then transmitted to humans via the milk.

Avoiding Plant Poisoning

Most plant poisonings involve children under three years of age. Therefore, potentially hazardous house plants such as Poinsettia, Philodendron, and Dieffenbachia should be kept out of reach of children. Children should be taught not to touch or eat house or garden plants, especially brightly colored berries or flowers, and to leave all mushrooms alone. Older children should be warned not to feed plant materials to friends or younger siblings. They must be told not to experiment with berries that they find in the yard or farther afield. Red and especially white berries are particularly suspect. Most white-fruited plants are toxic. Blue-colored berries are less likely to be poisonous. However, no general rule on berry color can assure edibility or toxicity. Children should be taught not to suck nectar from flowers or make "tea" from leaves.

Gardening supplies and plant materials such as bulbs, seeds, and roots should be properly labeled and stored out of the reach of children. Plant toxins are often most highly concentrated in roots and seeds. Many bulbs of the Iris family and tubers, corms, and rootstocks of other plants may be poisonous. Some seeds, such as those of the Castor plant *(Ricinus communis),* can be fatally poisonous. Remember that fertilizers and pesticides are poisonous. Some plant poisonings are actually due to fertilizers or pesticides applied to the plant's leaves.

Do not eat wild plants or mushrooms unless their identity has been positively confirmed.

Learn to identify Poison Ivy and related members of the genus *Toxicodendron* that occur in your vicinity. Children should also be taught to identify and avoid Poison Ivy.

When camping, avoid smoke from burning plants, especially logs that have "hairy rope" on them; these may be the vines of climbing Poison Ivy, with their numerous hairlike rootlets. The toxic components of Poison Ivy can become airborne in smoke. Also be careful not to use the twigs of poisonous shrubs, like mountain laurels (*Kalmia* species), as cooking skewers for meat or marshmallows. Don't use the twigs of any plant that has thick, shiny leaves or is evergreen in winter months.

If preparing wild edible plants, don't assume heating and cooking will destroy a toxic substance.

Be careful not to harvest from polluted areas. Some plants can accumulate and concentrate both natural and man-made toxins in their leaves.

Again, positive identification of wild edibles is essential. It is best to avoid eating wild-harvested members of the parsley or carrot family. Even professional botanists have a difficult time accurately determining the exact identity of members of this family, which contains some of our most deadly poisonous plants. The root of Water-hemlock (*Cicuta* species) has an apparently pleasant flavor. It is also deadly in relatively minute amounts.

Just because you see a plant being eaten by a pet, bird, or wild animal, do not assume it is safe for humans. Many gamebirds and songbirds are partial to the fruits of poisonous plants, even Poison Ivy. Deer, mice, bears, muskrats, and rabbits are known to browse on the stems and leaves of Poison Ivy and Poison Oak.

Assume plants with milky white sap, or especially yellow, orange, or red sap, are potentially poisonous.

Despite popular wisdom, there is no test that can determine whether a mushroom is edible or poisonous. Just because a silver spoon or coin tarnishes when placed in a dish with a mushroom, or a mushroom's cap is easily peeled, does not mean it is safe.

In Case of Plant Poisoning

As noted earlier, the best way to avoid accidental poisoning from plants is knowledge and prevention. Know which plants around your home or environs are poisonous and what problems they might cause. Keep jewelry made from tropical seeds away from children. Remove dangerous ornamental plants from your home or keep them out of reach of children, as you would any dangerous household chemical. While plants may be "natural," their toxic components are chemicals. Think of them as you would other chemicals, and take the appropriate precautions.

If you don't have a *positive* identification for a wild plant, don't eat it, especially bulbs or plants thought of as "wild carrots" or "wild parsnips," which may actually be deadly Poison-hemlock. Teach children not to eat wild plants, especially wild mushrooms. Don't eat wild mushrooms yourself, unless expert identification is available.

Serious plant poisoning is relatively rare. About 10 percent of calls to poison control centers involve plant ingestion. The majority of these callers report no symptoms. Serious adult poisoning usually involves ingestion of mushrooms, such as *Amanita* species (Deathcaps), Poison-hemlock (*Cicuta* species) that has been mistaken for edible wild members of the carrot family, or other wild poisonous plants that have been mistaken for wild edible plants. In some cases, the poisoning may be the result of contamination by pesticide residues on or within a plant.

In many cases, when an unknown or potentially poisonous plant has been ingested, especially by a child, symptoms may not be readily evident. Close, astute observation is important. If a child suddenly develops a stomachache, vomiting, headache, or just suddenly feels ill, the child should be questioned to find out whether he may have eaten something he shouldn't have.

What to do?

Despite the best intentions, accidents do happen. What steps should be taken in the event of accidental ingestion of a potentially poisonous plant?

First, don't panic. The first step is to make a call to your doctor, the nearest emergency room, or the nearest poison control center. Even if no symptoms are evident, go ahead and call. Better to be embarrassed now than sorry later. Be prepared to provide the medical care-givers with the following important information:

- If known, give the name of the plant, preferably the scientific name (spell it, or pronounce it phonetically). Scientific names are given in this field guide. Common names are subject to regional variations, and several common names may be used for a single plant in a locality. If you don't know the name of the plant, try to get a specimen that can be identified later. If the victim has vomited, keep the vomit, as it may contain plant parts that may be useful in determining the species involved, as well as the quantity ingested. If the emergency room personnel or doctors are unable to identify the plant, a botanist at a nearby college or university, high school biology teacher, agricultural extension agent, or other knowledgeable individual might be contacted for help. The poison control center will likely have information on identification of the plant, or have a list of persons who can identify poisonous plants.
- Try to find out what plant parts were eaten, and how much.
- Estimate how long ago the plant was eaten.
- Give the age and estimated weight of the individual. This is very important for determining the course of action for treatment, especially for children.
- Observe any symptoms. A clear description of the immediate symptoms will help your health care practitioner know precisely what to tell you to do.
- If a knowledgeable health care practitioner is unavailable

by phone, get the victim to the nearest emergency medical facility. Below, space is provided for you to pencil in the number of your doctor, the nearest hospital, and the nearest poison control center. Two spaces are provided for each—one for your permanent residence, the other for your summer home or for use during a long camping trip.

Doctor:________________	Doctor: ________________
Hospital: ________________	Hospital: ________________
Poison control: ___________	Poison control: ___________

Acting quickly is important. If a child has eaten an unknown mushroom, yet exhibits no immediate symptoms, that doesn't mean that poisoning will not occur. In the case of the Deathcap mushroom *(Amanita phalloides)*, there may be a latent period of anywhere from 6 to 24 hours before symptoms appear. The sooner treatment is obtained, the better. If a doctor or emergency health care provider cannot be contacted by phone, get the patient to an emergency medical facility.

It may also be important to remove the poison from the body before it is absorbed into the system. One method often suggested is to give the patient a glass of warm water and then induce vomiting by tickling the back of the throat. Another method is to use an emetic such as syrup of ipecac. When administering emetics, be sure to follow the directions on the label as well as the instructions of medical personnel. Ipecac syrup may be inappropriate in certain situations: if a victim is drowsy, unconscious, or has convulsions, do not try to give anything to the victim orally. In some cases, such as ingestion of roots or leaves of arum family members, plant material may cause as much harm coming up during vomiting as it would in the digestive tract. Here, the decisive clinical judgment of a doctor, possibly aided with information from a poison control center, is essential.

Antidotes to plant poisonings are rare or nonexistent. The required action to manage a case of plant poisoning depends on many variables including the severity of poisoning, the symptoms that result, the quantities and nature of toxins in the species ingested (which can vary depending upon plant part, time of year, time of day, and geographic location), as well as other factors.

The decision for treatment or management of symptoms, including whether or not to induce vomiting, is a medical decision that should be made by a physician or trained emergency medical personnel. Follow the advice given by the professionals you contact.

A snake can strike out and bite you. Poisonous plants become dangerous when humans, in essence, bite them. But there is no need to be overly fearful of poisonous plants. We eat parts of potentially poisonous plants every day—tomatoes, potatoes, rhubarb, and apricots. Knowledge is your best preventative measure against poisoning from plants or mushrooms. Remember that just because a plant is not found in this book does not mean it is safe to ingest. A simple rule: if you don't know a plant, don't eat it. And make sure to keep the number for the nearest poison control center by the phone.

—Steven Foster

POISONOUS PLANTS ILLUSTRATIONS

MISCELLANEOUS SHOWY FLOWERS

OPIUM POPPY *Papaver somniferum* L. **Whole plant**

Sometimes white, most often pink to red. See on page 116.

DUTCHMAN'S BREECHES **Pl. 17** **Whole plant**

Dicentra cucullaria (L.) Bernh. Bleedingheart Family

Delicate perennial herb; 5–9 in. Leaves lacy, dissected. Flowers white, yellow-tipped; dangling on arched stalk, each with 2 *inflated "pantlike" spurs;* April–May.

Where found: Rich woods. N.S. to Ala.; N.D. to Okla.

Related species: There are nine N. American *Dicentra* species. Some have pink/red flowers, such as garden Bleedingheart *D. spectabilis.*

Comments: *Dicentras* are known to contain toxic isoquinoline alkaloids in the leaves and, especially, the tubers. However, they are unlikely to be ingested, except for cattle nibbling on the emerging spring leaves or exposed tubers in shaded pastures. Handling may cause skin rash.

WILD CALLA, WATER ARUM **Whole plant**

Calla palustris L. Arum Family

Aquatic perennial with shiny, elongated, heart-shaped leaves to 6 in. Flower a *white spathe* clasping a *clublike* yellowish green spadix; May–Aug.

Where found: Circumpolar; North America, Eurasia. Marshes, swamps, ponds. Nfld. to N.J.; Ind., Wisc., Minn. to Alaska.

Comments: Needlelike, insoluble calcium oxalate crystals occur in whole plant, especially root. Eating any fresh plant part causes intense irritation and burning and swelling of the entire mouth and throat. Symptoms usually subside in a few hours (rarely several days). Generally does not produce internal poisoning since the crystals are not absorbed in the digestive tract. Touching plant may cause dermatitis. Completely dried root and seeds have been used in food and medicine.

FOXGLOVE *Digitalis purpurea* L. **Whole plant**

Sometimes white, most often violet. See on page 128.

PRICKLY POPPY **Pl. 25** **Whole plant, seeds**

Argemone albiflora Hornem. Poppy Family

Herb with blue-green stems, 2–3 ft. *Yellowish* (white to orange-red) latex in broken leaf stems. *Thistlelike,* lobed leaves; stems with *sharp spines or prickles.* Flower with 4–6 (12) petals, 2 in. wide, numerous *orange* stamens; May–Sept.

Where found: Waste places, ditches; scattered. Conn. to Fla.; Texas to Mo., Ill.

Related species: The 15 North American *Argemone* species are mainly found in the Southwest U.S.; flowers yellow or lavender-tinted.

Comments: Isoquinoline alkaloids are found in the whole plant and seeds. The sharp spines or prickles discourage ingestion. Seed-contaminated grains have caused poisoning in humans. Symptoms include vomiting, diarrhea, fainting, and coma.

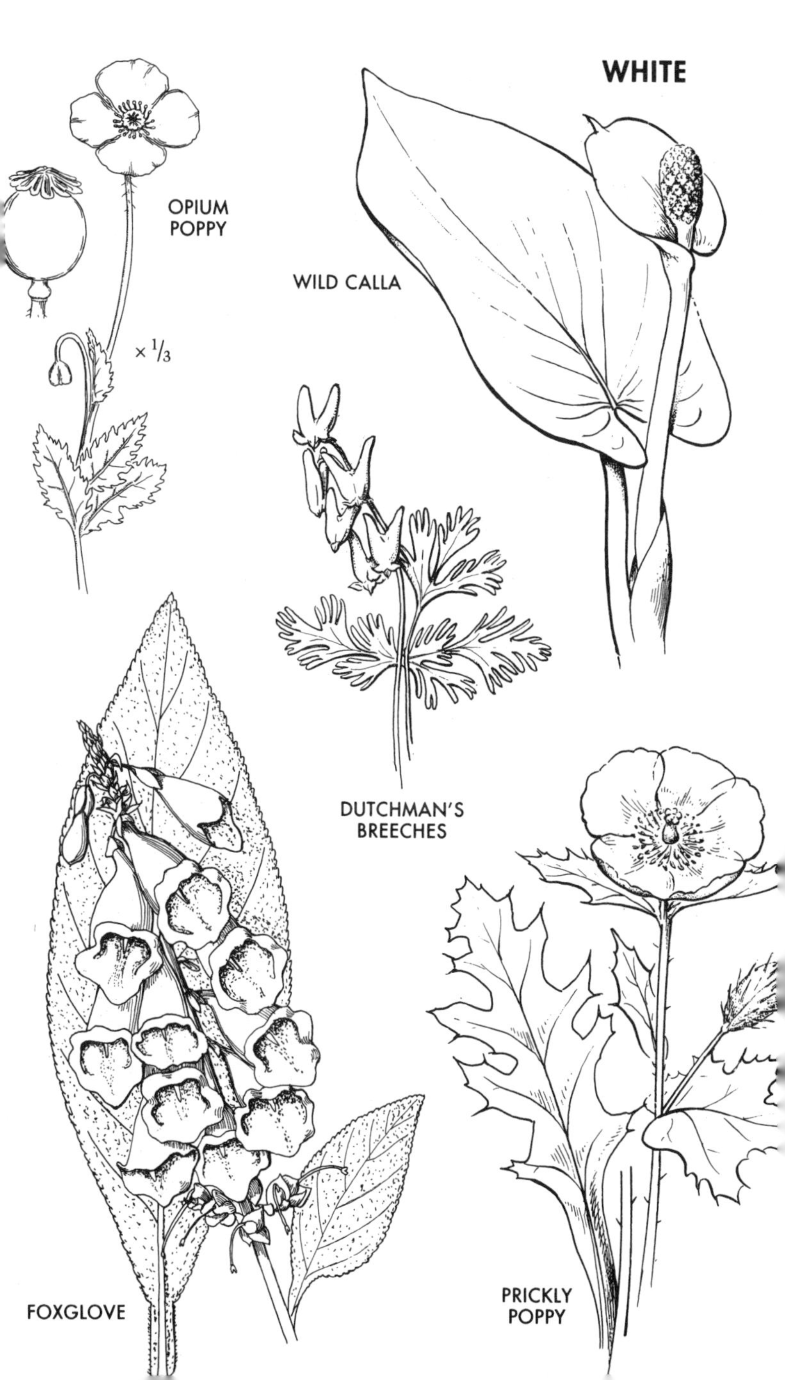
WHITE
OPIUM POPPY
× 1/3
WILD CALLA
DUTCHMAN'S BREECHES
FOXGLOVE
PRICKLY POPPY

FLOWERS IN GLOBULAR OR RADIATING CLUSTERS

RED BANEBERRY **Pl. 17** **Whole plant, especially berries**
Actaea rubra Ait. (Willd.) Buttercup Family

Perennial to 2–3 ft. Similar to White Baneberry, but the stalks are less stout. Flowers in globe-shaped or elongated clusters; April–May. Fruits lustrous, cherry red (sometimes white in *A. rubra* ssp. *arguta,* in the western U.S.); July–Oct.

Where found: Rich moist woods and thickets. S. Canada to n. N.J., W. Va.; west through Ohio and Iowa to S.D., Colo., Utah, Calif. to Ore.

Comments: See *A. pachypoda* below. Nervous system disturbances have been reported from ingesting Red Baneberry fruits.

INDIAN HEMP, DOGBANE HEMP **Pl. 25** **Whole plant**
Apocynum cannabinum L. Dogbane Family

Erect, usually smooth perennial; stem often reddish, with *milky juice;* 1–3 ft. Leaves opposite, oval to lance-shaped, without teeth, usually with a definite stalk to 1/2 in.; sometimes stalkless. Flowers *terminal,* small, *whitish green,* bell-like or urn-shaped, 5-parted; June–Aug. Seed pods paired, 4–8 in. long.

Where found: Roadsides, forest margins. W. Que., s. New England, to Fla., Miss., La., Texas, N.M., Calif. to Wash.

Comments: Poisoning is rare, even in livestock; the milky juice is exceedingly bitter, discouraging ingestion. Livestock have died after eating hay contaminated with Indian Hemp. As little as 1/2 ounce of the dried leaves may kill a large farm animal. Contains a cardiac glycoside, apocynamarine, which causes cardiac arrest. Human poisonings are unknown, but the plant is potentially dangerous, especially to children who may play with the milky-juiced stalks.

WHITE BANEBERRY, DOLL'S EYE **Pl. 17** **Whole plant, especially berries**
Actaea pachypoda L. Buttercup Family

Smooth-stemmed perennial; 1–2 ft. Leaves alternate, twice divided into sharp–toothed leaflets. Flowers white, with many stamens; in oblong cluster on a thick reddish stalk; April–May. Fleshy *ivory white* (rarely red) fruit with a conspicuous *dark dot* at tip; July–Oct.

Where found: Rich woods, thickets. S. Canada to Ga., La.; Okla. north to Minn.; sometimes cultivated.

Comments: The European Baneberry *Actaea spicata* L. is the source of most reports of toxicity from this genus. The acrid, bluish black berries of *A. spicata* have caused poisoning and death of children in Europe. An essential oil in the fruits may be responsible for poisoning. Symptoms include severe stomach cramps, headache, vomiting, and dizziness. As few as 5 or 6 fruits can produce toxic effects. The fruits are acrid-tasting, however; rarely are enough ingested to be fatal. Protoanemonin, a toxin widespread in the buttercup family, is reported from the genus *Actaea*. See Creeping Buttercup, page 104.

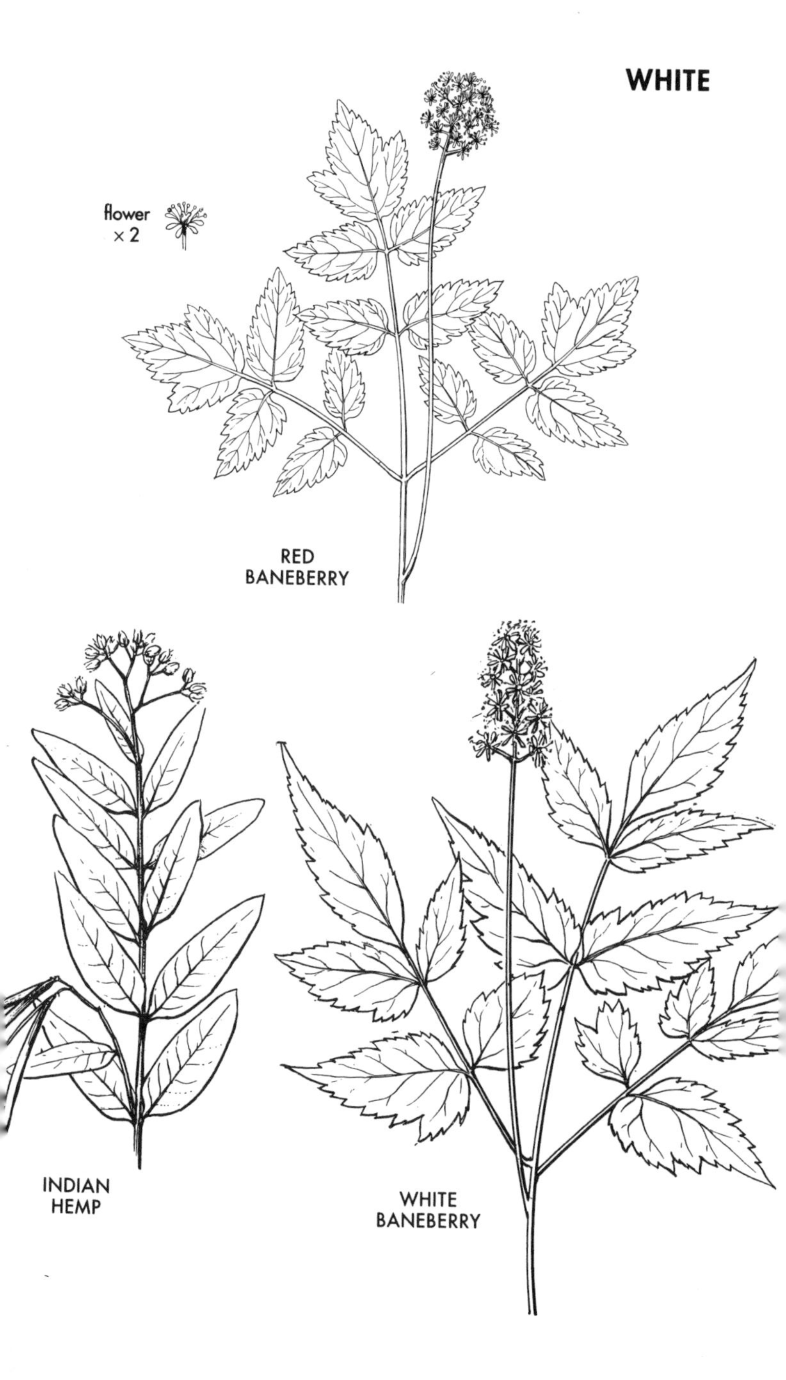
WHITE
flower
× 2
RED
BANEBERRY
INDIAN
HEMP
WHITE
BANEBERRY

3–6-PARTED FLOWERS IN TAPERING CLUSTERS; LARGE LEAVES

COMMON RHUBARB **Leaves, roots**

Rheum x *cultorum* Hort. [*R. rhabarbarum* L.] Buckwheat Family

Large, robust perennial to 8 ft. Leaves very large, broadly oval, wrinkled, with wavy margins; leaf stalks grooved, reddish. Flowers small, whitish, on large conspicuous branching panicles; June–July.

Where found: Old gardens and farm sites. Throughout, except South. Alien cultivar. Europe.

Comments: Leaf stalks (petioles) are a popular spring vegetable with a tart flavor produced by malic acid. The leaves contain soluble oxalates and anthraquinone glycosides. Even small amounts of the leaves can be lethal. Symptoms include burning mouth, stomach pains, vomiting, labored breathing, and internal bleeding.

CALIFORNIA CORN LILY **Whole plant**

Veratrum californicum Dur. Lily family

Tall perennial with *cornlike stalk;* 3–6 ft. Leaves alternate, large, untoothed, with *conspicuous parallel veins.* Flowers in a terminal panicle, 6-petaled; petals with V-shaped green mark at base.

Where found: Moist mountain meadows. Calif.

Comments: Like the European *Veratrum album,* the 8 North American "false hellebores" have highly toxic alkaloids that affect the heart and nervous system. Fringed Corn Lily *V. fimbriatum* (not shown), has *fringed flower petals;* it occurs in cen. Calif. to the Pacific Northwest. Siskiyou Corn Lily *V. insolitum* (not shown), found in openings and thickets in nw. Calif. and sw. Ore., is covered with *short velvety hairs.* Pregnant sheep feeding on early spring leaves of *V. californicum* have produced offspring known as monkey-faced lambs, with malformations of the face and cranium. *Veratrum* has been used in medicine, and its preparations are responsible for some reports of human toxicity. An Asian species *(V. japonicum)* poisoned Korean soldiers who ate soup made from the plant.

POKEWEED, POKE **Pl. 21** **Whole plant**

Phytolacca americana L. Pokeweed Family

Tall, stout, large-rooted perennial; 5–10 ft. Stem smooth, succulent, often reddish. Large leaves alternate, oval to lance-shaped, untoothed, to 10 in., progressively smaller toward top of plant. Flowers greenish white, petallike sepals; July–Sept. Fruits purple-black, many-seeded, inedible berries.

Where found: Waste places. New England to Fla.; Texas to Iowa, Neb. Locally, a West Coast weed.

Comments: Root is especially poisonous. In the South, young spring leaves gathered before stalks turn red are traditionally boiled in several waters as a pot herb. Poke poisoning can cause severe stomach pain with cramps, vomiting, diarrhea, labored breathing, convulsions, and death. Children have died from eating the fruits.

COMMON
RHUBARB
× 1/5
CALIFORNIA
CORN LILY
× 1/4
POKEWEED

FLAT-TOPPED CLUSTERS; LEAVES NOT STRONGLY DIVIDED

WHITE SNAKEROOT **Pl. 30** **Whole plant**
Eupatorium rugosum Houtt. Composite Family

Coarse, widespread, variable, rough perennial; 2–5 ft. Opposite leaves on slender, long stalks; oval, often slightly heart-shaped at base; margins coarsely toothed. Flowers white, tiny, 10–30 in small round heads on branched clusters; July–Oct.

Where found: Dry fields, thickets. Que. to Ga.; Texas to Sask.

Comments: Poisonous to livestock; the toxic element is transferred to humans via cow's milk, causing a condition known as "milk sickness," which claimed thousands of victims in the early 1800s, especially on the western frontier (the modern Midwest). Abraham Lincoln's mother died of the disease when Lincoln was seven years old. The disease mystified researchers for over 100 years, until its cause was finally isolated in 1928 when a USDA researcher discovered tremetol, an alcohol from White Snakeroot that caused "trembles" in laboratory animals. Symptoms include prostration, severe vomiting, tremors, liver dysfunction, constipation, delirium, and death. The disease is rarely encountered today either in livestock or humans, but the possibility exists that cows in the Midwest could eat the plant, imparting the toxins to milk.

FLOWERING SPURGE **Pl. 30** **Whole plant, sap**
Euphorbia corollata L. Spurge Family

Smooth-stemmed, blue-green perennial; 1–3 ft. The alternate *oval to linear* leaves are stalkless. Flowers in branched umbels, white, 5 "petals" (actually bracts) with rounded tips. Note *whorl of reduced leaves* beneath flower heads; June–Aug.

Where found: Dry fields. N.Y. to Fla.; Texas to Minn.

Comments: The milky sap of spurges *(Euphorbias)* contains a toxic component that can cause dermatitis and, if ingested, severe internal poisoning. Symptoms include burning and irritation of the mouth, throat, and stomach.

SNOW-ON-THE-MOUNTAIN **Pl. 30** **Whole plant**
Euphorbia marginata Pursh Spurge Family

Annual, usually *unbranched* herb, with *milky juice;* to 3 ft. Leaves opposite, oblong-oval to linear-oblong, to 5/8 in. long, margins minutely serrated. The *white bracts beneath the flowers* are more showy than the flowers themselves; June–Oct.

Where found: Limey prairies, roadsides, pastures, waste land; native to prairies. Minn. to Texas; N.M. to Mont. Widely cultivated and escaped elsewhere.

Comments: The milky latex (sap) is highly caustic, causing dermatitis. It has even been used to brand cattle. A decoction (tea) of the plant used to induce an abortion resulted in a fatality.

WHITE SNAKEROOT

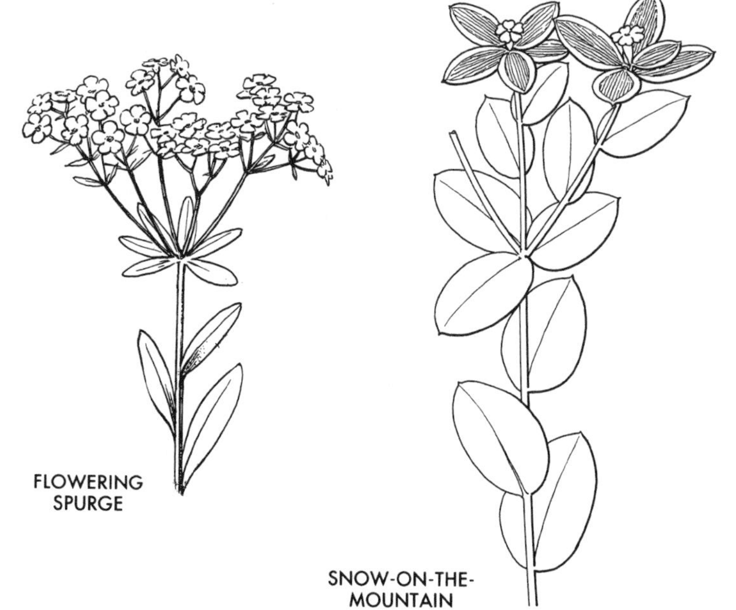

FLOWERING SPURGE

SNOW-ON-THE-MOUNTAIN

UMBRELLALIKE CLUSTERS; LEAVES FINELY DIVIDED

POISON-HEMLOCK **Pl. 21** **Whole plant**
Conium maculatum L. Parsley Family
Smooth-stemmed biennial; purple-streaked or spotted; 4–8 ft. Leaves *carrotlike,* divided into 3–4 segments; up to 4 ft.; leaflets finely divided, tiny. *Foul-scented* when crushed. Flowers white, in umbels; May–Aug.
Where found: Waste ground. Throughout, except deserts. Alien.
Comments: Young plant resembles wild carrot. This is the poison immortalized by Socrates. Seeds and root are especially toxic.

WATER-HEMLOCK **Whole plant**
Cicuta maculata L. Parsley Family
Stout, smooth, *purple-streaked or spotted, rank-smelling* biennial; 3–10 ft. Leaves divided into 2–3 segments; leaflets lance-shaped, coarsely toothed. Flowers in loose, flat umbels; May–Sept.
Where found: Wet meadows, swamps. Throughout. Me. to s. Mexico; Calif. to Alaska.
Comments: One bite of root, which may be mistaken for wild parsnip, may kill an adult. Can cause violent tremors, intense stomach pain, salivation, delirium, and death.

BULB-BEARING WATER-HEMLOCK **Whole plant**
Cicuta bulbifera L. Parsley Family
Smooth stem; 1–3 ft. Fingerlike tubers spread outward from central stalk. Leaves divided into 2–3 segments; leaflets *linear,* slender-toothed, 1/2–2 1/2 in.; upper leaf axils bearing *clustered bulblets.* Flowers in 1 to few umbels, 1–3 in. across; July–Sept.
Where found: Swamps, wetlands. Nfld. to Va.; Ill., Neb., Minn., west to Mont. and Ore.
Comments: See Water-hemlock above.

FOOL'S PARSLEY **Whole plant**
Aethusa cynapium L. Parsley Family
Carrotlike, foul-scented annual; 1 1/2–2 ft. Leaves divided into 2–3 *fine segments.* Flowers white, umbel rays of unequal length, *bracts absent beneath main umbel; present under secondary umbels;* May–Aug.
Where found: Waste places. N.S. to Del., Penn. to Ohio. Alien.
Comments: Mistaken for parsley. May cause vomiting, headache.

WESTERN POISON WATER-HEMLOCK **Whole plant**
Cicuta douglasii (DC) Coult. & Rose Parsley Family
Stout, often purplish herb; 2–6 ft. *Hollow horizontal chambers* in thick base of *mature* stem. Leaves divided into 1–3 segments; leaflets lance-shaped, *saw-toothed,* 1 1/2–4 in. Flowers in flat-topped umbel. Fruits subrounded, with *narrow intervals between corky ribs; reddish brown.*
Where found: Freshwater stream banks, ditches, below 8,000 ft. Nev., e. Mont. to cen. and n. Calif., north to Alaska.
Comments: See Water-hemlock above.

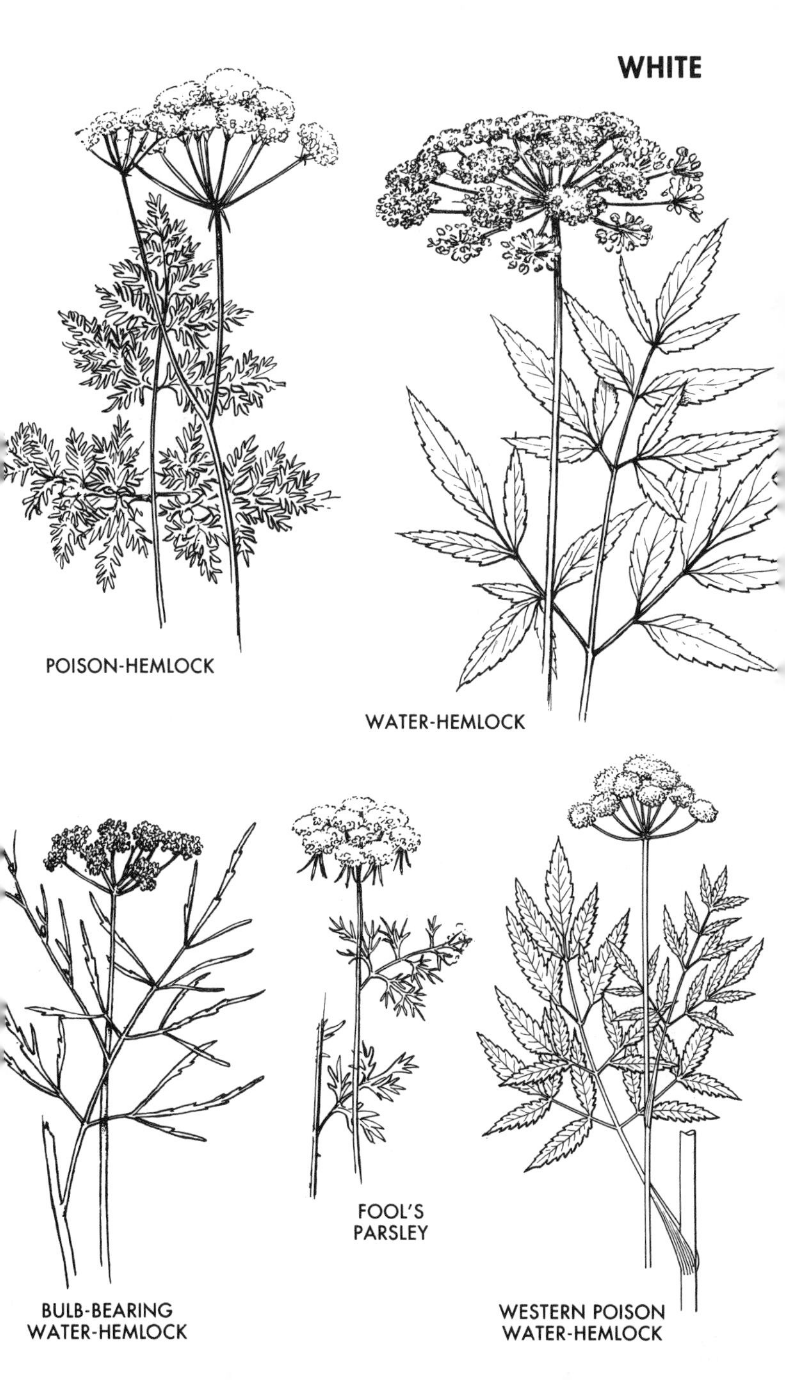
WHITE
POISON-HEMLOCK
WATER-HEMLOCK
BULB-BEARING
WATER-HEMLOCK
FOOL'S
PARSLEY
WESTERN POISON
WATER-HEMLOCK

5-PARTED FLOWERS
HOODED, SPURRED, OR TRUMPETLIKE

JIMSONWEED **Pl. 29** **Whole plant**
Datura wrightii Nightshade Family
(*D. inoxia* P. Mill., *D. meteloides* DC.)
Rank, hairy, branched perennial; to 5 ft. Leaves oval, margins untoothed or *wavy.* Flowers white (or violet-tinged), trumpet-shaped; 6–8 in.; May–Oct. Capsules *nodding, densely prickly.*
Where found: Dry open ground. New England to N.C. (rare), sw. Okla., w. Texas, to Calif. Alien. South America.
Comments: All parts poisonous; see *D. stramonium* below.

JIMSONWEED, DATURA **Pl. 29** **Whole plant**
Datura stramonium L. Nightshade Family
Smooth-stemmed annual; 2–5 ft. Leaves *coarse-toothed, angled.* Flowers white to pale violet, trumpet-shaped, *3–5 in.;* May–Sept. Capsules sharp-spined.
Where found: Waste ground. Throughout. Native to North America. A widespread weed throughout the world.
Comments: A very dangerous weed. Causes severe hallucinations, rapid heartbeat, dry mouth, etc. Most poisonings are the result of intentional ingestion. Livestock avoid it.

TRAILING MONKSHOOD, WOLFSBANE **Whole plant**
Aconitum reclinatum Gray Buttercup Family
Reclining perennial; 3–4 ft. Leaves alternate, palmate, with *3–7 deep lobes.* Flowers white (or yellowish); 5 petallike sepals, the uppermost helmetlike, *twice as wide as long;* axis of flower stalk *downy;* June–Sept.
Where found: Rare. Mountains of W. Va. to w. Va., south to Ga.
Comments: Human poisonings, mostly historical overdoses of medicinal extracts, have caused cardiac arrhythmias. Has caused livestock deaths. Contains the alkaloid aconitine.

CHRISTMAS ROSE, BLACK HELLEBORE **Pl. 31** **Whole Plant**
Helleborus niger L. Nightshade Family
Leathery, semievergreen perennial; 1–2 ft. Divided leaves in 7–9 segments originating from horseshoe-shaped base leaflets oval to lance-shaped, toothed at apex. Flower white (or pinkish), 5-parted, 2–3 in. across. Fruits black capsules.
Where found: Cultivated; sometimes escaped. Alien. Europe.
Comments: Poisoning is rare; may include salivation, tingling of throat and mouth, diarrhea, and vomiting.

PLAINS DELPHINIUM **Whole plant**
Delphinium virescens Nutt. Buttercup Family
Erect herb, $1^1/_2$–3 ft.; *glandular hairs on upper stalk.* Leaves deeply palmate, segments *forked.* Flowers white (or blue-tinted), 5-parted, spur $1^1/_2$ times the length of top sepal; May–July.
Where found: Prairies, open woods. Plains states; Texas, s. N.M., to se. Ariz.
Comments: Young leaves and seeds especially toxic. Often affects grazing cattle. All *Delphiniums* considered toxic. Contains the alkaloids delphinine, ajacine, etc. May cause nervous symptoms, nausea, depression, or death in large doses. See related species on Plate 26.

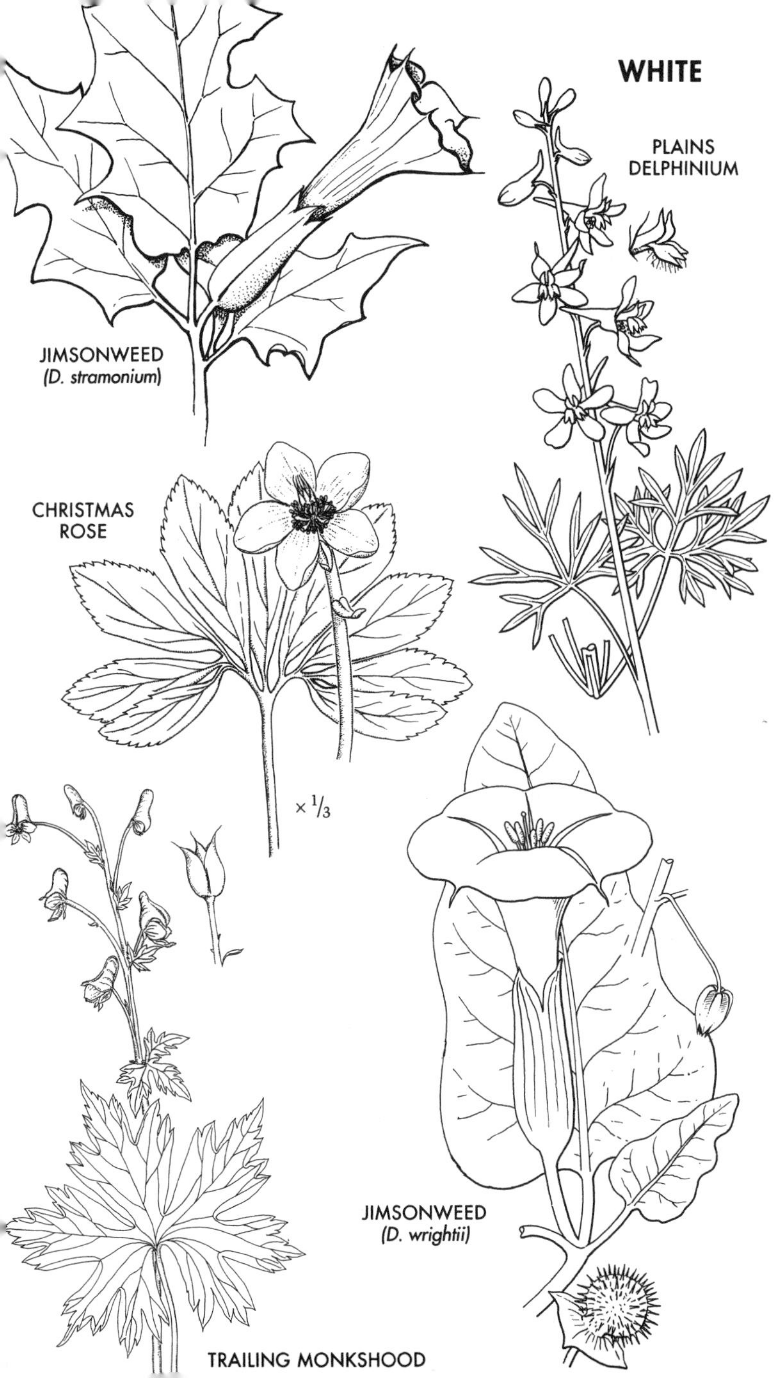
WHITE
PLAINS DELPHINIUM
JIMSONWEED
(D. stramonium)
CHRISTMAS ROSE
× 1/3
JIMSONWEED
(D. wrightii)
TRAILING MONKSHOOD

5-PARTED FLOWERS; NIGHTSHADES

HORSE-NETTLE, CAROLINA NIGHTSHADE **Pl. 29** **Whole plant**
Solanum carolinense L. Nightshade Family

Perennial; 1–4 ft. Stem with *yellow, flattish, sharp spines.* Leaves variable, oval to elliptical, lobed to coarse-toothed. Flowers violet to white; May–Oct. Berries yellow-orange; Aug.–Sept.

Where found: Light soils; waste places, pastures. N.Y. to Fla.; Texas and n. Mexico to S.D.

Comments: Considered a noxious poisonous weed, especially a problem in pastures in S.D. Livestock poisoning common. In 1963, a child died from eating the berries.

GROUND CHERRIES, CHINESE LANTERN *Physalis* spp. **Whole plant**

Usually with yellow flowers, sometimes white. See page 102.

BLACK NIGHTSHADE **Whole plant**
Solanum nigrum L. Nightshade Family

Widely naturalized European species, similar to *S. ptycanthum. S. nigrum* is coarser, leaves thick. Berries dull black.

Where found: Waste places. Throughout.

Comments: Leaves of *Solanum* species are highly variable in alkaloid content. Cases of poisoning of this and many other *Solanum* species, however, are reported. The alkaloid solanidine may cause digestive irritation, trembling, restlessness, etc.

BLACK NIGHTSHADE **Whole plant**
Solanum ptycanthum Dun. ex DC Nightshade Family
(*S. americanum* P. Mill, *S. nigrum* L. var. *virginicum* L.)

Not shown. Annual, tap-rooted herb; 12–24 in. Leaves alternate, smooth or sparsely hairy; shape and size highly variable; margins untoothed to toothed. Flowers white (with yellow star); May–Oct. Berries lustrous purple or black.

Where found: Waste places, fields, and gardens. S. Que. to Fla., Texas, north to Man.

Comments: See *S. nigrum* above.

POTATO **Pl. 33** **Whole plant**
Solanum tuberosum L. Nightshade Family

Familiar perennial widely cultivated as an annual for its tubers; spineless, weak-stemmed; 1–3 ft. Leaves pinnate, to 10 in. long; leaflets oval. Flowers white (or bluish) to 1 in. wide; May–Oct. Berries yellow-greenish.

Where found: Cultivated throughout. Alien. Andes.

Comments: While the tubers are edible, all green parts of the plant, which contain the highest concentrations of the alkaloid solanine (including the green skin of some potatoes), are potentially toxic. Sprouted or green-skinned potatoes have caused most reported poisoning cases.

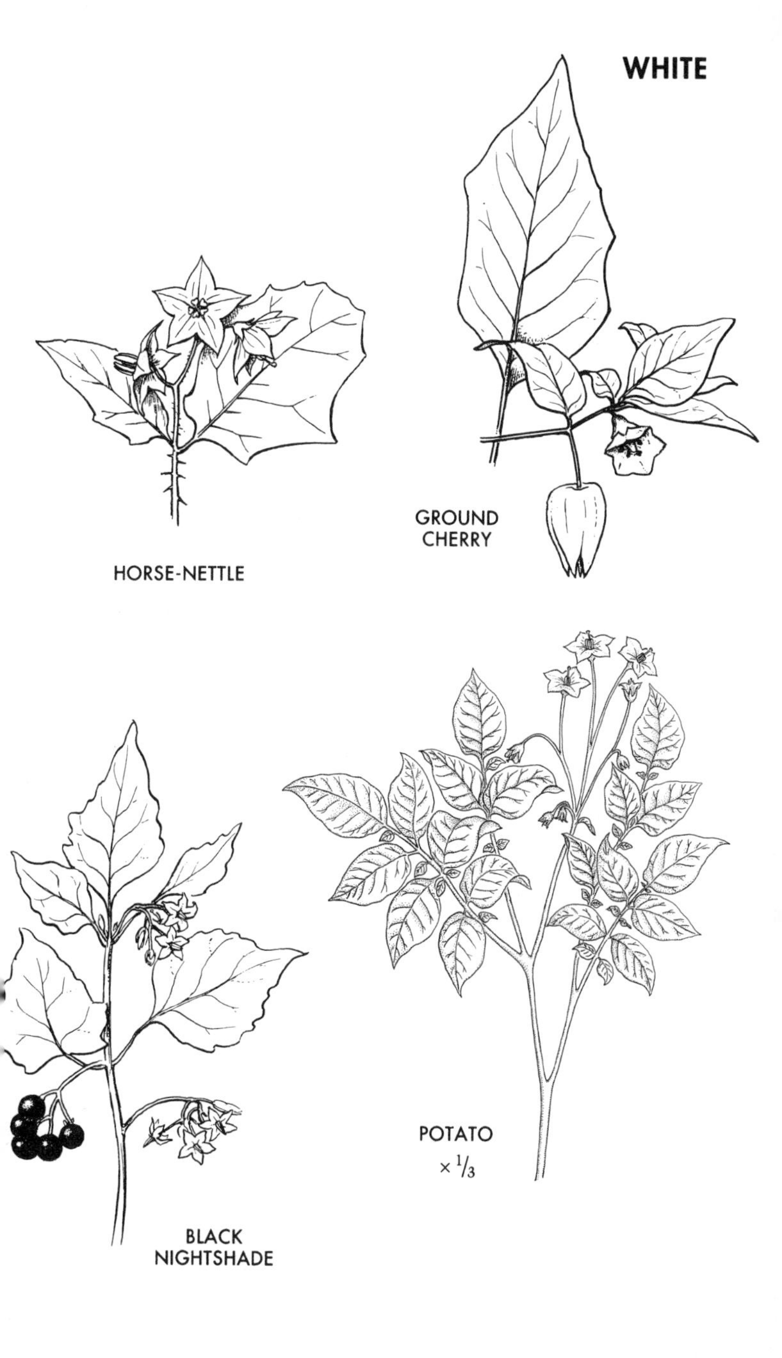
WHITE
GROUND
CHERRY
HORSE-NETTLE
POTATO
× 1/3
BLACK
NIGHTSHADE

MILKWEEDS
5 PETALS, FLOWERS IN UMBEL, MILKY SAP

Milkweeds, which generally have milky sap, have distinctive flowers in loose to crowded umbels. The 5 petallike parts of the flower curve downward beneath the extra floral parts (5 inflated hoods with horns) that make up the corona, which surrounds a central column with the stamens and pistil. It sits atop a short pedestal. Minute sepals are hidden beneath the downcurved petals. Some species contain poisonous cardiac glycosides. When Monarch butterfly larvae feed on milkweed leaves containing glycosides, the toxins remain concentrated in the bodies of the mature butterflies, making them unpalatable or poisonous to potential predators such as birds.

WHORLED MILKWEED **Whole plant**
Asclepias verticillata L. Milkweed Family
Slender perennial; 15–34 in. Leaves whorled (3–6), linear or nearly threadlike. Flowers in upper leaf axils, 6–20 blooms per group; petals white to greenish white (purple-tinged); June–Sept.
Where found: Dry slopes, open woods. Mass., N.Y., south to Fla., west to Ariz., north to Mont. and s. Man.
Comments: Ingestion of the green plant equal to 2 percent of an animal's weight can be fatal. Galitoxin, a resinous substance isolated from the plant, is thought to cause symptoms such as weakness, spasms, and seizures induced by milkweed poisoning. It is probably the primary toxin, though other components, such as glycosides, may be involved in cases of livestock poisoning. Most species of milkweeds can be expected to be more or less toxic, with reactions varying from minor to severe, depending upon the amount consumed.

PLAINS MILKWEED **Whole plant**
Asclepias pumila (A. Gray) Vail Milkweed Family
Slender perennial; 3–12 (15) in. tall. Leaves in whorls (3–5), strongly spreading or upward-curved. Flowers white (tinted rose to yellow-green), with short greenish white pedestal; horns curve over central column; July–Sept.
Where found: Sandy or limey clay soils. S. Great Plains to nw. Texas, e. N.M.
Comments: Plains Milkweed has been known to cause deaths of livestock. Ingesting enough of the green plant to equal 1–2 percent of the animal's body weight can be fatal.

SHOWY MILKWEED **Whole plant**
Asclepias speciosa Torr. Milkweed Family
Stout perennial, upper stalk densely hairy; 20–40 in. Leaves usually opposite, broad lance-shaped to oval-oblong, smooth to velvety; 3–8 in. long, 1–3 in. wide; apex variable, acute to broadly rounded. Flowers usually pale pink, sometimes whitish, few to many in upper leaf axils, hoods *toothlike* and strongly *outward curving;* May–Aug.
Where found: Moist sandy loam to rocky soils; along waterways. S. Canada, prairie states; west to Wash., cen. Calif., to s. N.M.
Comments: Rarely implicated in poisoning, but potentially toxic.

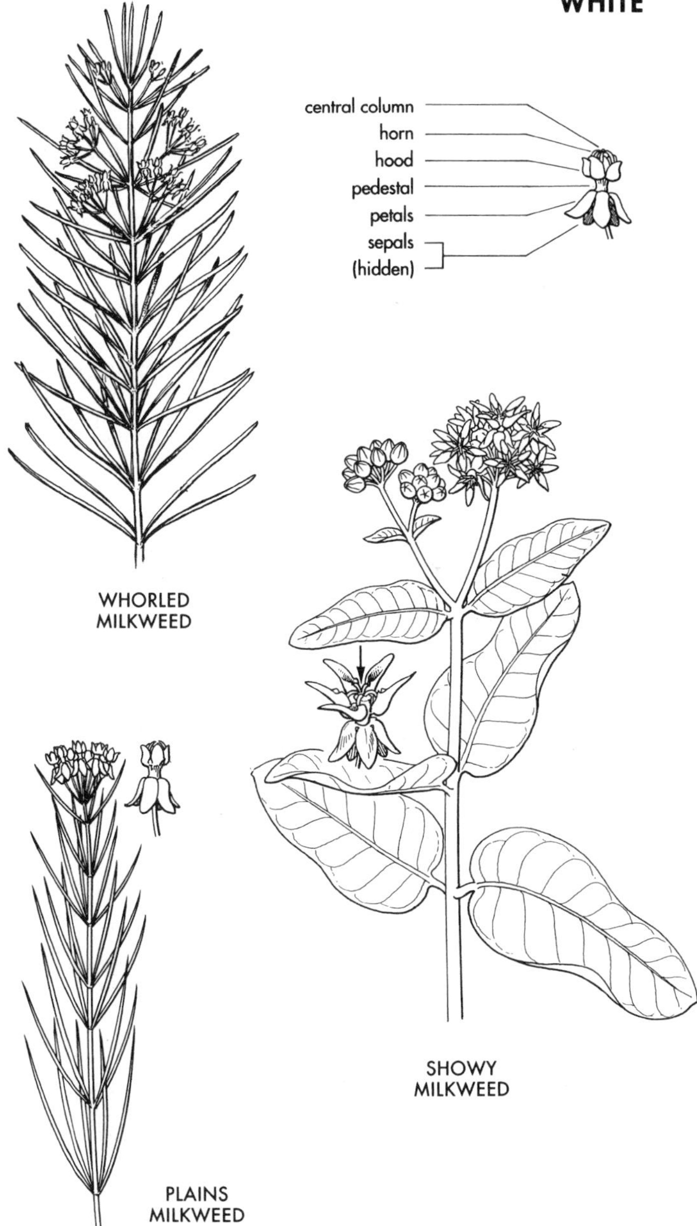
central column
horn
hood
pedestal
petals
sepals
(hidden)
WHORLED
MILKWEED
SHOWY
MILKWEED
PLAINS
MILKWEED

ASTRAGALUS
PEALIKE FLOWERS, INFLATED SEED PODS

WOOTON LOCO, GARBANCILLO **Whole plant**
Astragalus wootonii Sheldon Pea Family
Coarse, leafy, hairy, sprawling or prostrate annual or biennial; 3–20 in. Leaves greenish yellow; 11–23 leaflets linear–oblong, smooth, *usually folded.* Loose racemes of 2–10 white flowers; March–July. Pods inflated.
Where found: Deserts. W. Texas, N. M., Ariz., Se. Calif.
Comments: See White Dwarf Locoweed below.

WHITE DWARF LOCOWEED **Whole plant**
Astragalus didymocarpus Hook. & Arn. Pea Family
Variable, slender, often bending or prostrate herb separated into four varieties. Leaves pinnate, with (9) 11–17 linear-oblong to wedge-shaped leaflets, with *squared* apex. Flowers white (lavender-tinged), in dense racemes of 7–30 flowers; calyx densely white- or black-hairy; Feb.–May. Pods plumply oval, squat, *wrinkled,* usually hairy.
Where found: Grassy hillsides, rolling plains, gravel washes, deserts. E. Texas to s. Ariz., Baja Calif. to cen. Calif.
Comments: Locoweeds poison livestock in Rocky Mountain states and westward, affecting animals in different ways. Locoism makes animals act in an irrational manner; they may seek out locoweeds over palatable forage. Locoism brings about structural changes in nerve cells. It has caused many livestock deaths.

SPECKLED LOCO **Whole plant**
Astragalus lentiginosus Dougl. ex Hook. Pea Family
Highly variable; over 30 varieties. A genetic jigsaw puzzle that includes annuals, biennials, and perennials, with white to pink-purple flowers. Typical white-flowered plants are spreading or prostrate. Leaves pinnate; 7–27 leaflets. Flowers short, in small racemes; not elongated when in fruit; May–July. Pods *spotted or freckled.*
Where found: Dry plains, deserts, sagebrush; in mountains below the 40th parallel. Wash. to Idaho; Calif., Utah to B.C.
Comments: See White Dwarf Locoweed above.

NARROW-LEAVED POISON VETCH **Whole plant**
Astragalus pectinatus Dougl. ex. G. Don. Pea Family
Stout, mostly smooth, sprawling or prostrate herb, forming *bushy depressed mats;* 6–24 in. Leaves with 7–15 opposite, stiff, pale *green-gray, linear* leaflets; 1½–2 in. long. Flowers cream-white in dense racemes of 12–30 flowers; May–June. Pods drooping, plumply oblong, to 1 in. long, sides somewhat flattened, smooth.
Where found: Clay or sandy soils, gullies, barren hilltops, prairies. Sw. Man. to w. Kans., north through e. Colo., Wyo., Mont., to s. Alta.
Comments: Accumulated selenium in plant causes chronic syndromes in livestock such as "blind staggers" and "alkali disease."

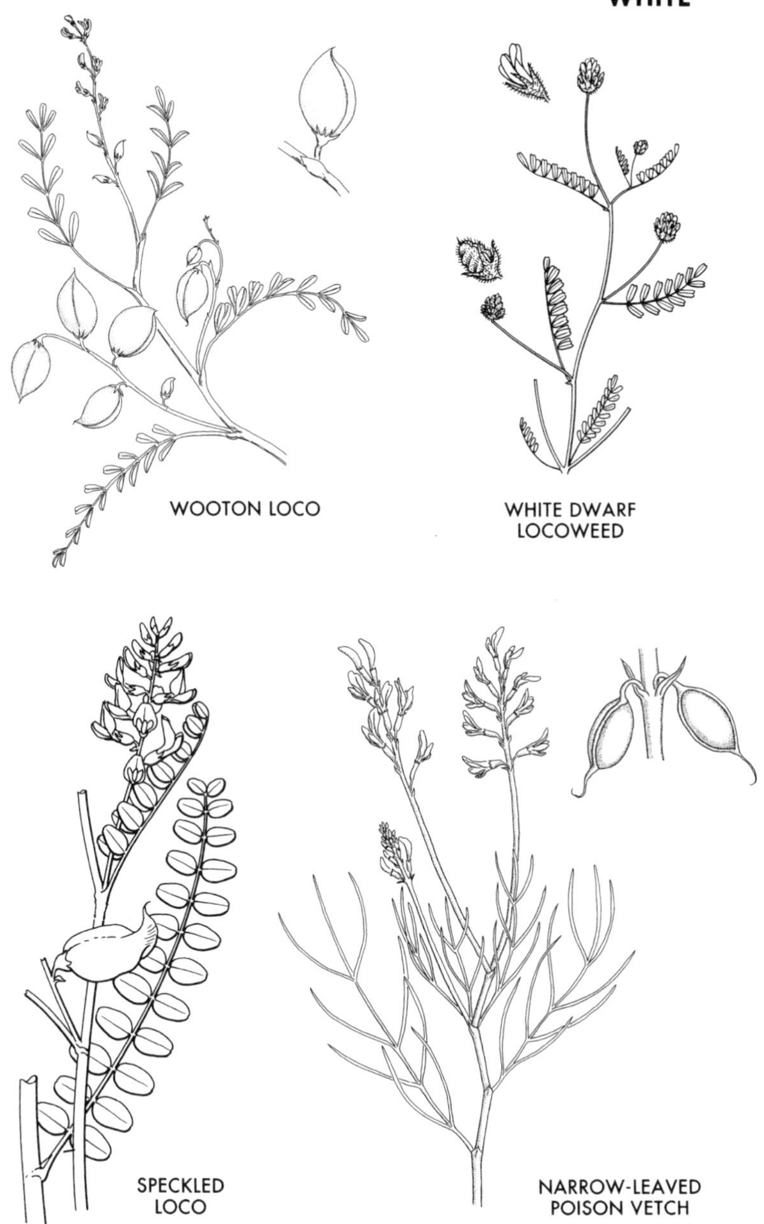
WOOTON LOCO
WHITE DWARF
LOCOWEED
SPECKLED
LOCO
NARROW-LEAVED
POISON VETCH

PINE LUPINE **Whole plant**
Lupinus albicaulis Dougl. ex. Hook. Pea Family
Stout, branched perennial with sparse hairs lying flat to stem; 20–36 in. Leaves palmate; 5–9 lance-shaped leaflets broadest at apex. Flowers *dirty white* to purplish, fading to brown, in *loose racemes;* April–Aug.
Where found: Dry grassy hillsides and openings, 2,000–8,500 ft., in yellow pine woods. Cen. Calif. to w. Wash.
Comments: See Whitewhorl Lupine below.

WHITE WILD INDIGO **Pl. 26** **Whole plant**
Baptisia alba (L.) Vent. Pea Family
[*B. lactea* (Raf.) Theriet; *B. leucantha* T. & G.]
Perennial, stems and leaves with grayish white, waxy film, main stem solitary, branching above; 3–5 ft. Leaves 3-compound. Flowers white, on a single stout raceme; May–July.
Where found: Ravines, streambanks, roadsides. Ont., Ohio to Miss.; Texas to Minn., se. Neb., Mich.
Comments: White Wild Indigo is cited as causing death of cows that ate hay containing the plant. May cause diarrhea and loss of appetite. Contains quinolizidine alkaloids.

FAVA BEAN, HORSE BEAN **Beans, pollen**
Vicia faba L. Pea Family
Coarse, erect, annual vine, without tendrils; to 6 ft. Leaves alternate, compound; 2–6 oval leaflets. Flowers in leaf axils, 1 in. long, off-white, with *dull purple dot;* June–Aug.
Where found: Widely cultivated as forage crop, especially in South. Alien. N. Africa, sw. Asia.
Comments: While the beans are sold canned and frozen, they are dangerous to individuals with an inherited red blood cell enzyme deficiency called favism. Within 5–24 hours after eating beans, some may experience headache, dizziness, vomiting, fever, jaundice, anemia, and death. Inhaling pollen may affect the same individuals. Most serious in children; most common in boys. The active components are cyanogenic glycosides.

WHITEWHORL LUPINE **Whole plant**
Lupinus densiflorus Benth. Pea Family
Hairy-stemmed annual; 8–16 in. Leaves palmate, with 7–9 smooth, lance-shaped leaflets broadest at apex. Flowers white (or tinted with rose to violet veins), with 5–12 blossoms in *whorls;* April–June.
Where found: Grassy fields, open hillsides, mixed evergreen woods. Calif.
Comments: Contains anagyrine, a toxic alkaloid. Ingestion of large amounts (especially seeds or seed pods) may be dangerous to livestock, although this species has not been involved in livestock poisoning. Cows eating toxic lupines may give birth to deformed calves. One case of human birth defects resulting from the mother drinking lupine-contaminated milk has been reported.

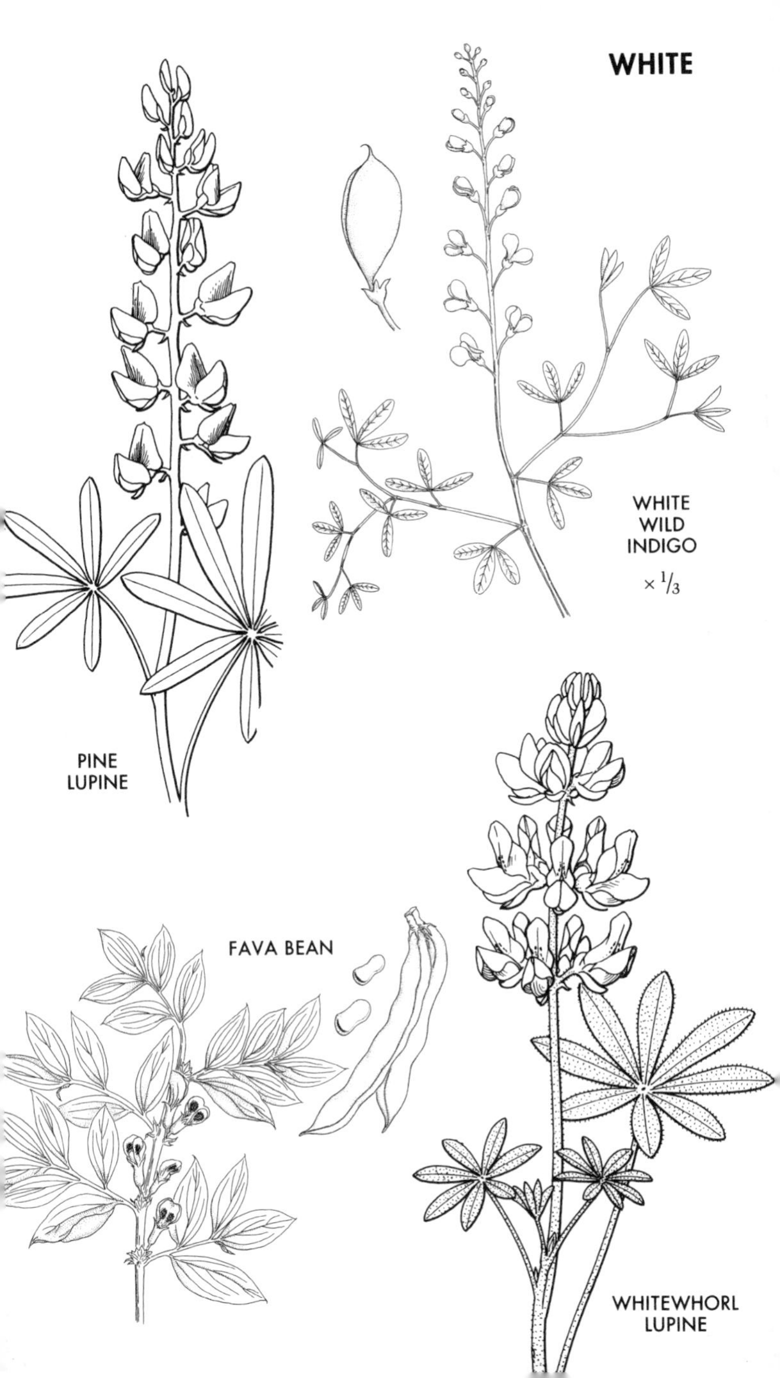
WHITE
WHITE
WILD
INDIGO
× 1/3
PINE
LUPINE
FAVA BEAN
WHITEWHORL
LUPINE

6-PARTED FLOWERS; LEAVES GRASSLIKE OR WITHOUT TEETH

FLY POISON **Bulb**

Amianthium muscaetoxicum (Walt.) Gray Lily Family
[*Zigadenus muscaetoxicum* (Walt.) Zimm.]

Perennial; 1½–4 ft. Leaves linear, mostly basal, up to 2 ft. long, 1 in. wide, much reduced on flower stalk. Flowers white, in terminal raceme to 5 in. long; May–July.

Where found: Low sandy soils, open woods. S. N.Y., Penn. to Va., Fla., west to Ark., Mo., Okla.

Comments: All parts, especially the bulb, contain poisonous alkaloids including zygadenine and zygacine. Poisoning same as in Death Camases (see next page).

STAR-OF-BETHLEHEM **Whole plant, flowers, bulbs**

Ornithogalum umbellatum L. Lily Family

Smooth perennial; to 10 in. Leaves basal, linear, *channeled*. Flowers white, 6-parted stars on *leafless* corymb rising from bulb; petals with *green stripe*; March–June.

Where found: Roadsides, lawns, waste places. Nfld. to N.C., Miss.; Mo., Kans. to Ont. Alien. Europe.

Comments: Consumption of the flowers and the small onionlike bulbs has caused human poisoning, especially in children. The bulbs may also kill livestock, especially sheep, after frost heaves the bulbs to the soil surface. May cause gastrointestinal disorders and nausea. Tulipalin A, which causes dermatitis from handling tulips, is also reported from this plant.

LILY OF THE VALLEY **Whole plant**

Convallaria majalis L. Lily Family

Perennial, spreading by root runners; 4–8 in. Leaves 2–3, basal, oblong-oval, without teeth, *parallel veined*; to 1 ft. Flowers white, aromatic, drooping, somewhat bell-shaped; May–June.

Where found: Widely cultivated; naturalized. Alien. Europe.

Comments: Ingestion of any plant part may cause digestive disturbance, purging, irregular heartbeat, coma, and death. Contains cardiac glycosides such as convallarin, convallamarin, and convallatoxin.

NARCISSUS, JONQUILS, DAFFODILS **Pl. 31** **Bulb, leaves**

Narcissus spp. L. (*N. poeticus* L). Amaryllis Family

The genus *Narcissus* contains about 26 species with hundreds of cultivars with white, yellow, or orange blooms. Perennials with onion-shaped bulbs; 8–18 in. Leaves basal. Flowers one to several on a scape. The 6-parted petals sit beneath a long or short, large or small, tubular corona. Typically flowers in early spring.

Where found: Commonly cultivated; escaped. Alien. W. and cen. Europe.

Comments: A short time after ingesting small amounts of the plant, humans have developed symptoms such as diarrhea, vomiting, and sweating. Fatalities reported. Contains toxic alkaloids and needlelike calcium oxylate crystals, which may cause external irritation from handling the plant. Bulbs have been mistaken for onions *(Allium cepa)*, causing poisoning.

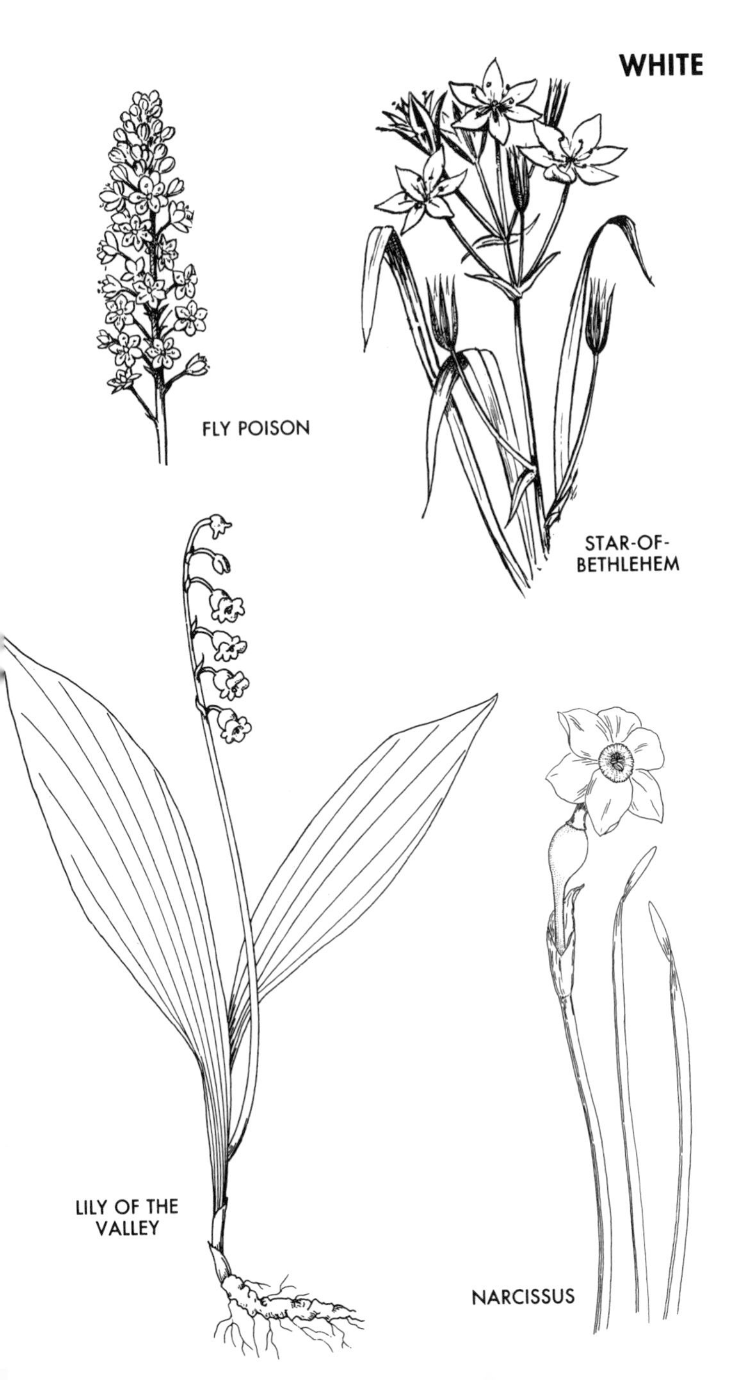
FLY POISON
STAR-OF-BETHLEHEM
LILY OF THE VALLEY
NARCISSUS

DEATH CAMAS
6-PARTED STARS IN RACEMES OR PANICLES

Zigadenus species (sometimes misspelled *Zygadenus*) are smooth, onionlike perennials with linear, mostly basal leaves and *no onion scent*. Flowers have 6 segments often fused at the base. One or two yellow-green glands are often present on lower bases of petals. Bulbs and new foliage are especially poisonous. Human poisonings result from confusing Death Camases with wild onions (*Allium* spp.) or edible Camas (Wild Hyacinth *Camassia scilloides* and related species). Symptoms include dry, burning mouth, thirst, headache, dizziness, severe vomiting, cardiac irregularities, loss of muscle control, and in severe cases coma and death. Famous source of sheep deaths in the West. Contains zygadenine, a toxic steroidal alkaloid

DESERT CAMAS **Whole plant**
Zigadenus brevibracteatus (M.E. Jones) Hall Lily Family
To 20 in.; basal leaves to 1 ft. long, margins rough. Flowers cream-white to yellowish; petal gland flattened, U-shaped; April–May.
Where found: Sandy soil. Mountains. Mojave Desert. S. Calif.

ELEGANT CAMAS, WHITE CAMAS **Whole plant**
Zigadenus elegans Pursh Lily Family
Slender herb; 10–34 in. Leaves thin, with *very narrow tip.* Flowers in *slender* loose raceme; may have small dark gland; June–July.
Where found: Limey prairies. Mo., w. Texas, Ariz. to Alaska.

NUTTALL'S CAMAS **Whole plant**
Zigadenus nuttallii Gray ex S. Wats. Lily Family
Stout; 12–30 in. Leaves papery. Flowers in *thick, cylindrical raceme,* petals 1/4 in. long, *narrowed* at base, with *round yellow* gland; March–May.
Where found: Prairies, rocky soils. Tenn. to Kans.; Texas.

DEATH CAMAS **Whole plant**
Zigadenus venenosus S. Wats. Lily Family
Slender to stout; 4–15 in. Leaves up to 12 in. long. Flowers numerous in a thick spike, *stamens exceeding petals;* petals obtuse, gland globe-shaped; March–June.
Where found: Sask. to s. Neb., Colo., Calif., to B.C.

CROW POISON **Whole plant**
Zigadenus densus (Desr.) Fern. Lily Family
To 40 in.; bracts persistent. Flowers in dense simple raceme; sepals and petals 3/8 in. long; very small gland at base of fresh petals; May–June.
Where found: Pine barrens, coastal plains. Va., Fla., to e. Texas.

FREMONT'S CAMAS **Whole plant**
Zigadenus fremontii (Torr.) Torr. ex S. Wats. Lily Family
Smooth herb; 12–40 in. Leaves 8–24 in. long, folded and arching, margins rough. Flowers cream-white, in loose spike; stamens half the length of petals; glands present; March–May.
Where found: Grassland, chaparral, thickets. Coast Ranges. S. Calif. to s. Ore.

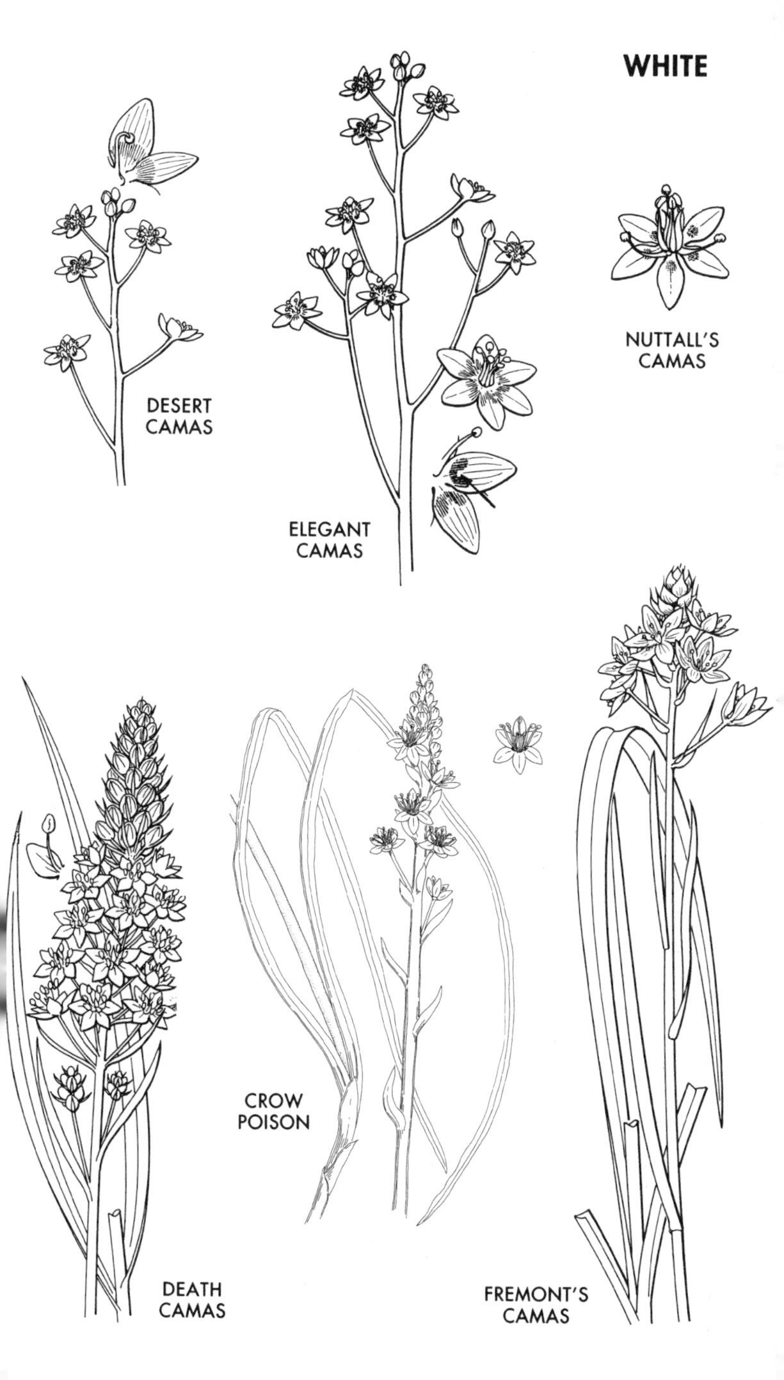
WHITE
DESERT CAMAS
ELEGANT CAMAS
NUTTALL'S CAMAS
DEATH CAMAS
CROW POISON
FREMONT'S CAMAS

SHOWY FLOWERS WITH 6 OR MORE PETALS

WESTERN PASQUE FLOWER **Whole Plant**
Anemone occidentalis S. Wats. Buttercup Family
[*Pulsatilla occidentalis* (S. Wats.) Freyn.]

Perennial; 6–20 in. Finely divided leaves with *lacy segments.* Flower white to yellow-green, strongly cup-shaped; May–Sept. Seeds with wispy feathery plumes.

Where found: Alpine and subalpine meadows. Sierra Nevada in Calif. to B.C.

Comments: See Pasque Flower below.

BLOODROOT **Pl. 19** **Root, leaves**
Sanguinaria canadensis L. Poppy Family

Perennial; 6–12 in. *Orange juice* in stems; roots *blood red* within. Leaves with deep or shallow sinuses. Flowers white, 8–10 petals, to 2 in., *usually appearing before leaves;* March–June.

Where found: Rich woods. N.S. to Fla.; e. Texas to s. Man.

Comments: Contains alkaloids similar to those of the Opium Poppy, including sanguinarine, which can depress the central nervous system. Overdoses cause vomiting, irritation of mucous membranes, diarrhea, fainting, shock, and coma. Most poisonings reported were from medicinal preparations.

MAYAPPLE **Pl. 19** **Root**
Podophyllum peltatum L. Mayapple Family

Perennial; 10–18 in. Leaves paired, *umbrellalike,* stem attached to leaf center from beneath. Flowers white, *waxy,* solitary, 8-petaled, *nodding from division of leaf stems;* April–May. Fruit like a small crabapple; edible when fully ripe; June–July.

Where found: Woods, clearings. Often in large clumps. S. Me. to Fla.; Texas to Minn.

Comments: The root components produce a toxic effect on cell division, and are used as anticancer agents in chemotherapy. Improperly administered, however, the root can be a deadly poison, causing respiratory stimulation, vomiting, catharsis, coma, or death. Unripe fruits have caused painful digestive disturbances. Handling roots may cause dermatitis.

PASQUE FLOWER **Whole Plant**
Anemone patens L. Buttercup Family
[*Pulsatilla patens* (L.) P. Mill.]

Small perennial; 2–16 in. *Silky* leaves basal, divided into 3–7 *linear* segments. Flowers white to purple; to 1½ in. wide; petals arising from *cup-shaped receptacle;* March–June. Seed heads with feathery plumes.

Where found: Open prairies. Wisc. to s. Ill., Mo., N.M. to Wash. and Alaska. Also in Eurasia.

Comments: Touching or eating any plant part may produce toxic effects. External symptoms include acute inflammation, blistering; if ingested, may cause irritation of the mucous membranes, ulceration, and severe digestive irritation, including bloody vomiting and diarrhea. Toxins are protoanemonins, found in many buttercup family members (Ranunculaceae). See Creeping Buttercup, page 104.

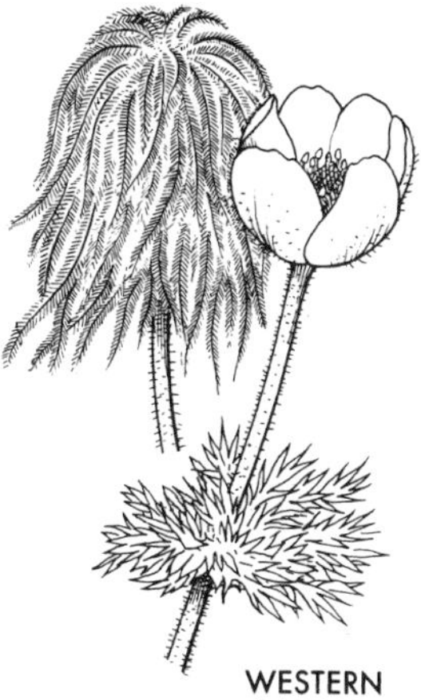

WESTERN PASQUE FLOWER

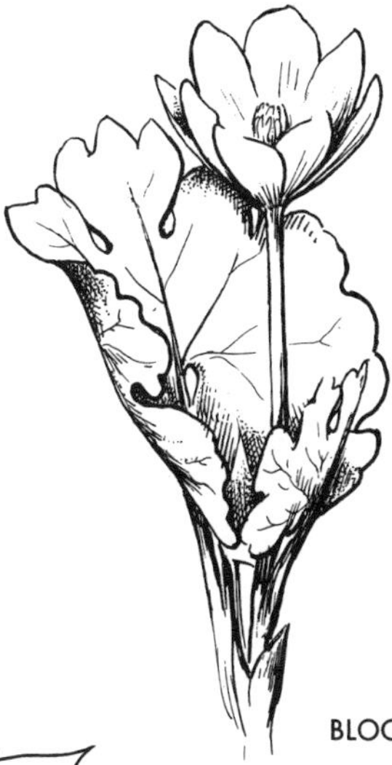

BLOODROOT

PASQUE FLOWER

MAYAPPLE

MISCELLANEOUS SHOWY FLOWERS

DAFFODIL **Pl. 31** **Bulb, leaves**

Narcissus pseudonarcissus L. Amaryllis Family

A commonly grown, yellow-flowered *Narcissus.* Onion-shaped bulbs; leaves basal, *broad, grasslike.* Flowers solitary; 6-parted petals beneath a prominent, tubular corona; March–April.

Where found: Cultivated; escaped. Alien. W. and cen. Europe.

Comments: See Narcissus, page 94.

PHEASANT'S EYE **Whole plant**

Adonis vernalis L. Buttercup Family

Small perennial; 1½ ft. Leaves *feathery,* with linear lobes. Flowers yellow, solitary, terminal, *daisylike,* to 3 in.; April–June.

Where found: Cultivated. Rarely escaped. Alien. Cen., se. Europe.

Comments: Few cases of poisoning reported, none in North America. Contains more than 20 poorly absorbed cardiac glycosides, especially adonitoxin. A European drug plant.

MARSH MARIGOLD **Whole Plant**

Caltha palustris L. Buttercup Family

Hollow-stemmed, succulent perennial; 10–24 in. *Kidney-shaped, glossy leaves,* to 7 in. wide. Flowers bright yellow, with 5–9 "petals" (sepals), to 2 in. across; stamens numerous; April–June.

Where found: Swamps, wet ditches. Nfld. to N.C., west to Neb., north to Alaska.

Comments: Bruised leaves develop the irritant protoanemonin. May cause inflammation or ulceration of the mouth and throat, gastroenteritis. See Creeping Buttercup, page 104.

CELANDINE **Pl. 22** **Whole plant**

Chelidonium majus L. Poppy Family

Brittle-stemmed biennial with *orange-yellow juice;* 1–2 ft. Leaves divided, with *irregular round lobes.* Flowers yellow, 4-petaled, to ¾ in.; April–Aug. Seed pods smooth, erect, to 1½ in.

Where found: Waste ground. Que. to Ga.; Mo. to Iowa. Alien. Eurasia.

Comments: Used medicinally. Stem juice may irritate skin. Plant contains numerous alkaloids, especially chelidonine. Overdoses said to cause vomiting, bloody diarrhea, circulatory disorders, etc., but such poisoning is apparently rare.

YELLOW PRICKLY POPPY **Whole plant**

Argemone mexicana L. Poppy Family

Annual with *bright yellow latex;* to 30 in. Leaves with whitish film that rubs off, prickly on veins above; clasping, lobed; margins spine-tipped. Flowers *bright yellow;* to 2¾ in. wide, 20–75 stamens; May–Sept.

Where found: Waste places, roadsides, fields. New England to Ill., to Texas.

Comments: Isoquinoline alkaloids found in whole plant, seeds. Prickles discourage ingestion. See Prickly Poppy, page 74.

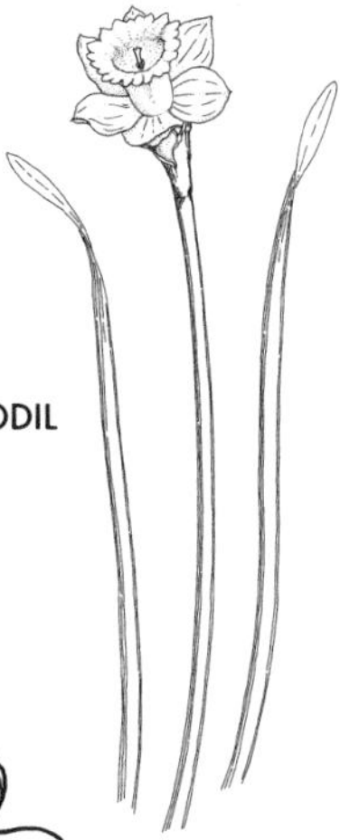

DAFFODIL

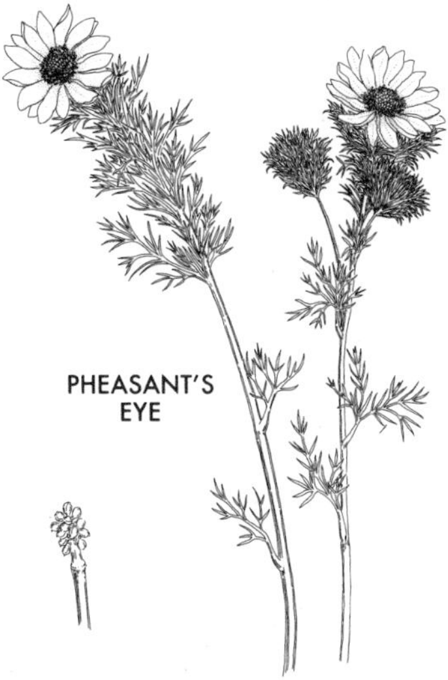

PHEASANT'S EYE

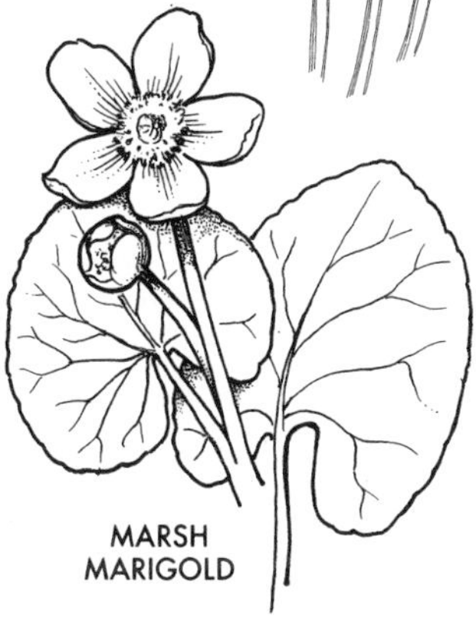

MARSH MARIGOLD

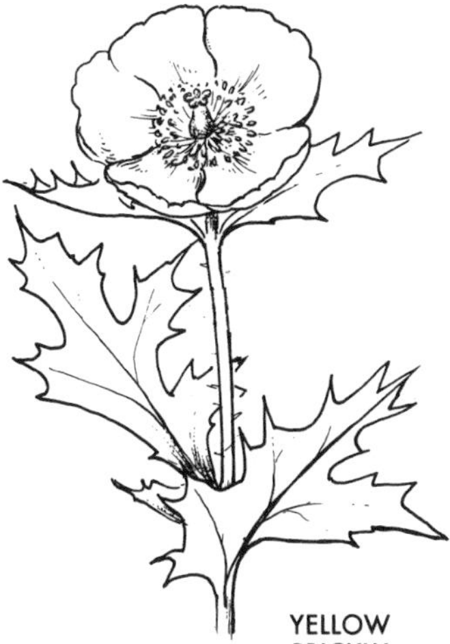

YELLOW PRICKLY POPPY

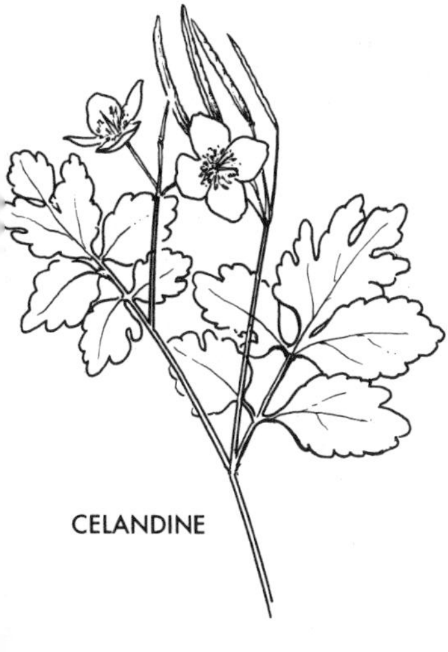

CELANDINE

BUFFALO BUR NIGHTSHADE, KANSAS THISTLE **Whole plant**
Solanum rostratum Dun. Nightshade Family
[*S. cornutum* Lam., *Androcera rostrata* (Dunal) Rydb.]
Coarse annual; 12–28 in. with radiating stem hairs. Leaves alternate, 1–2 pinnate-lobed, with spines *from central vein.* Flowers bright yellow, 5–15 on each raceme, petals *widely flared;* May–Oct. Fruits *yellow-spined.*
Where found: Waste places. Great Plains; naturalized elsewhere.
Comments: Spines may cause mechanical injury. Generally not ingested because of spines. Hogs have been poisoned from dried leaves in feed or from unearthing the roots. Causes severe enteritis (inflammation of the intestines) and hemorrhage.

TOMATO **Leaves**
Lycopersicon esculentum P. Mill. Nightshade Family
[*Lycopersicon lycopersium* (L.) Karst. ex Farw.]
The familiar garden tomato. Clammy, downy, sprawling, spineless annual. Leaves variable, compound, toothed or lobed. Flowers deeply 5-lobed. Fruit red, succulent.
Where found: Cultivated throughout. Alien. S. America.
Comments: Dried or fresh leaves of immature plants contain a steroidal glyco-alkaloid, and have produced symptoms of solanine poisoning in livestock. Alkaloids disappear after fruits ripen. Leaves cause dermatitis. Children have been poisoned from drinking leaf tea.

HENBANE **Whole Plant**
Hyoscyamus niger L. Nightshade Family
Annual or biennial; 8–30 in. Leaves oval-oblong, strongly wavy; lower leaves short-stalked, upper ones somewhat clasping. Flowers dirty yellow, with *violet veins;* inner base *dark purple;* June–Sept. Globular capsule with flared persistent calyx, many black seeds.
Where found: Infrequently cultivated; sporadically escaped throughout. Alien. Europe.
Comments: Long used in medicine. Contains the toxic alkaloids hyoscyamine, scopolamine, and atropine. Ingestion may cause salivation, headache, nausea, hallucinations, increased cardiac output, coma, and death. Poisoning now rare.

GROUND CHERRIES, CHINESE LANTERN **Pl. 33** **Whole plant**
Physalis spp. Nightshade Family
About 80 species, annual or perennial, with alternate, simple, soft-textured, coarse-toothed leaves. Flowers mostly single, in leaf axils, usually yellow (blue or white), dark and hairy in center. Fruits yellowish to red globular berries *enclosed in yellowish to red inflated bladders* (the calyx). We have 28 native or naturalized species.
Comments: Fruits or leaves may cause gastroenteritis, burning in throat, fever, and diarrhea. Unripe fruits are more dangerous than ripe fruits; ripe fruits are considered edible in some species.

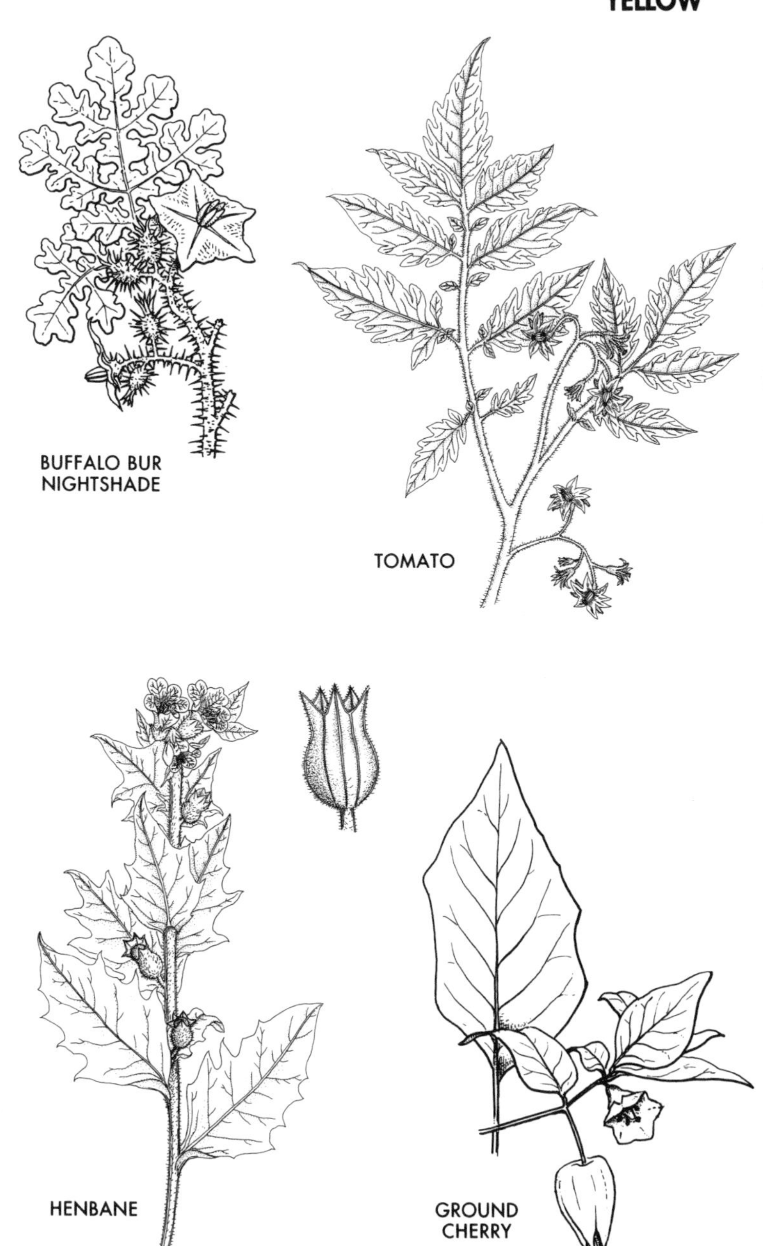
BUFFALO BUR
NIGHTSHADE
TOMATO
HENBANE
GROUND
CHERRY

BUTTERCUPS
5 OR MORE PETALS, STAMENS NUMEROUS

There are over 90 indigenous or naturalized species of *Ranunculus* (buttercups) in North America. Buttercups include annual or perennial aquatic plants, forest species, and cosmopolitan weeds of meadows, pastures, lawns, and waste places, with about 250 species in temperate climates worldwide. These herbs have alternate simple or compound leaves with yellow, white, or sometimes red sepals and petals. Normally there are five (or more) petals and numerous stamens. Fruits are a head of achenes (small, dry, nonsplitting, one-seeded fruits).

COMMON BUTTERCUP **Fresh plant juice**

Ranunculus acris L. Buttercup Family

Perennial with erect branched stems; 11–30 in. Lower leaves palmate, 5–7 main segments, divided further into narrow, toothed lobes. Flowers golden yellow, usually 5 petals (or as many as 20); May–Aug. Fruits, a head of one-seeded achenes, *smooth, flattish,* with erect beak and discernible margins.

Where found: Roadsides, fields, clearings. Cosmopolitan weed throughout. Alien. Eurasia.

Comments: See Creeping Buttercup below.

CREEPING BUTTERCUP **Fresh plant juice**

Ranunculus repens L. Buttercup Family

Highly variable, hairy, trailing perennial, flowering stems erect; 2–3 ft. Leaves dark, sometimes white-mottled, strongly 3-divided; leaflets irregularly divided and toothed, with obvious stalks. Flowers yellow, 5–9 petals (more in double-flowered varieties); sepals hairy and spreading; Feb.–Sept. Seed heads globe-shaped, seed oval with somewhat triangular erect beak, hooked at tip.

Where found: Moist disturbed ground, old gardens, roadsides, ditches. Found in much of our area; common in Pacific Northwest. Alien. Europe.

Comments: All buttercup species are thought to be toxic to some degree. The sap of the fresh plants contains protoanemonin, an acrid skin irritant. When dry it is converted into an inactive component, anemonin. Amount of protoanemonin in various *Ranunculus* species is highly variable. Fresh sap is highly irritating to the skin and mucous membranes, causing blistering or ulceration. Livestock have been poisoned by grazing on large amounts of the fresh plant, but such cases are rare, given the acrid flavor of the plants. Symptoms of fresh plant poisoning may include gastrointestinal irritation, colic, diarrhea, or nephritis; in severe cases it may cause paralysis of the central nervous system. Beggars in Europe are said to have applied the juice of buttercups to their feet, producing blisters to gain deeper sympathies. The poisonous roots have been eaten by children, producing toxic reactions.

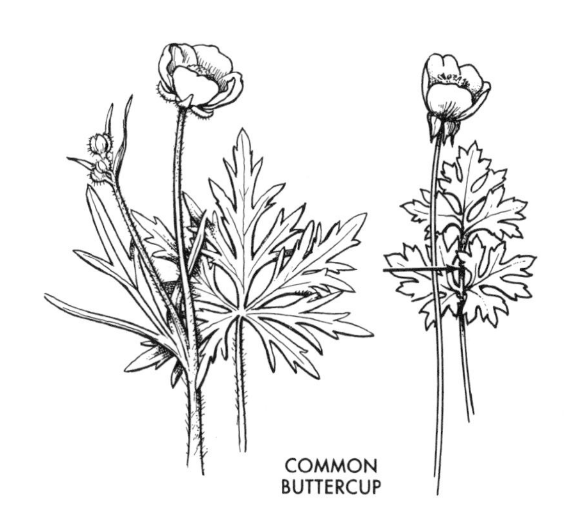
COMMON
BUTTERCUP

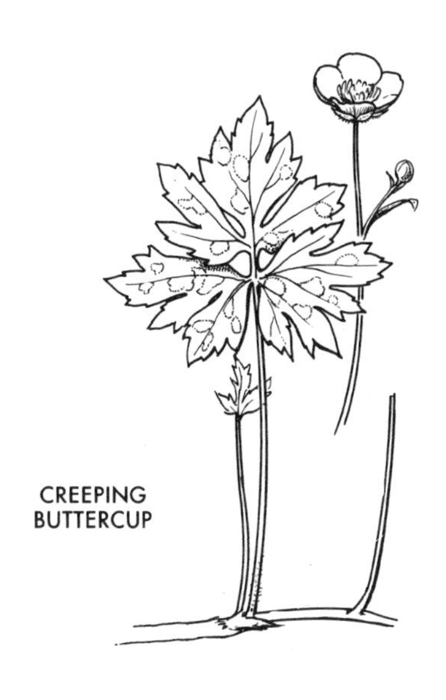
CREEPING
BUTTERCUP

SPECKLED LOCO **Whole plant**

Astragalus lentiginosus Dougl. ex. Hook. Pea Family

Highly variable; with more than 30 varieties. A genetic jigsaw puzzle; includes biennials, perennials, and annuals, with white to pink-purple flowers.

Where found: Dry plains, deserts, sagebrush; in mountains below the 40th parallel. Wash. to Idaho; Calif., Utah to B.C.

Comments: Causes locoism. See White Dwarf Locoweed, page 90.

WILD INDIGO **Whole plant**

Baptisia tinctoria (L.) R. Br. Pea Family

Smooth, blue-green perennial; 1–3 ft. Leaves nearly stalkless, cloverlike. Flowers yellow, pealike, loose, in racemes on upper branchlets; May–Sept.

Where found: Dry clearings. S. Me. to Va.; Fla. to se. Minn.

Comments: Various *Baptisia* species have been implicated in livestock poisoning. Human poisonings usually from overdoses of medicinal preparations. Used in plant medicines in modern Europe.

CREAMY POISON MILKVETCH **Whole plant**

Astragalus racemosus Pursh Pea Family

Stout, coarse, leafy herb, with three varieties; 10–28 in. Leaves to 7 in.; 11–29 lance-shaped to elliptical leaflets, flat or loosely folded, tips notched. Flowers 20–70 in dense racemes; whitish yellow with lilac-veined tips; May–July. Peapodlike fruits.

Where found: Hills, barren limestone. Plains states. N.D. to Texas; N.M., w. Utah, and Wyo.

Comments: Accumulates selenium from soil. Acute selenium poisoning (rare) may cause breathing difficulties, depression, coma, and death. Chronic poisoning in livestock causes "blind staggers" and "alkali disease," including emaciation, diarrhea, rapid and weak pulse, hoof loss, and eventual death from thirst, starvation, or failure of heart and lungs.

LAYNE'S LOCOWEED, LAYNE MILKVETCH **Whole plant**

Astragalus layneae Greene Pea Family

Low, coarse, *gray-hairy* perennial; 2–12 in. Primary root woody with branching, twinelike rhizomes, forming colonies. Leaves pinnate; leaflets 13–21, variable, mostly oval, to 1 in. Flowers 15–45 in loose racemes, light yellow, wing petals with *purple tips;* March–May. Pods *purple-spotted, curved,* to 2½ in. long.

Where found: Sandy desert flats, Creosote scrub. Common in Mojave and Colorado deserts. S. Calif., w. Nev., and nw. Ariz.

Comments: Contains aliphatic nitro compounds (a third group of *Astragalus,* in addition to selenium-concentrating species and locoweeds). Poisoning is acute, appearing in less than a week or, in cattle, in 4–20 hours. Symptoms in livestock include labored breathing with wheezing ("roaring") and loss of motor control, causing knocking of hind feet ("cracker heel"). Animals become so weak they may die if forced to move.

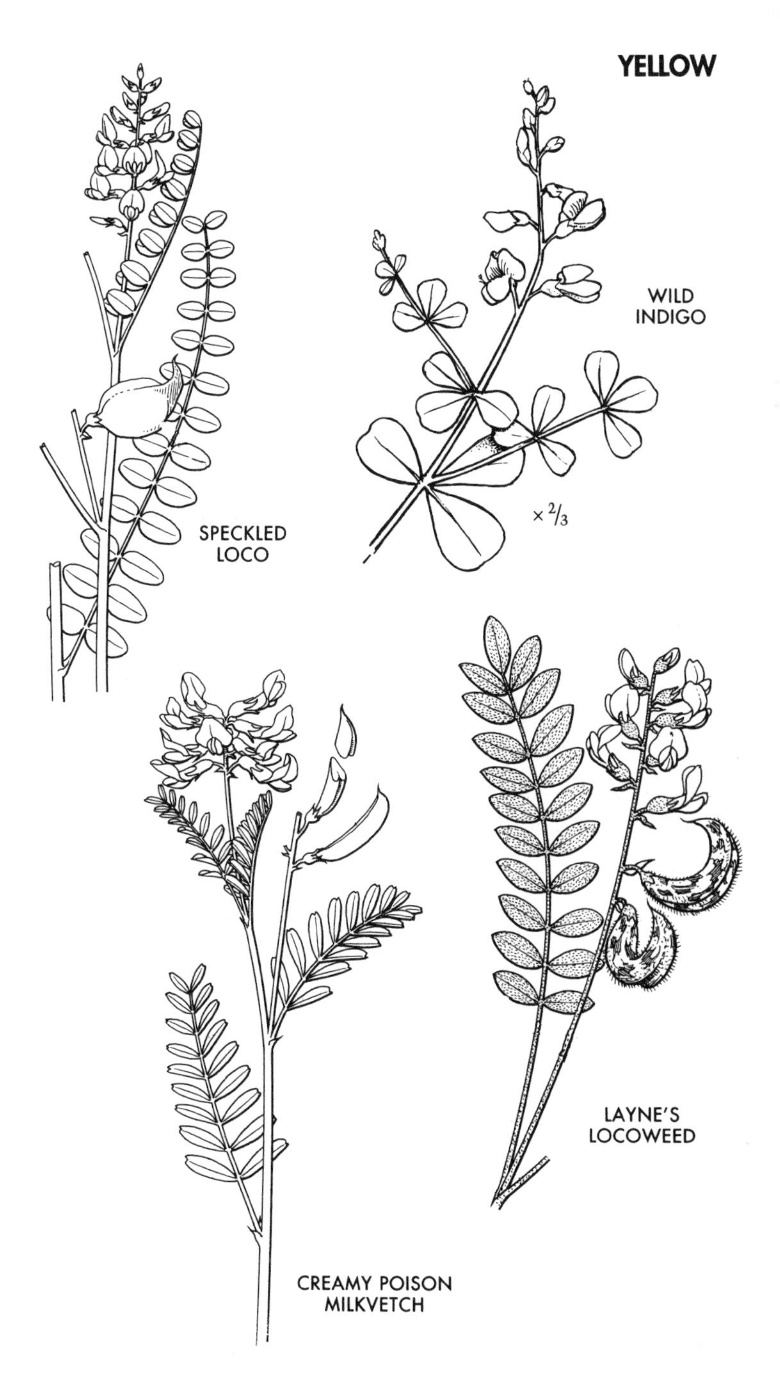
YELLOW
WILD
INDIGO
× 2/3
SPECKLED
LOCO
LAYNE'S
LOCOWEED
CREAMY POISON
MILKVETCH

PEALIKE FLOWERS

SCOTCH BROOM **Whole plant**
Cytisus scoparius (L.) Link Pea Family
Evergreen shrub, *stems angled;* 3–10 ft. Leaves 3-divided; leaflets tiny, $^1/_4$–$^1/_2$ in. long. Flowers yellow, to 1 in., single or paired in axils, *style strongly curved;* May–June. Fruits are flat legumes to 2 in. long.
Where found: Pine barrens, sandy soil, roadsides. Eastern and western North America. Alien. Europe. Invasive weed in Calif.
Comments: Contains the alkaloids cytisine (causes abdominal pain and diarrhea) and sparteine (impairs heart function). Cytisine is rapidly absorbed in the digestive tract, including the mouth, causing abdominal pain and diarrhea. Reports of human poisoning are rare and suspect.

RATTLEBOX **Whole plant**
Crotalaria sagittalis L. Pea Family
Bushy annual or short-lived perennial with *downward pointing arrowlike stipules;* 4–15 in. Leaves alternate, simple, lance-shaped to elliptical. Flowers yellow (fading to white), in 1- or 2–4-flowered racemes; May–Sept. Black pods rattle when dry.
Where found: Dry rocky prairies, fields. Eastern U.S. to e. Great Plains, southward.
Comments: Growing in field bottoms, this plant is the probable source of "bottom disease," which killed horses in the 19th century. All livestock, including poultry, are susceptible to poisoning. Contains pyrrolizidine alkaloids. May cause liver lesions (veno-occlusive disease of the liver). Human poisoning has resulted from seed-contaminated grain or folk remedies.

BUTTER LUPINE **Whole Plant**
Lupinus luteolus Kellogg Pea Family
Widely branched, hairy annual; 1–3 ft. Leaves bright green, palmate with 7–9 lance-shaped leaflets broadest at apex. Flowers pale or light yellow in *whorls on long raceme;* May–Aug. Pods to $^1/_2$ in. long.
Where found: Dry flats, hillsides. Yellow pine and juniper thickets. Calif. to s. Ore.
Comments: Ingestion of large amounts of any of the 200 or so species of *Lupinus* (80 in California alone), eaten over a short period of time, can kill livestock. Human poisoning rare, mostly resulting from seed or pod ingestion by children. Symptoms include nervousness and excitability, followed by depression, difficult breathing, muscle spasms, convulsion, and coma. All Lupines are potentially toxic.

SULPHUR LUPINE **Whole Plant**
Lupinus sulphureus Dougl. ex Hook. Pea Family
Perennial with *silver-gray* stems and leaves; 1–3 ft. Leaves palmate, with 7–9 lance-shaped leaflets. Flowers *sulphur colored,* in irregular, loose raceme; April–June.
Where found: Dry soils. Great Basin.
Comments: See Butter Lupine above.

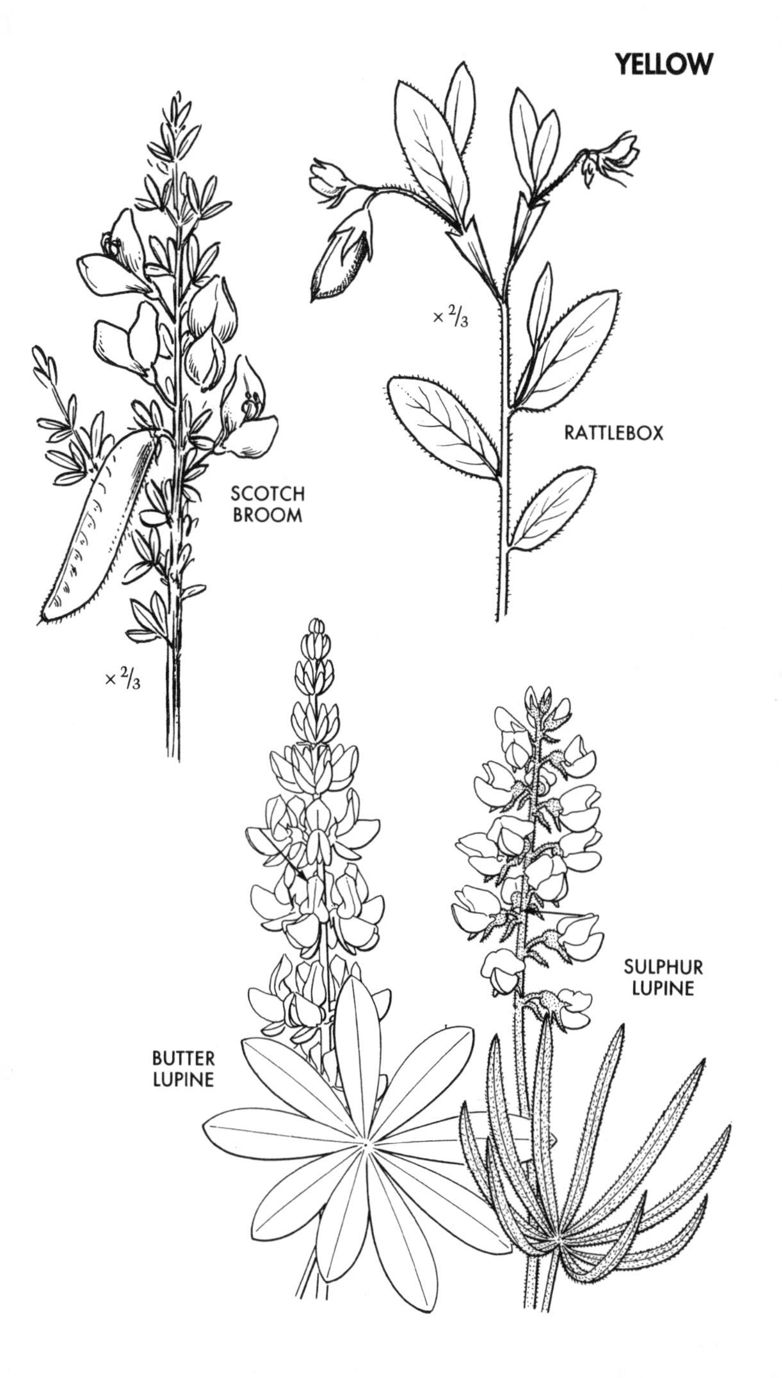
YELLOW
× 2/3
RATTLEBOX
SCOTCH
BROOM
× 2/3
SULPHUR
LUPINE
BUTTER
LUPINE

COMPOSITES WITHOUT SHOWY RAY FLOWERS

COMMON TANSY **Pl. 24** **Whole plant**
Tanacetum vulgare L. Composite Family

Strongly aromatic perennial; to 4 ft. Leaves finely divided into *fern-like* segments. Flowers yellow *buttons* in flat-topped clusters; July–Sept.

Where found: Roadsides, waste places. Ne. U.S., se. Canada. Alien. Europe.

Comments: Ingestion of the highly concentrated essential oil or a strong extract from Tansy, formerly used as an abortifacient and worm expellant, has been implicated in human fatality. The essential oil of Tansy contains a monoterpene called thujone, which is a toxic constituent of several plants including Wormwood *(Artemisia absinthium).* Thujone may stimulate the central nervous system partially by stimulating the reflexes of the respiratory system. Some distinct chemical types of Tansy are apparently free of thujone. It is theorized that thujone may interact with the same brain receptor sites affected by the active chemical compounds of marijuana.

GROUNDSEL **Whole plant**
Senecio vulgaris L. Composite Family

Annual, 4–12 in. Leaves lance-shaped overall with deep round or sharp lobes; lowermost without stalks, uppermost clasping. Flowers tufted, in loose clusters, without rays, *bracts with black tips;* May–Oct.

Where found: Weed of rich soils. Throughout. Alien. Eurasia.

Comments: Liver-damaging pyrrolizidine alkaloids widely present in *Senecio* species. Causes liver lesions (veno-occlusive disease of the liver) in animals or in humans who ingest the plants in the form of folk medicines. The alkaloids are also known to be carcinogenic. They have been identified in more than 100 of the 1,300 species of *Senecio* that grow throughout the world. Symptoms of *Senecio* poisoning appear only after several weeks or months and include loss of appetite, exhaustion, abdominal pain and swelling, and eventually liver enlargement, hardening, and cirrhosis.

RAYLESS GOLDENROD, JIMMY WEED **Whole plant**
Haplopappus heterophyllus (Gray) Blake Composite Family
[*Isocoma wrightii* (Gray) Rydb.]

Perennial, usually smooth-stemmed, branching subshrub; 12–30 in. Leaves on stems linear, erect, $^{1}/_{12}$–3 in., with stalks covered with resin dots. Flowers goldenrodlike, in loose clusters, with few to numerous heads; June–Sept.

Where found: Disturbed sandy soils. Cen. Texas panhandle west to Ariz.

Comments: Contains an alcohol, tremetol. Ingestion of the green plant over a period of several weeks may cause symptoms and finally death in livestock. Obvious trembling is a primary symptom. Poisoning is most common in fall and winter, when other plants are not available. Human poisoning is unknown.

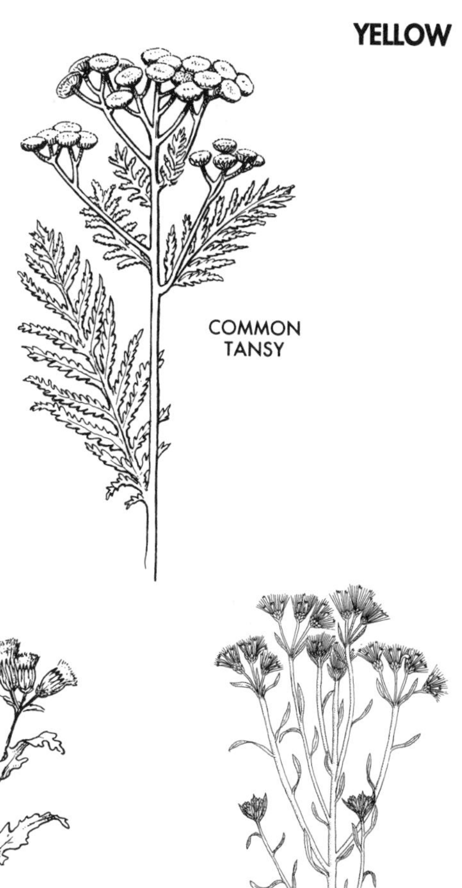
COMMON
TANSY

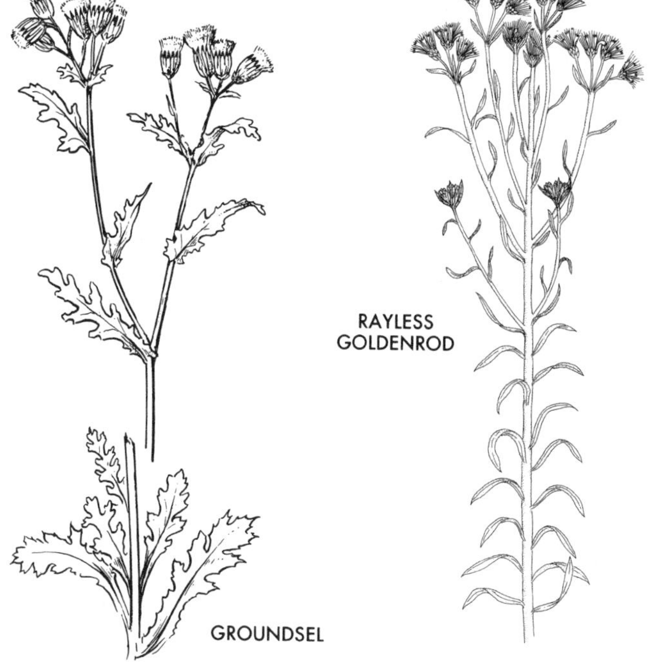
RAYLESS
GOLDENROD
GROUNDSEL

ORANGE SNEEZEWEED, MOUNTAIN HELENIUM **Whole plant**
Helenium hoopesii Gray Composite Family
[*Dugaldia hoopesii* (Gray) Rydb.]

Stout perennial; 15–36 in. Leaves thickish, lance-shaped and broadest at apex, 5–10 in. long, untoothed, stalked. Flower heads showy, to 3 in., *flat* (spreading); rays *narrow*; disk to 3/4 in. high; June–Sept.

Where found: Mountain meadows. Wyo. to N.M.; cen. Calif to Ore.

Comments: Usually avoided by animals. Causes vomiting ("spewing sickness") in sheep. Hymenovin, a sesquiterpene lactone, is the toxic component. May cause depression and tremors.

TANSY RAGWORT, TANSY BUTTERWEED **Pl. 22** **Whole plant**
Senecio jacobaea L. Composite Family

Biennial or perennial; 3–4 ft. Leaves smooth, divided, with large rounded terminal lobe and smaller lobes beneath. Flowers all-yellow, daisylike, to 1 in. across, in broad corymbs; July–Oct.

Where found: Dry soils. North of N.J. Alien. Eurasia.

Comments: Known to cause livestock poisoning. Contains at least six pyrrolizidine alkaloids, especially jacobine.

GOLDEN RAGWORT **Pl. 24** **Whole plant**
Senecio aureus L. Composite Family

Highly variable perennial; 2–4 ft. Leaves of two shapes: basal leaves *heart-shaped;* stem leaves lance-shaped, incised. Flowers yellow, daisylike, in flat-topped clusters; March–July.

Where found: Stream banks, moist soils. P.E.I. to Fla.; west to Texas, Iowa, north to Minn.

Comments: *Senecio* species often contain pyrrolizidine alkaloids. Consumption of the plant or tea may cause acute veno-occlusive liver lesions, which can lead to cirrhosis and, in severe cases, death.

BITTER SNEEZEWEED **Pl. 24** **Whole plant**
Helenium amarum (Raf.) H. Rock Composite Family

Strongly bitter-scented annual; 8–12 in., with *ribbed stems.* Leaves *stringlike.* Yellow flower heads to 3/4 in. across, with globe-shaped cone, petals triangular, strongly toothed at tip; June–Nov.

Where found: Dry poor pastures, roadsides. Va. to Fla.; Texas to e. Kans.

Comments: Livestock may die from eating the plant, but they usually avoid it. Tenulin (a sesquiterpene lactone) is the primary toxic component.

ARNICA **Pl. 24** **Whole plant**
Arnica spp. L. Composite Family

The genus includes 32 species with *all-yellow flowers, opposite leaves,* and hollow, minutely barbed or feathery hairs atop seeds.

Where found: Primarily mountains of western North America.

Comments: The best-known species, European *A. montana,* is a medicinal plant. Toxic element is helenalin. May cause a toxic-allergic skin reaction. Overdoses have caused rapid pulse, heart palpitation, shortness of breath, and death. Not eaten by livestock, as it occurs in alpine habitats where livestock is not pastured.

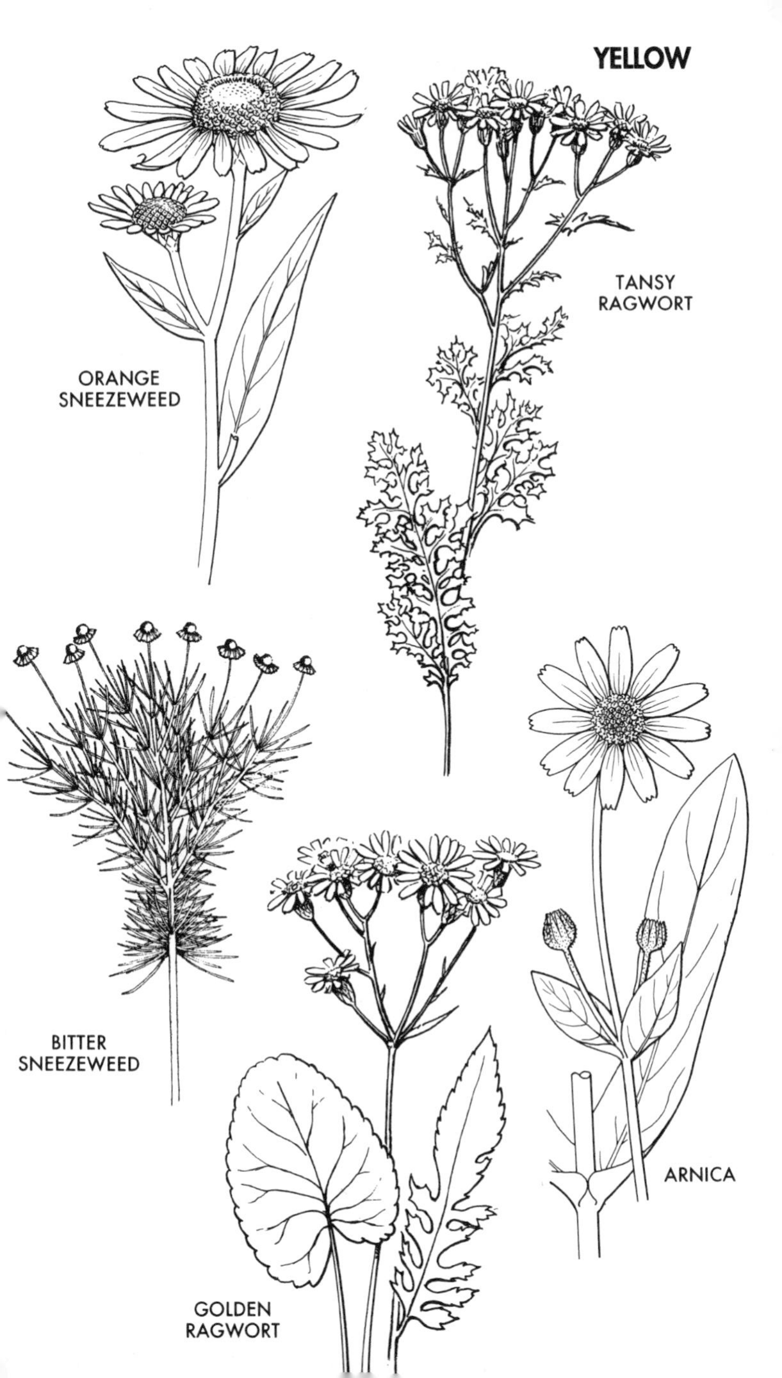
YELLOW
TANSY
RAGWORT
ORANGE
SNEEZEWEED
BITTER
SNEEZEWEED
ARNICA
GOLDEN
RAGWORT

MISCELLANEOUS ORANGE FLOWERS

CANON DELPHINIUM **Whole plant**

Delphinium nudicaule Torr. & A. Gray Buttercup Family

Erect, smooth perennial; to 2 ft. Basal leaves 3–5 palmate-divided; lobes shallow; stem leaves reduced. Flowers orange-red, *cornucopia-shaped,* in few-flowered spikes; *spurs straight;* March–June. See related species on Plate 26.

Where found: Open mountain woods and foothills. Calif. to sw. Ore.

Comments: Larkspurs (*Delphinium* species) are second to locoweeds in causing poisoning of grazing animals in the West. Entire plant, especially young leaves and seeds, is toxic. All *Delphiniums* are considered potentially toxic to varying degrees.

LANTANA **Pl. 31** **Leaves, unripe fruits**

Lantana camara L. Vervain Family

Hairy herb or subshrub, stems *square,* unarmed or prickly; 1–4 ft. Leaves opposite or whorled, oval to oblong-oval, to 5 in. long, with rounded teeth; rough above; fragrant. Flowers *orange-yellow* (sometimes changing to red or white), in *flat-topped clusters* to 2 in. across; May–Oct. Fruits clusters of small black berries.

Where found: Cultivated throughout; naturalized in South from Fla. to Texas. Invasive in Hawaii and tropics. Alien. Tropical America.

Comments: In the wild, Lantana is poisonous to grazing animals. May produce jaundice, photosensitivity, gastrointestinal disturbances, and constipation. Lantadene A and B are considered the major toxic components. Ingestion of the unripe (green) fruits has been reported in the poisoning of two small children in Tampa, Florida. Ripe fruits are questionably edible. Leaves may cause dermatitis.

BUTTERFLY WEED, PLEURISY ROOT **Pl. 27** **Leaves**

Asclepias tuberosa L. Milkweed Family

Familiar midwestern and eastern wildflower. Hairy perennial, *without milky juice;* to 3 ft. Leaves crowded, *surrounding stem,* lance-shaped to oblong-lance-shaped, up to 5 in. long. Orange flowers (rarely yellow) in showy, flat-topped, or somewhat rounded clusters; May–Sept.

Where found: Dry soils. S. Me. to Fla.; Texas, Kans., to Minn. Often cultivated as an ornamental.

Comments: Root used as a medicinal plant, primarily for lung ailments. Overdoses of the root are reportedly toxic. Humans are not known to ingest the leaves. Nineteenth-century medical literature generally refers to the safety of root in medicinal preparations rather than noting any toxicity. Nevertheless, most milkweeds are considered at least potentially toxic to some degree.

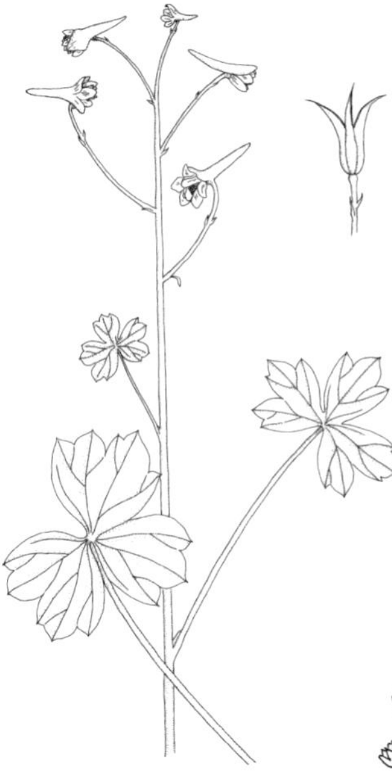

CANON DELPHINIUM

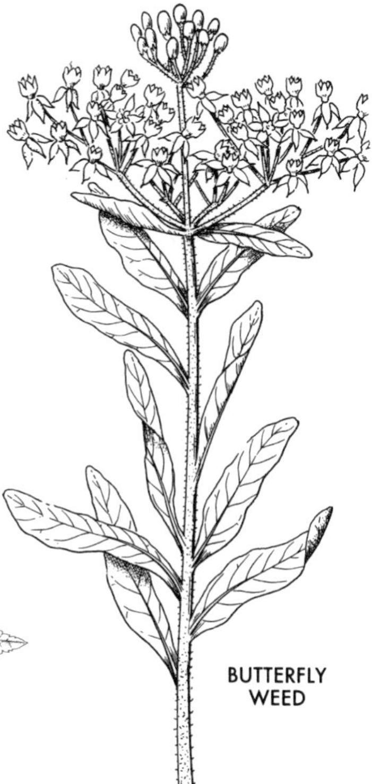

BUTTERFLY WEED

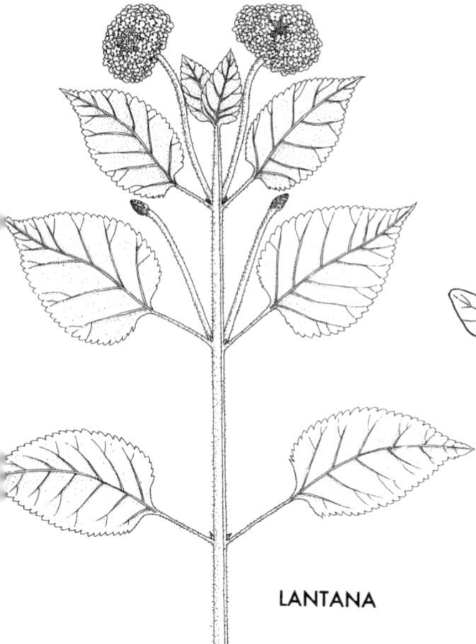

LANTANA

MISCELLANEOUS SHOWY FLOWERS

CLIMBING LILY **Whole plant**
Gloriosa superba L. Lily Family
Climbing perennial; to 5 ft. Leaves broad, lance-shaped, with tendril-like tips; 4–6 in. long. Flowers reddish (yellow at first), segments *wavy-edged, strongly back-curving.*
Where found: Potted plant, grown outdoors in Florida and Hawaii. Alien. Tropical Africa, Asia.
Comments: Roots and seeds contain colchicine. See Autumn Crocus below.

CROWN-OF-THORNS **Whole plant**
Euphorbia milii Desmoul. Spurge Family
Shrubby, *very spiny;* 1–4 ft. Leaves few, mostly on new growth, 1–2½ in. Floral parts above 2 broad *red bracts* to ½ in.
Where found: House plant; common outdoors in South and Hawaii.
Comments: Contains toxic diterpenes. Can cause dermatitis. If ingested, may cause severe internal poisoning.

PEYOTE **Pl. 33** **Whole plant**
Lophophora williamsii (Lem.) J. Coult. Cactus Family
Hassock-shaped, spineless cactus; forming "buttons" with 7–10 ribs, *tufts of silky hairs;* to 3 in. Flowers pink; March–April.
Where found: Deserts. S. N.M., s. Texas to n. Mexico. Rare because of overcollecting.
Comments: Dried buttons contain alkaloids such as mescaline and lophophorine; symptoms include vomiting, stomach pain, diarrhea, vivid hallucinations, etc. Used in religious rites of Indians.

OPIUM POPPY **Pl. 32** **Whole plant**
Papaver somniferum L. Poppy Family
Annual, 2–3 ft.; with bluish film that rubs off. Leaves irregularly lobed. Flowers showy, pink (white to violet); stamens many; stigma large; May–Aug. Fruits urn-shaped capsules.
Where found: Cultivated. Escaped; uncommon.
Comments: Dried latex from unripe pods is crude opium. Unripe fruits can cause symptoms of morphine poisoning, depressing the central nervous system and the respiratory and circulatory systems.

AUTUMN CROCUS **Whole plant**
Colchicum autumnale L. Lily Family
Perennial to 1 ft. Leaves 3–8, linear. Flowers pink to violet, to 4 in. across; *autumn* (rarely spring). Rattlelike pods appear with leaves in spring.
Where found: Cultivated. Escaped. Alien. Europe.
Comments: Contains the highly toxic alkaloid colchicine, which causes dilation and serious damage to blood vessels. Effects have been likened to arsenic poisoning. Long latent period of 2–6 hours before symptoms appear. The highly toxic seeds in the rattlelike pods may be attractive to children. Once used to treat gout.

POINSETTIA *Euphorbia pulcherrima* Willd. ex Klotzsch **Whole plant**
House plant. Floral parts subtended by *showy red* (or yellowish green) leaflike bracts. See on page 142.

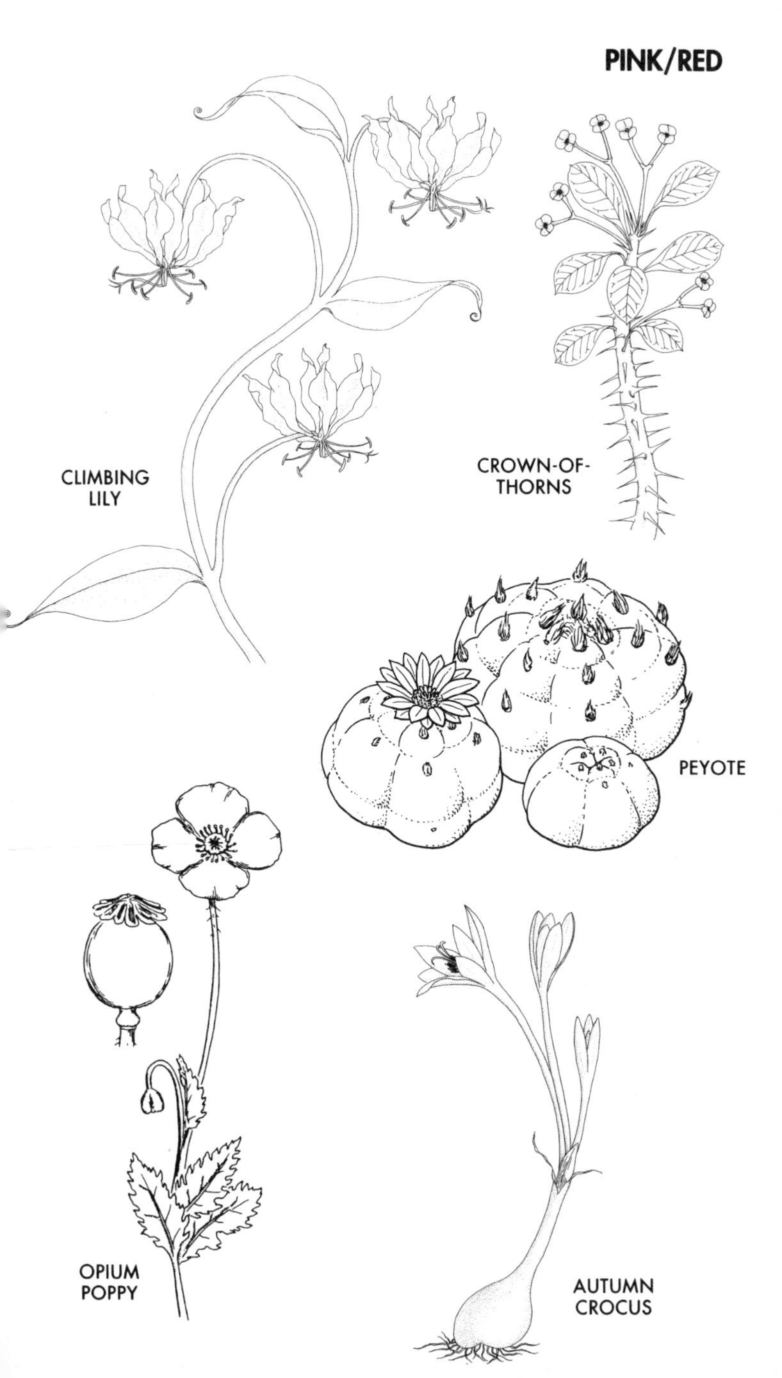
CLIMBING
LILY
CROWN-OF-
THORNS
PEYOTE
OPIUM
POPPY
AUTUMN
CROCUS

MISCELLANEOUS 5–PARTED FLOWERS

DOGBANE **Pl. 25** **Whole plant**
Apocynum androsaemifolium L. Dogbane Family
Perennial; 1–4 ft. *Milky juice within.* Leaves opposite, oval; dark green, smooth above. Flowers drooping pink bells, *red-striped within;* both in leaf axils and terminal; June–Aug.
Where found: Dry thickets. Much of North America, except Kans. and south of N.C. mountains.
Comments: Contains a toxic cardiac glycoside, apocynamarin, which can cause cardiac arrest. The milky juice in the stems and leaves is exceedingly bitter, usually discouraging ingestion. Human poisonings are unknown, but the plant is potentially dangerous, especially to children who may play with the milky-juiced stalks.

PHEASANT'S EYE **Whole plant**
Adonis aestivalis L. Buttercup Family
Taprooted herb; 4–16 in. Leaves alternate, 3-*pinnate,* segments linear. Flowers crimson, to 1 in. across, *sepals tightly hugging the spreading petals;* summer.
Where found: Cultivated, sometimes escaped. Calif. Alien. Europe. *A. annua* (not shown), also with red-orange flowers, occasionally escapes in the eastern United States.
Comments: No poisonings reported in North America. Contains glycosides but in smaller amounts than *A. vernalis,* page 100.

JAGGED JATROPHA **Whole plant**
Jatropha macrorhiza Benth. Spurge Family
Stout, thick-rooted perennial; 12–30 in. Leaves alternate, with *3–5 palmate lobes,* 6–7 in. long or broad. Flowers whitish pink, star-shaped, in clusters; summer. Fruit is calico-colored, 3-lobed capsule.
Where found: Mesas, plains. W. Texas, s. N.M. to Ariz., cen. Mexico. Other native and introduced species may have yellow, purple, or scarlet flowers.
Comments: The attractive fruits have poisoned children. As few as three seeds may cause severe poisoning. Sap in all plant parts contains curcin (a toxalbumin). May produce burning in throat, violent vomiting, stomach pain, muscle cramps, bloody diarrhea, coma, and death. All *Jatropha* species are considered potentially toxic. Some species are used medicinally.

SCARLET DELPHINIUM **Whole plant**
Delphinium cardinale Hook. Buttercup Family
Hollow-stemmed; 3–6 ft. Leaves 5–7-parted, the divisions *twisted.* Flowers scarlet, *single-spurred;* May–July. See related species on Plate 26.
Where found: Dry openings. South Coast Ranges to Baja.
Comments: All *Delphiniums* are considered toxic to some degree. Contains the alkaloids delphinine, ajacine, and others. May cause nervous symptoms, nausea, depression, or death in larger doses.

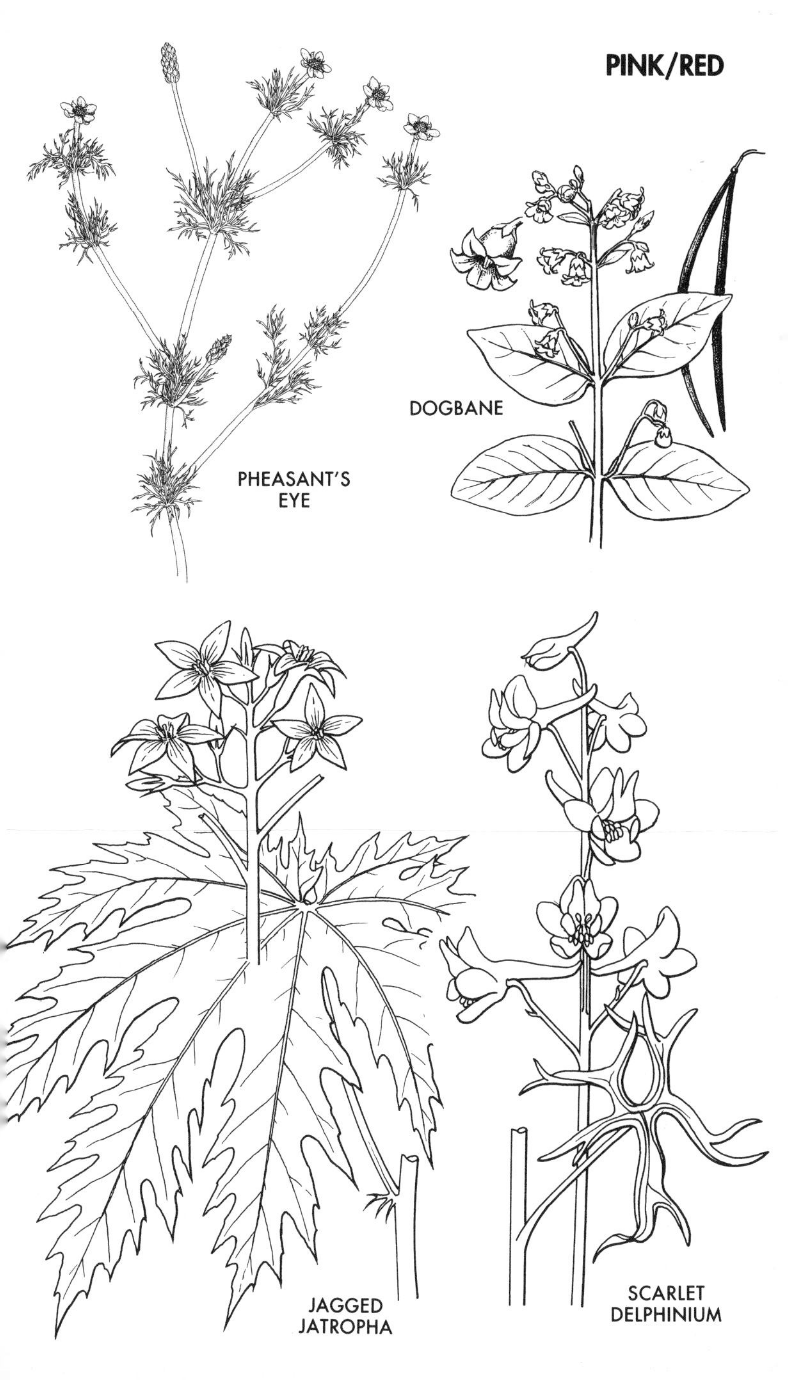
PHEASANT'S EYE
DOGBANE
JAGGED JATROPHA
SCARLET DELPHINIUM

5-PARTED TUBULAR COROLLA OR CALYX

CORNCOCKLE **Seeds**

Agrostemma githago L. Pink Family

Silky-hairy, winter annual; 1–3 ft. Leaves opposite, grasslike. Flowers pink, with noticeable veins. Calyx *10-ribbed,* inflated at base. Linear sepals longer than petals; June–Sept.

Where found: Waste places. Much of our area. Alien. Europe.

Comments: All parts, but especially the seed embryos, contain a toxic saponin, githagenin. Considered hazardous to grazing livestock. Seeds may cause gastroenteritis, abdominal pain, vomiting, respiratory failure, and death. Once a troublesome contaminant of European and American grain supplies, but modern cleaning equipment and weed control have largely eliminated it as a potential toxin.

FOUR-O'CLOCK **Roots and seeds**

Mirabilis jalapa L. Four-o'Clock Family

Smooth perennial; to 3 ft. Grown as an annual. Leaves opposite, oval, often heart-shaped at base. Flowers narrowly tubular, flaring at end; 1–2 in. Usually red (also pink, yellow, or white); often striped. Opening in late afternoon; summer.

Where found: Cultivated. Escaped in the South and Southwest. Alien. Tropical America.

Comments: Ingesting the seeds or roots has caused poisoning in children, with acute stomach pain, vomiting, and diarrhea, plus skin and mucous membrane irritation. Used medicinally. Toxic reactions are generally managed by treating the symptoms. The native Heart-leaved Four-o'Clock *M. nyctaginea* (not shown), in the Prairie states and locally eastward, is also potentially poisonous.

TOBACCO **Leaves**

Nicotiana tabacum L. Nightshade Family

Sticky-hairy, rank, stout annual; 3–9 ft. Leaves large, broadly oval-lance-shaped. Flowers pink to greenish, 5-flared trumpets; Aug.–Sept.

Where found: Cultivated, escaped. Alien. Tropical America.

Comments: A well-known, widely used poisonous plant. Smoked or chewed, it can produce nausea, giddiness, and sweating; long-term use causes lung disease, cancer, and death. The succulent seeds of Tree Tobacco *N. glauca* (not shown) have been eaten, causing several deaths in California. *N. glauca* contains the toxic alkaloid anabasine.

INDIAN PINK **Pl. 18** **Whole plant**

Spigelia marilandica L. Logania Family

Perennial; 12–24 in. Leaves opposite, oval. Flowers *scarlet,* 5-lobed flaring trumpets with yellowish interior; May–June.

Where found: Rich woods openings. Md. to Fla.; Texas, Okla., Mo., to Ind.

Comments: Historically used to expel intestinal worms. Scattered historical reports of poisoning generally result from overdoses of medicinal preparations. Physicians of 19th century reported narcotic effects, including dilated pupils, spasms, or convulsions.

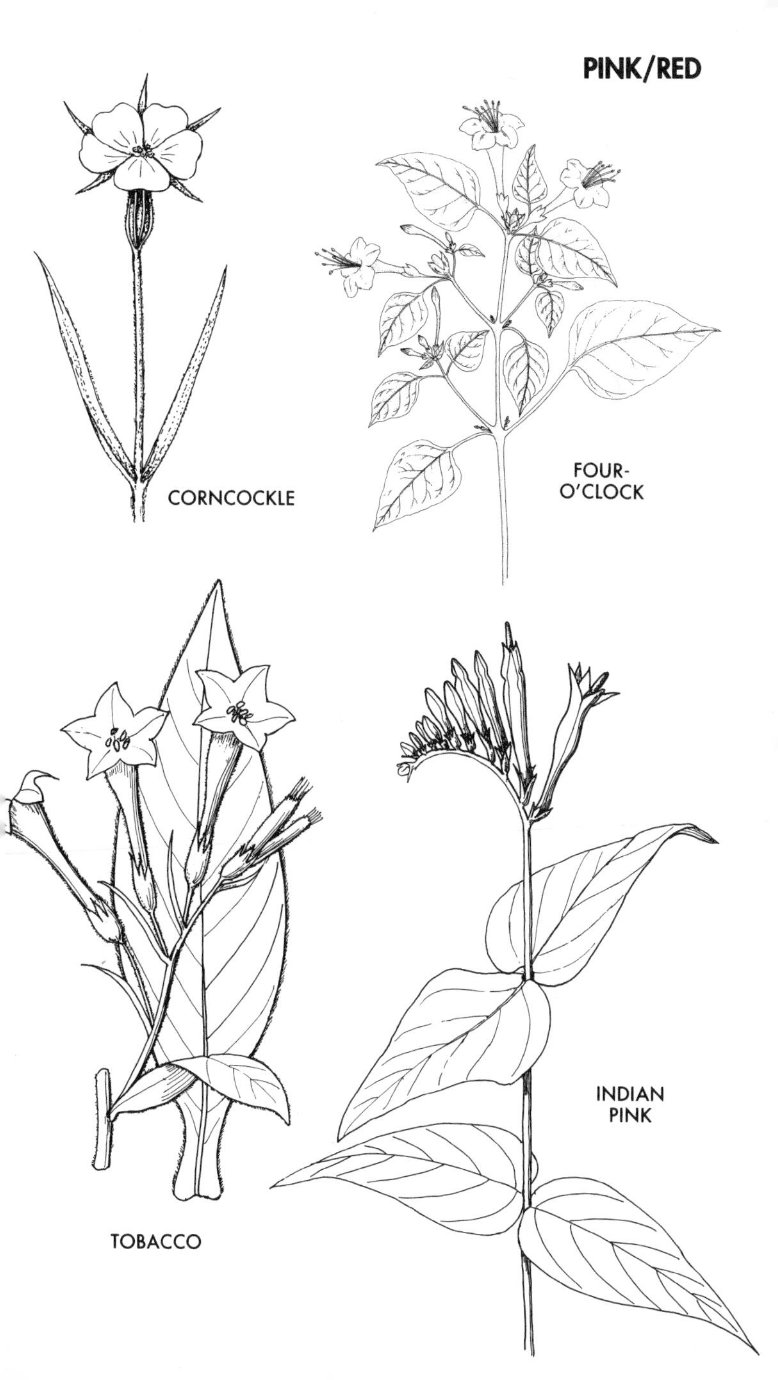
PINK/RED
CORNCOCKLE
FOUR-
O'CLOCK
TOBACCO
INDIAN
PINK

MILKWEEDS

SWAMP MILKWEED **Pl. 27** **Whole plant**
Asclepias incarnata L. Milkweed Family

Mostly smooth, branched perennial; 2–4 ft. Leaves *opposite* (or apparently so), *lance-shaped* or broader at base; soft-hairy beneath. Flowers in small umbels, deep rose or *bright pink;* June–Sept.

Where found: Moist meadows, marshes, pond edges, and stream banks. S. Canada to Fla.; N.M. to e. N.D.

Comments: Reported to have poisoned sheep in Indiana. Root historically used in folk medicine, reported to be strongly laxative.

SHOWY MILKWEED **Whole plant**
Asclepias speciosa Torr. Milkweed Family

Stout perennial, upper stalk densely hairy; 20–40 in. Leaves usually opposite, broad lance-shaped to oval-oblong, smooth to velvety; 3–8 in. long, 1–3 in. wide. Flowers usually pale pink (sometimes whitish), few to many in upper leaf axils, hoods *toothlike,* strongly *outward-curving;* May–Aug.

Where found: Moist sandy loam to rocky soils along waterways. S. Canada, prairie states, s. N.M. to w. Wash.

Comments: Not commonly implicated in poisoning, but experimental ingestion of large amounts of the plant has produced toxicity. All milkweeds are considered potentially toxic to some degree.

COMMON MILKWEED **Pl. 27** **Leaves, sap**
Asclepias syriaca L. Milkweed Family

Downy perennial *with milky sap;* 2–4 ft. Leaves *opposite,* widely elliptical, to 8 in. long. Flowers in *showy globe-shaped clusters* from leaf axils, often drooping under their own weight; June–Aug. Pods *warty.*

Where found: Pastures, roadsides. S. Canada to Fla., Ala.; Okla., Kans. to N.D. Common in the eastern United States.

Comments: The pods and young shoots are included as wild foods in many edible plant guides. While it has been circumstantially implicated in livestock poisoning, the bitter sap in leaves and stems would probably be distasteful to most animals. Sap may cause dermatitis. Not known to cause poisoning in humans, probably because the fresh plant is unlikely to be ingested. Contains cardiac glycosides.

SPIDER MILKWEED **Leaves, stems**
Asclepias viridis Walt. Milkweed Family

Solitary or paired-stemmed perennial; 10–25 in. Leaves alternate or slightly opposite, lance-shaped to oval, 2–4 in. long, to 2 in. wide; *margins strongly upturned.* Flowers in *airy globe-shaped clusters* (hence the common name); *mostly greenish,* but calyx tips and center of flower pale rose-purplish; April–Aug.

Where found: Sandy or rocky soil. Ohio to Fla.; Texas to Neb.

Comments: Suspected of poisoning livestock. Unlikely to be consumed.

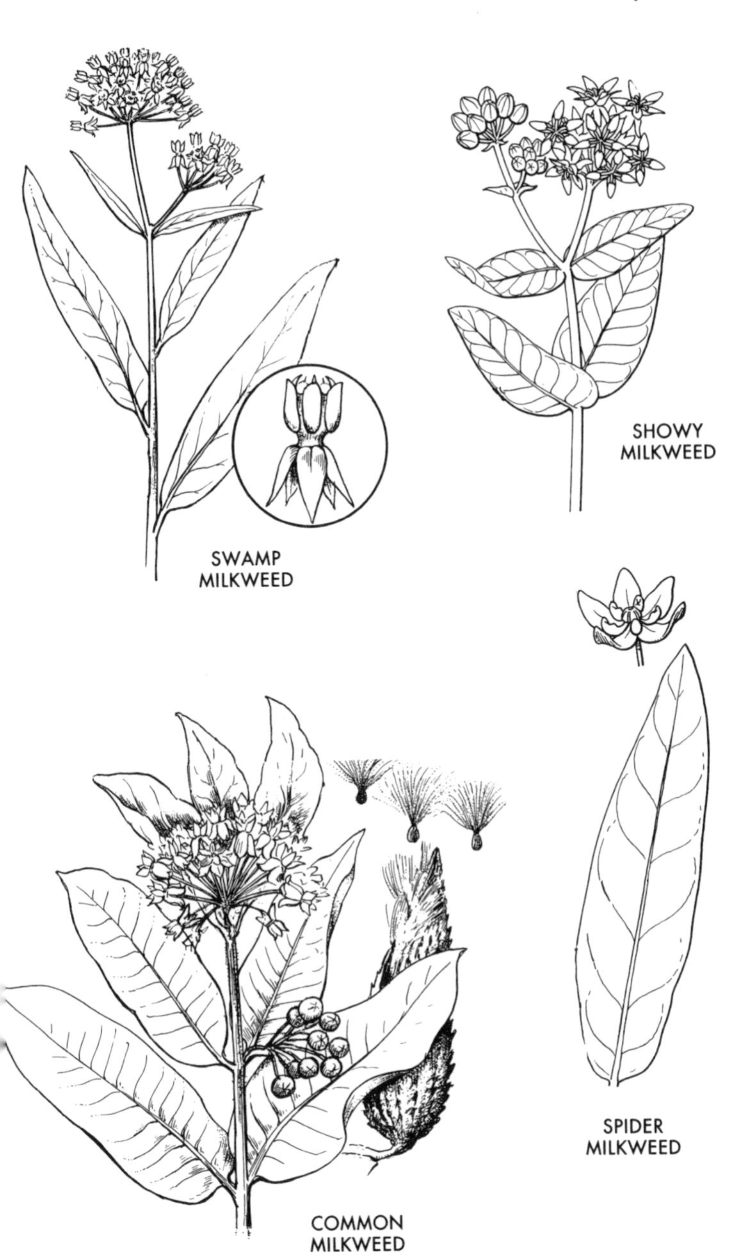
SHOWY
MILKWEED
SWAMP
MILKWEED
SPIDER
MILKWEED
COMMON
MILKWEED

LAYNE'S LOCOWEED *Astragalus layneae* Greene **Whole plant**
Wing petals with pink–purple tips. See on page 106.

KING'S LOCOWEED **Whole plant**
Astragalus calycosus Torr. ex. S. Wats. Pea Family
Low, short-stemmed, tufted, silvery perennial. Leaves to 4 in. long with 1–13 (commonly 3) leaflets. Flowers 2–6, pinkish to dark red, with *narrow 2-lobed wing petals;* April–June.
Where found: Arid woods. S. Wyo. and Idaho, Utah, nw. N.M., Ariz., Nev. to Calif.
Comments: Causes locoism. See comments under Woolly Loco *A. mollissimus* below.

SPECKLED LOCO *Astragalus lentiginosus* Dougl. ex. Hook. **Whole plant**
Highly variable; flowers white to pink-purple. See on page 106.

LAMBERT'S CRAZYWEED **Whole plant**
Oxytropis lambertii Pursh Pea Family
Stemless perennial, leaf and flower stalks arising from root; 12–22 in. Leaves pinnate, with 7–23 linear to lance-shaped, silver-hairy leaflets. Flowers in compact or loose spikes, *brilliant red-purple;* April–Sept.
Where found: Minn. to Texas, Ariz. to Mont., Man.
Comments: A true locoweed, causing locoism: trembling, listlessness, lack of appetite (except for a craving for the crazyweed), impaired vision, paralysis, prostration, and death in as little as one month after an animal first begins eating the plant.

WOOLLY LOCO, PURPLE LOCO **Whole plant**
Astragalus mollissimus Torr. Pea Family
Low, leafy, bushy, short-lived, soft, *silver-woolly-hairy* perennial; 8–12 in. Leaves feathery-pinnate, with 11–35 *roundish* leaflets. One of the earliest plants to produce spring foliage, when other succulent forage is unavailable. Flowers *pink-purple* to dull pinkish lavender, 7–45 on dense racemes (elongating with age) rising above leaves; April–July. Seed pods short, sharply curved at tip, and hairy.
Where found: Prairies, mesas. S.D., Neb., to Texas; e. Ariz., N.M.; e. Colo., to se. Wyo. Abundant in N.M. and the Texas panhandle.
Comments: *Loco* is Spanish for crazy or foolish. Animals can develop narcoticlike cravings for loco once they begin to graze it, refusing to feed on other forage. Poisoning occurs in two stages. The first, lasting several months, produces depression, laziness, weight loss, lack of coordination, and mania, with bizarre antics. In the second stage, the animal refuses to eat other plants and becomes emaciated, with sunken eyeballs, dull hair, and weakness. The animal may die from starvation in one to two years. Bees and prairie dogs are also affected. Horses are most severely affected.

WOOTON LOCO, GARBANCILLO **Whole plant**
Astragalus wootonii Sheldon
Flowers sometimes pinkish. See on page 90.

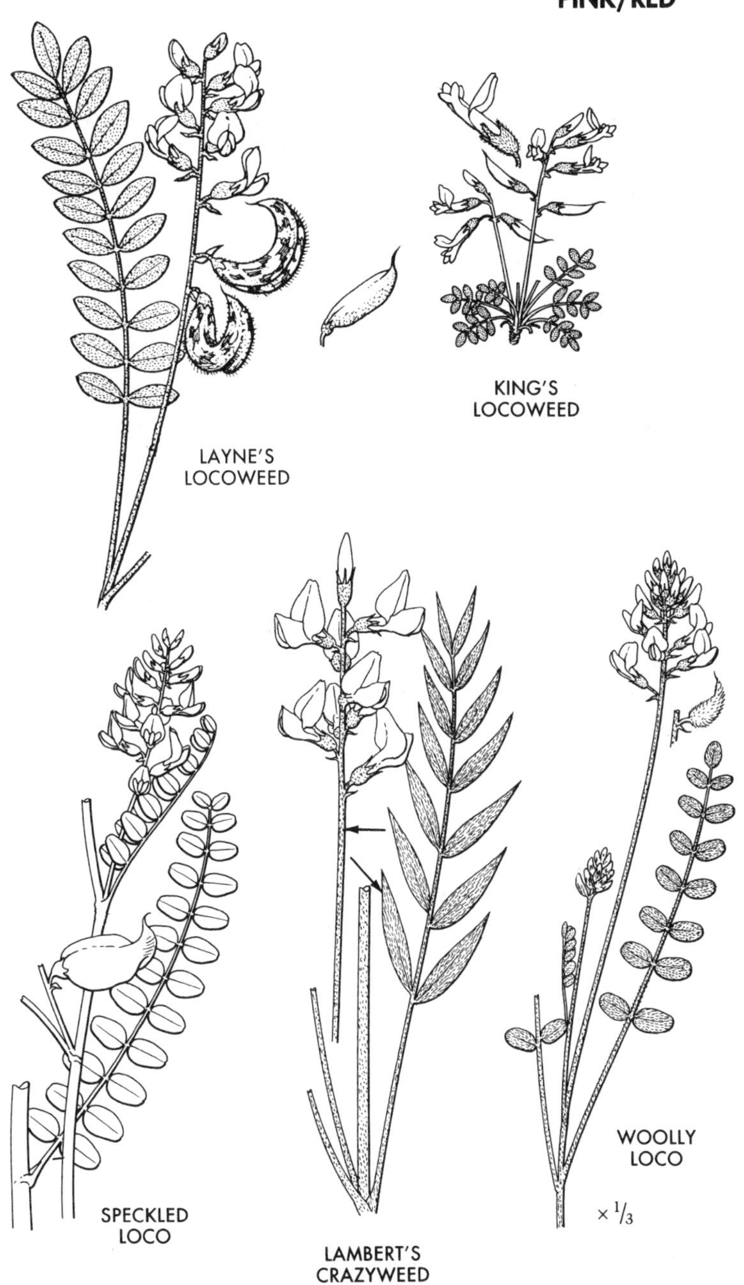
KING'S
LOCOWEED
LAYNE'S
LOCOWEED
SPECKLED
LOCO
LAMBERT'S
CRAZYWEED
WOOLLY
LOCO
× 1/3

SINGLETARY PEA — **Whole plant**
Lathyrus hirsutus L. — Pea Family

Weak-stemmed, annual legume; 1–4 ft. Stems *with prominent narrow wings.* Leaves compound, with *paired leaflets,* 1–3 in. long, major veins parallel, with terminal branching tendril. Flowers pealike, 2–3 per stalk, red-violet, *calyx lobes equal in size and shape;* March–early April. Seed pod a broad, flat, many-seeded hairy legume, *twisting tightly upon drying.*

Where found: Naturalized and scattered in U.S., south of Va., also in Calif. and Ore. Cultivated as a winter annual cover legume. Alien. Europe.

Comments: Closely related to the Sweet Pea *Lathyrus odoratus* L. (not shown). There are over 40 native or naturalized *Lathyrus* species in North America. Numerous species are known to produce lathyrism if large quantities of the seeds are eaten. The condition has been recognized in humans and livestock since the day of Hippocrates. Historically, when only seeds of *Lathyrus sativus* (not shown) were available to populations in times of famine, epidemics of lathyrism have been recorded, especially in India. When used as a staple, poisoning occurs in 4–8 weeks. Symptoms include spastic paralysis of the legs and, in severe cases, the arms, from nerve tissue lesions and degeneration of the spinal cord (neurolathyrism). Secondary symptoms disappear after the seeds are removed from the diet, but the paralysis is permanent. *Lathyrus odoratus* produces skeletal deformities in animals (osteolathyrism). *Lathyrus hirsutus* has been implicated in livestock poisoning in the U.S. Though it is used as a forage crop, livestock are removed from the pasture when the pods begin to form in spring because of the toxicity of the seeds.

EVERLASTING PEA — **Pl. 34** — **Whole plant**
Lathyrus latifolius L. — Pea Family

Climbing perennial to 9 ft. Stems *broadly winged.* Leaves divided into paired lance-shaped to elliptical leaflets, terminating in branched tendril. Flowers pealike, pink, blue or white, in 4–10-flowered raceme; *lower calyx lobes twice as long as the upper calyx lobes;* June–Sept.

Where found: Cultivated. Widely escaped in the U.S. Alien. S. Europe.

Comments: Implicated in livestock poisoning from nerve degeneration, including symptoms such as hyperexcitability, convulsions, and death. The active toxin in this species does not produce skeletal deformities as does *L. odoratus.*

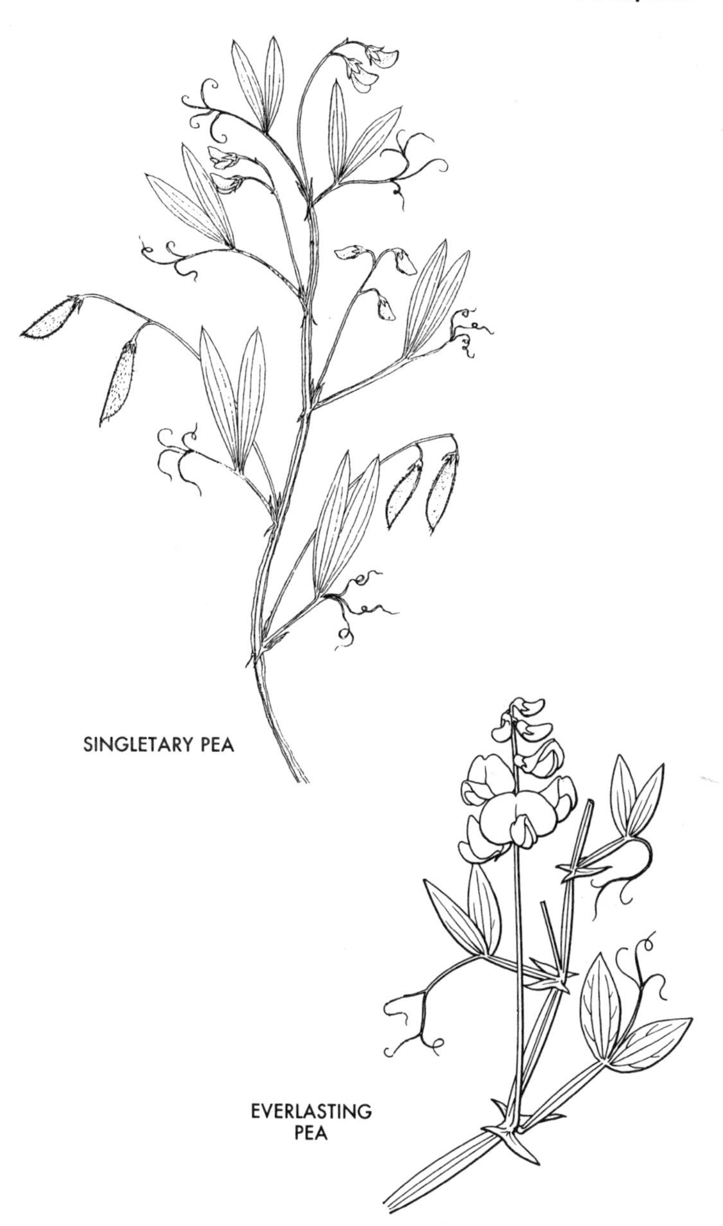
SINGLETARY PEA
EVERLASTING PEA

MISCELLANEOUS FLOWERS; MOSTLY SHOWY

PASQUE FLOWER **Whole plant**
Anemone patens L. Buttercup Family
[*Pulsatilla patens* (L.) P. Mill.]

Small perennial; 2–16 in. Silky leaves basal, divided into 3–7 *linear* segments. Flowers white to purple, to 1½ in. wide, petals arising from *cup-shaped receptacle;* March–June.

Where found: Open prairies. Wisc. to s. Ill., Mo., N.M. to Wash. and Alaska. Also in Eurasia.

Comments: Contact or ingestion may cause corrosive irritation of the mucous membranes. See on page 98.

VIPER'S BUGLOSS **Pl. 22** **Leaves**
Echium vulgare L. Borage Family

Bristly biennial; 12–24 in. Leaves lance-shaped. Flowers azure blue with *protruding red-tipped stamens* on curled branches. Flowers bloom singly on *curled stalks;* June–Sept.

Where found: Waste places; much of our area.

Comments: Irritant hairs may cause contact dermatitis. Contains pyrrolizidine alkaloids, which may cause veno-occlusive liver disease if ingested over a long period of time.

FOXGLOVE **Pl. 32** **Whole plant**
Digitalis purpurea L. Figwort Family

Rough, felty biennial; 2–4 ft. Oval to lance-shaped, 1-ft.-long, round-toothed leaves in basal rosette the first year; reduced on flowering stem in the second year. Flowers violet (pink to white) *spotted thimbles,* to 1¼ in. long on showy spike in second year; June–Aug.

Where found: Widely cultivated and escaped. Alien. Europe.

Comments: Contains cardiac glycosides used in modern medicine. Leaves are intensely bitter, discouraging their ingestion. In recent years, however, the leaves have been mistaken for Comfrey *Symphytum* species (not shown), resulting in fatalities. Symptoms of poisoning include dizziness, vomiting, irregular heartbeat, delirium, or hallucinations.

LOBELIA **Pl. 30** **Leaves, seeds**
Lobelia inflata L. Bluebell Family

Much-branched annual; 6–18 in. Leaves oval, toothed; hairy beneath. Small, pale blue flowers, to ½ in. long, in raceme. Upper *corolla split;* June–Oct. Seed pods *inflated.*

Where found: Fields, waste places, open woods. N.S. to Ga.; La., Ark., e. Kans. to Sask.

Comments: Contains 14 pyridine alkaloids, especially lobeline, similar in structure and effects to nicotine. Ingesting small amounts of leaves or seeds produces burning in throat and nausea; larger doses produce vomiting, sweating, rapid pulse, possibly coma and death. There have been reports of fatalities, but they are not well substantiated. All *Lobelia* species are believed to have similar compounds and activity. About 48 species of *Lobelia* are found in the wild in North America.

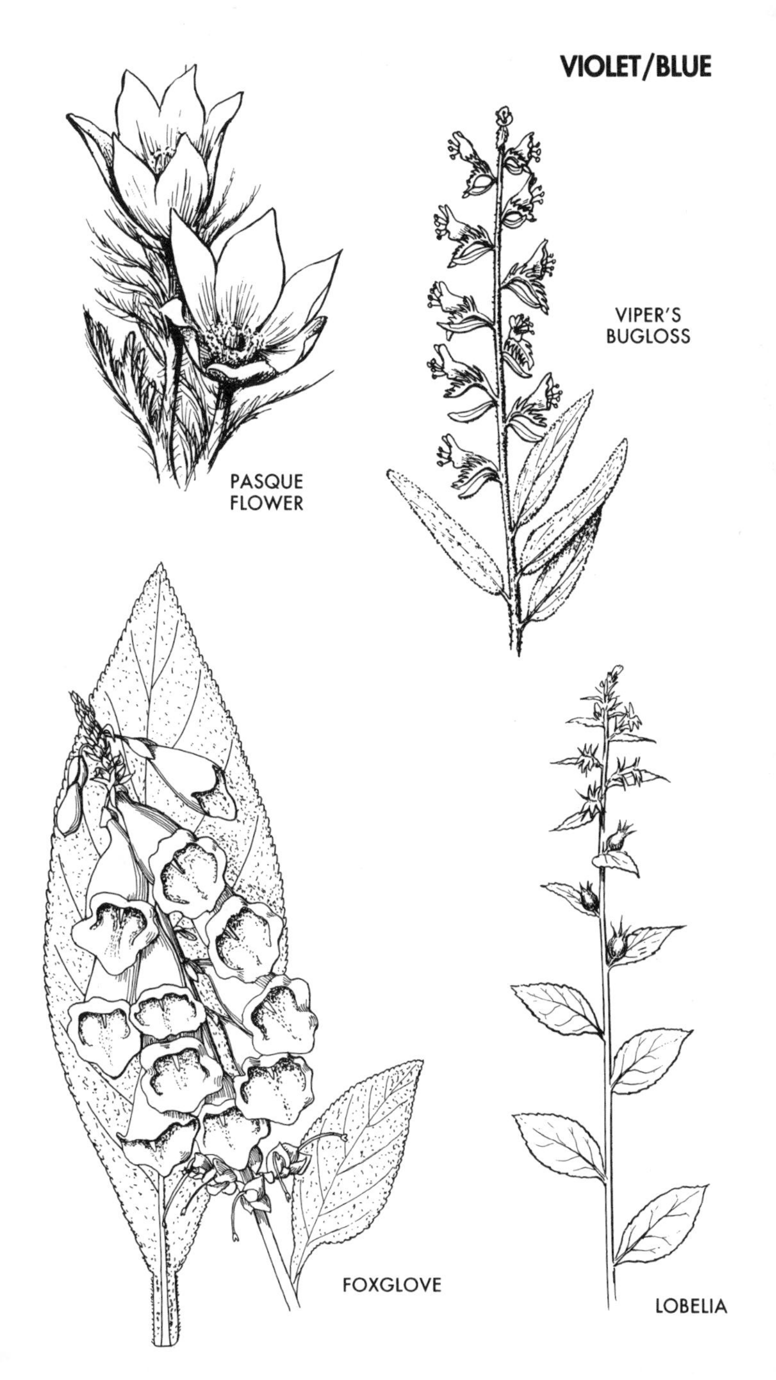
VIOLET/BLUE
PASQUE FLOWER
VIPER'S BUGLOSS
FOXGLOVE
LOBELIA

DEADLY NIGHTSHADE, BELLADONNA **Pl. 28** **Whole plant**
Atropa belladonna L. Nightshade Family

Rank perennial; 1–4 ft. Leaves *paired, one larger leaf always with a smaller leaf,* stalk *slightly winged.* Flowers single, drooping, bell-shaped, violet-brownish; June–Aug. Fruits *glossy black berries,* to $1/2$ in., seated on *star-shaped calyx.*

Where found: Sometimes cultivated; rarely seen in the U.S. Alien. Europe.

Comments: A famous European poisonous and medicinal plant. Important source of the medicinal but highly toxic alkaloids scopolamine, atropine, and *L*–hyoscyamine. Symptoms of poisoning include reddening of face, dry mucous membranes, rapid pulse, and dilated pupils. In larger doses it produces hallucinations, fits of frenzies and crying, talkativeness, respiratory paralysis, coma, and death. Atropine is used in eye surgery and as an antidote to nerve gas poisoning. Scopolamine is used in skin patches for motion sickness.

HORSE–NETTLE, CAROLINA NIGHTSHADE **Whole plant**
Solanum carolinense L.

Flowers violet to white. See description on page 86.

Comments: A noxious weed, present in pastures in S.D. Livestock poisoning common. At least one child has died from eating the berries.

JIMSONWEED *Datura wrightii* **Pl. 29** **Whole plant**

Flowers sometimes violet-tinged. Contains meteloidine, hyoscyamine, norscopolamine, and other toxic alkaloids. Taxonomy confused. See on page 84.

JIMSONWEED, DATURA **Pl. 29** **Whole plant**
Datura stramonium L. Nightshade Family

Flowers are often pale violet. See description on page 84.

Comments: Very dangerous weed. All parts toxic, even the nectar. Causes severe hallucinations, rapid heartbeat, dry mouth, etc. Most poisonings result from intentional ingestion. Livestock avoid it. Contains numerous alkaloids, including hyoscine, hyoscyamine, and atropine.

SILVERLEAF NIGHTSHADE **Whole plant**
Solanum elaeagnifolium Cav. Nightshade Family

Perennial to $1\frac{1}{2}$ ft., covered with *orange spines* and densely covered with *silvery, radiating* hairs. Leaves oblong to lance-shaped with *wavy margins.* Flowers violet stars; March–Oct. Fruits mottled yellow berries turning brownish with age.

Where found: Disturbed soils. Mo. to Texas; Kans. to Ariz. Spreading eastward and westward.

Comments: Cattle have died from eating mature berries. Unripe berries, possibly attractive to children, may cause nausea, vomiting, diarrhea, and depression of circulatory and respiratory systems. Contains the toxin solanine.

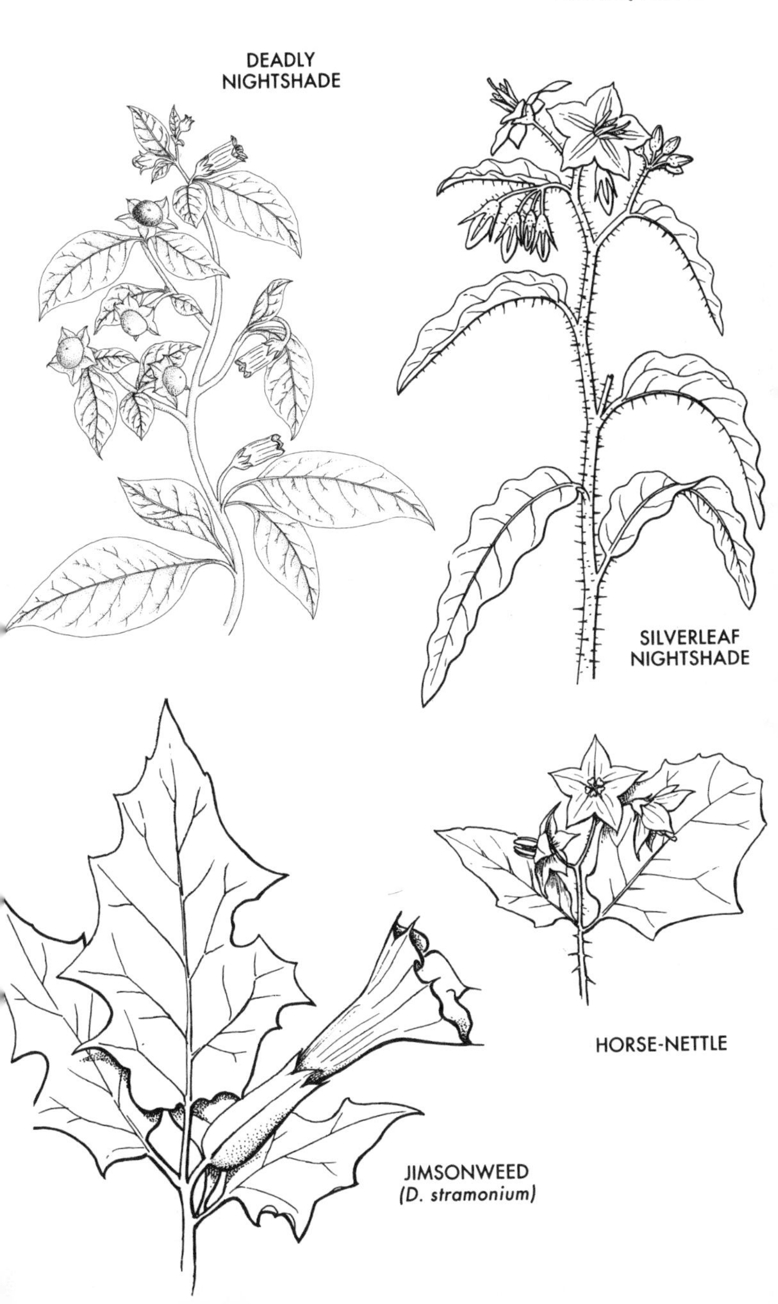
DEADLY
NIGHTSHADE
SILVERLEAF
NIGHTSHADE
HORSE-NETTLE
JIMSONWEED
(D. stramonium)

TALL LARKSPUR **Whole plant**
Delphinium exaltatum Ait. Buttercup Family
Slender, smooth-stemmed perennial; 2–6 ft. Leaves divided into 3–5 *long wedge-shaped* segments. Flowers blue (or white); May–June. See related species on Plate 26.
Where found: Thickets, rich woods. S. Pa. to Ala., north to Ohio.
Comments: See Carolina Delphinium below.

MONKSHOOD, WOLFSBANE **Pl. 28** **Whole plant**
Aconitum napellus L. Buttercup Family
Erect perennial; 2–6 ft. Leaves numerous, divided into many segments, the ultimate divisions linear. Flowers blue-violet, in many-flowered raceme; *helmet-shaped* sepal with visorlike beak, to 1 in. tall; July–Sept.
Where found: Roadsides, fields. Nfld. to Ont.; N.Y. Cultivated elsewhere. Alien. Europe.
Comments: See Western Monkshood below.

WILD MONKSHOOD **Whole plant**
Aconitum uncinatum L. Buttercup Family
Slender, erect to leaning, mostly smooth perennial; 4–6 ft. Leaves in 3–5 lobes or cleft, to 6 in. long and wide. Flowers *blue,* hood *rounded* and *conical;* Aug.–Oct.
Where found: Rich woods. Va. to Ga.; Tenn., W. Va.
Comments: See Western Monkshood below.

CAROLINA DELPHINIUM **Whole plant**
Delphinium carolinianum Walt. Buttercup Family
Downy perennial; 30 in. Leaves dissected into 3–5 linear lobes. Flowers *deep blue* to violet on slender raceme, spur arched slightly upward; May–June. See related species on Plate 26.
Where found: Fields, openings. Va. to Fla.; Texas to Mo.
Comments: Young leaves and seeds are especially toxic. All *Delphiniums* are considered toxic to some degree. Alkaloids include delphinine, delphineidine, ajacine, etc. May cause nervous symptoms, nausea or depression, and death in larger doses.

WESTERN MONKSHOOD **Whole plant**
Aconitum columbianum Nutt. Buttercup Family
Weak-stemmed perennial; 1–7 ft. Leaves palmate, deeply incised in 3–5 parts; 2–3 in. across. Flowers purple, with *hoodlike* sepals suspended over rounded petals; June–Aug.
Where found: Mountain meadows, open woods. S. S.D. to N.M., Calif., north to B.C. and Alaska.
Comments: All *Aconitum* species are poisonous. Causes livestock poisoning and death. Contains aconitine. Symptoms in humans include tingling, then numbing or burning of the mouth, tightening of the throat and speech difficulties, salivation, nausea, and vomiting. May produce blurred vision, anxiety, an oppressive feeling in the chest, weakness, dizziness, and lack of coordination. Many fatalities have resulted from overdoses of once widely used medicinal preparations of the European *A. napellus.* Roots and leaves are particulately toxic.

VIOLET/BLUE

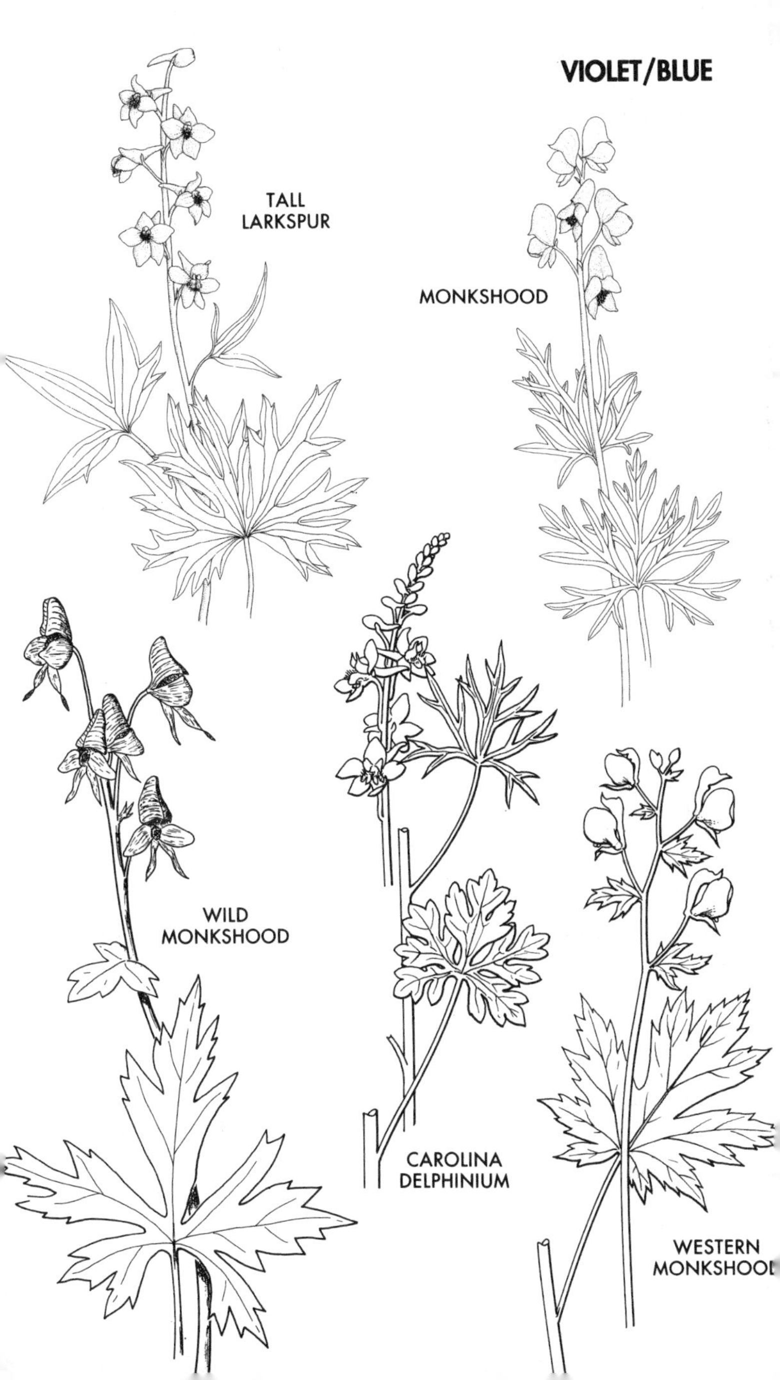
TALL
LARKSPUR
MONKSHOOD
WILD
MONKSHOOD
CAROLINA
DELPHINIUM
WESTERN
MONKSHOOD

WHITE LOCO, WHITE POINT LOCO **Whole plant**
Oxytropis sericea Nutt. ex. Torr. & A. Gray Pea Family

Variable, silky-hairy, robust perennial. Leaves to 12 in. long, with 11–19 lance-shaped to oval-oblong leaflets. Flowers in racemes of 10–30 flowers, whitish to pale orchid, keel petal *purple-tipped;* April–June. Pods oblong, to 1 in.

Where found: Rocky prairies, valleys, hillsides. Sw. S.D. to Okla.; cen. N.M., ne. Utah, s. Idaho, to w. Mont.

Comments: A source of loco poisoning in the western prairies and mountains. See description of locoism under *Astragalus mollissimus* on page 124.

POISON MILKVETCH **Whole plant**
Astragalus bisulcatus (Hook.) Gray Pea Family

Unpleasantly scented, *reddish-stemmed* perennial; 6–20 in. Leaves to 5 in. long, with 15–35 oval-oblong leaflets. Flowers in drooping, cylindrical raceme with 25–80 flowers; purple (or whitish); June–Aug. Legumes with *2 deep grooves,* pendulous, to 1 in.

Where found: Prairies, roadsides. Man., Sask. to N.D.; N.M. to Colo.

Comments: Poisons livestock by virtue of its ability to accumulate selenium, causing chronic syndromes such as "blind staggers" and "alkali disease."

BLUE FALSE INDIGO **Pl. 26** **Whole plant**
Baptisia australis (L.) R. Br. Pea Family

Smooth perennial; 3–5 ft. Leaves divided into 3 cloverlike leaflets, oval, broadest at apex. Flowers in erect racemes, blue to violet, 1 in. long; April–June.

Where found: Open woods, roadsides. Pa. to Ga.; Texas to Neb., s. Ind.

Comments: Various *Baptisia* species have been implicated in livestock poisoning. Human poisonings are usually the result of overdoses of medicinal preparations. May cause loss of appetite and diarrhea. The genus contains active quinolizidine alkaloids.

FAVA BEAN, HORSE BEAN **Beans, pollen**
Vicia faba L. Pea Family

Coarse, erect, annual vine, *without tendrils;* to 6 ft. Leaves alternate, compound, with 2–6 oval leaflets. Flowers in leaf axils, 1 in. long, off-white, with *dull purple dot;* June–Aug.

Where found: Widely cultivated as a forage crop, especially in South. Alien. N. Africa, sw. Asia.

Comments: Eating beans or inhaling pollen may cause favism in individuals with an inherited red blood cell enzyme deficiency. Within 5–24 hours after eating fava beans, these individuals may experience headache, dizziness, vomiting, fever, jaundice, profound anemia, and death. Most serious in children; most common in boys. The toxic components are cyanogenic glycosides.

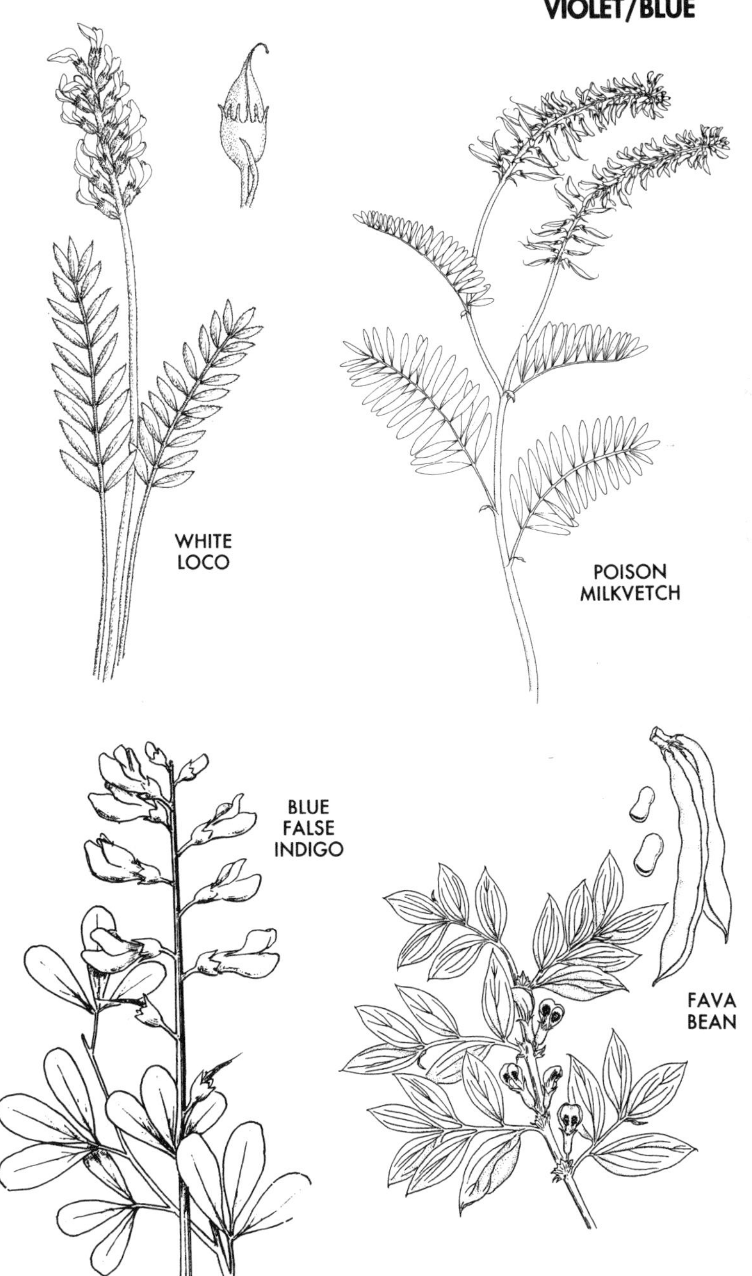
WHITE
LOCO
POISON
MILKVETCH
BLUE
FALSE
INDIGO
FAVA
BEAN

LUPINES: PEALIKE FLOWERS IN SPIKES

TEXAS BLUEBONNET — **Whole plant**

Lupinus texensis Hook. — Pea Family

The famous Texas Bluebonnet, widely planted by garden clubs along roads in Texas. A Texas endemic. Annual; 1½–2 ft. Leaflets usually 5, tips acute. Flowers rich dark blue on dense raceme; spring.

Where found: Limey soils in Texas. Cultivated elsewhere.

Comments: Livestock avoid eating this plant because it is unpalatable. Not generally implicated in livestock poisoning.

ELEGANT LUPINE — **Whole plant**

Lupinus lepidus Dougl. ex Lindl. — Pea Family

Low, spreading herb, usually less than 1 ft. high. Leaves palmate, with 5–9 lance-shaped, densely hairy leaflets. Flowers in dense racemes, *rich blue,* banner petal with *square* white spot; June–Sept.

Where found: Prairies, meadows. Mont. to Colo.; Calif. to B.C.

Comments: Seeds and young leaves are the most dangerous parts of lupines, which contain alkaloids such as lupinine and anagyrine. Symptoms may include weak pulse and nervousness, then depression, labored breathing, convulsions, and paralysis. Lupines have caused birth defects of bone in cattle, termed "crooked calf disease," in which calves are born with bent legs or arched backs. Bone deformity was reported in a human baby whose mother drank goat's milk during the pregnancy. Apparently the goat had fed on *Lupinus latifolius* (not shown).

WILD LUPINE — **Pl. 25** — **Whole plant**

Lupinus perennis L. — Pea Family

Perennial; 1–2 ft. Leaves radiating from a single point, with 7–9 segments lance-shaped, broadest at apex. Flowers *pealike,* in showy raceme; April–July.

Where found: Open woods, dry fields, roadsides. Sw. Me., N.Y. to Fla.; W. Va., Ohio, Ind. to Ill.

Comments: Not all lupines are toxic, but it is very difficult to distinguish between nontoxic and poisonous species. The seeds of this species are considered toxic. Lupines are commonly cultivated in gardens. They are not often involved in human poisoning.

STINGING LUPINE — **Whole plant**

Lupinus hirsutissimus Benth. — Pea Family

Perennial covered with long *stiff yellow stinging hairs.* Palmate leaves with rounded segments. Flowers *red-violet* to *magenta.* Banner petal often with yellow blotch; March–May.

Where found: Thickets, woods. S. Calif. Coast Ranges to Baja Calif.

Comments: The stiff hairs will sting, which usually prevents ingestion of the plant.

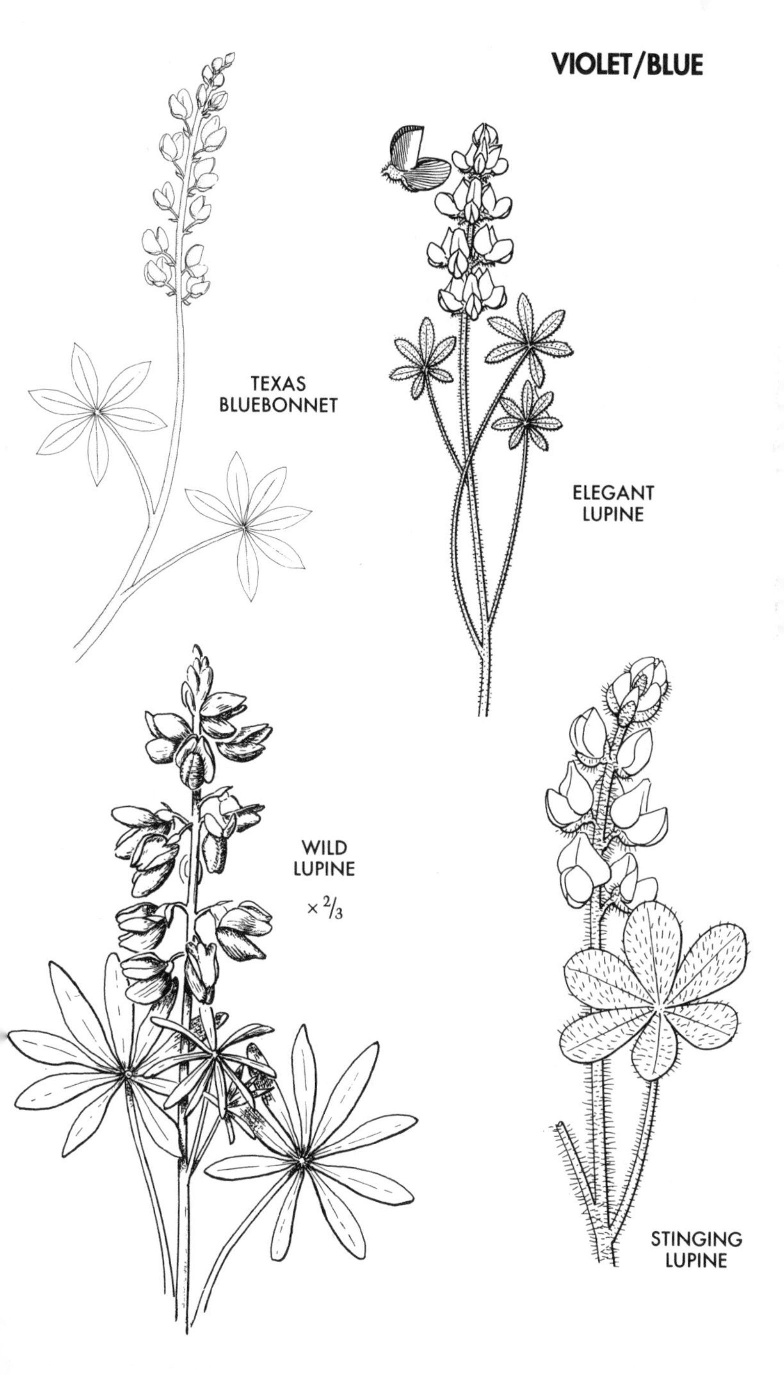

TEXAS
BLUEBONNET
ELEGANT
LUPINE
WILD
LUPINE
× 2/3
STINGING
LUPINE

BLUE COHOSH **Pl. 19** **Berries**
Caulophyllum thalictroides (L.) Michx. Barberry Family
Smooth-stemmed perennial with bluish film that rubs off; 1–2 ft. Leaves compound with 3 (to 5) 2–3-lobed leaflets. Flowers greenish yellow (or darker) in terminal clusters, with *6 petallike sepals* and 6 stamens; April–June, before expansion of leaves. Fruits blue berries.
Where found: Moist rich woods. N.B. to S.C.; Ark., N.D. to Man.
Comments: The raw blue berries have been reported to poison children, apparently irritating the gastrointestinal tract, especially the mucous membranes of the intestines. Fresh berries, leaves, and roots may cause contact dermatitis. Contains a number of alkaloids and saponins. Root used medicinally.

STINGING NETTLE **Pl. 28** **Stinging hairs**
Urtica dioica L. Nettle Family
Perennial with 4-angled stems; 1–4 ft. Stems and lower leaf veins with *stiff stinging hairs.* Leaves opposite, oval, bases somewhat *heart-shaped.* Flowers greenish, in branched clusters from leaf axils; June–Sept. The American plant, *U. dioica* subsp. *gracilis* (Ait.) Seland., with 6 varieties, differs from the European *U. dioica* subsp. *dioica* primarily in that the latter has male and female flowers on the same plant. Some botanists treat the varieties of *U. dioica* subsp. *gracilis* as separate species. All the *Urticas* that occur in North America have stinging hairs.
Where found: Waste places, moist thickets. Most of N. America. Lab. to Alaska, southward. The European subspecies is nturalized in some areas.
Comments: Fresh plants sting on contact. Each stinging hair is like a tiny syringe with a bladderlike base. Upon contact, the hair tip breaks off, injecting a burning chemical mixture into the skin. The burning sensation may last for up to one hour. The chemical mixture contains histamine, acetocholine, 5–hydroxytryptamine, and small amounts of formic acid. Dried plant is used medicinally and for tea; cooked leaves are edible. A recent study shows the freeze–dried plant may be useful in the treatment of asthma.

SPIDER MILKWEED **Leaves, stems**
Asclepias viridis Walt. Milkweed Family
Perennial with solitary or paired stems; 10–25 in. Leaves alternate or slightly opposite, lance-shaped to oval, 2–4 in. long, to 2 in. wide, margins *strongly upturned.* Flowers in *airy globe-shaped clusters* (hence the name Spider Milkweed); *mostly greenish,* but calyx tips and center of flower pale rose-purplish; April–Aug.
Where found: Sandy or rocky soils. Ohio to Fla.; Texas to se. Neb.
Comments: Suspected of poisoning livestock, but unlikely to be consumed.

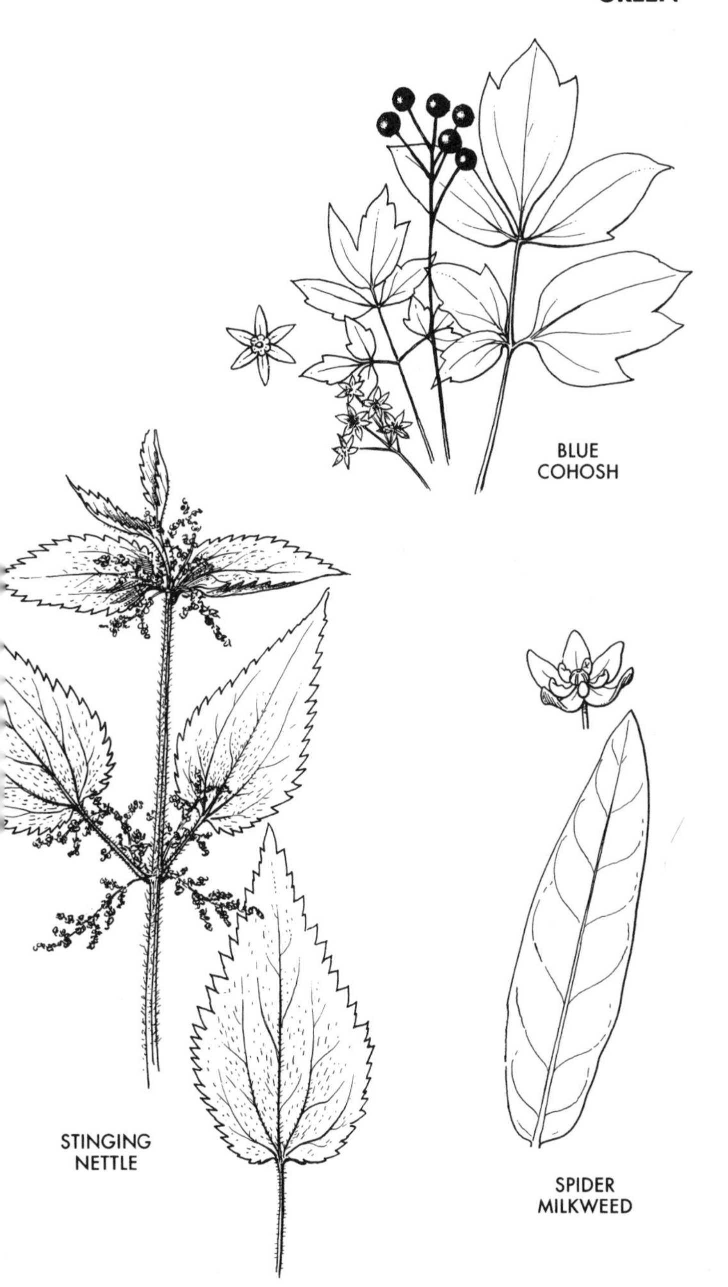
BLUE
COHOSH
STINGING
NETTLE
SPIDER
MILKWEED

ARUM FAMILY
HOODLIKE FLOWERS OR MOTTLED LARGE LEAVES

GREEN DRAGON **Pl. 18** **Whole plant**
Arisaema dracontium (L.) Schott Arum Family

Perennial, 1–3 ft. high. Single leaf, divided into 5–15 lance-shaped leaflets arranged in a horseshoe shape. Spathe sheathlike, *narrow, spadix much longer.* The unusual sheathed flowers are green. May–July.

Where found: Rich, moist woods. Sw. Que., Vt., s. N.H. to Fla.. Texas, e. Kans., Neb., Wisc., Mich.

Comments: Ingesting any part of fresh plant may cause intense burning from irritating calcium oxalates.

CALADIUM **Whole plant**
Caladium bicolor (Ait.) Vent. and other species Arum Family

Stemless perennial. Leaves with stalk attached to center of leaf, oval, to 14 in. long, multicolored, *red to white,* margins green, veins often *bright red.* Flowers green outside, white-green within; spathes up to 4 in. long.

Where found: Tropical America. Widely grown as house plant and garden annual. Naturalized in subtropical areas.

Comments: Contains calcium oxalate raphides (needles). If ingested, causes intense irritation and burning of mouth, lips, and throat. See Dieffenbachia below.

DIEFFENBACHIA, DUMBCANE **Whole plant**
Dieffenbachia seguine (Jacq.) Schott Arum Family

Highly variable, large, smooth perennial. Leaves large; to 15 in., broad-oval, *mottled with various shades of green-yellow.* Stems are sheathed (often to more than half their length). Flower a greenish spathe.

Where found: Tropical America. A popular house and greenhouse plant, grown for its foliage. Reportedly naturalized in subtropical climates.

Comments: Contains calcium oxalate raphides (needles) located in specialized ejector cells. Slight pressure on these ampoule-shaped cells forces open a cap, causing swelling of structures inside, instantaneously ejecting the needlelike raphides, which can penetrate the mucous membranes of the mouth and throat. A release of histamine can cause swelling. Severe cases may cause speech to become slurred and unintelligible (thus the common name Dumbcane). Chewing the plant or coming in contact with the juice invokes the reaction. May also produce contact dermatitis. The calcium oxalates are insoluble, so poisoning is localized, not systemic. Other toxins are also present.

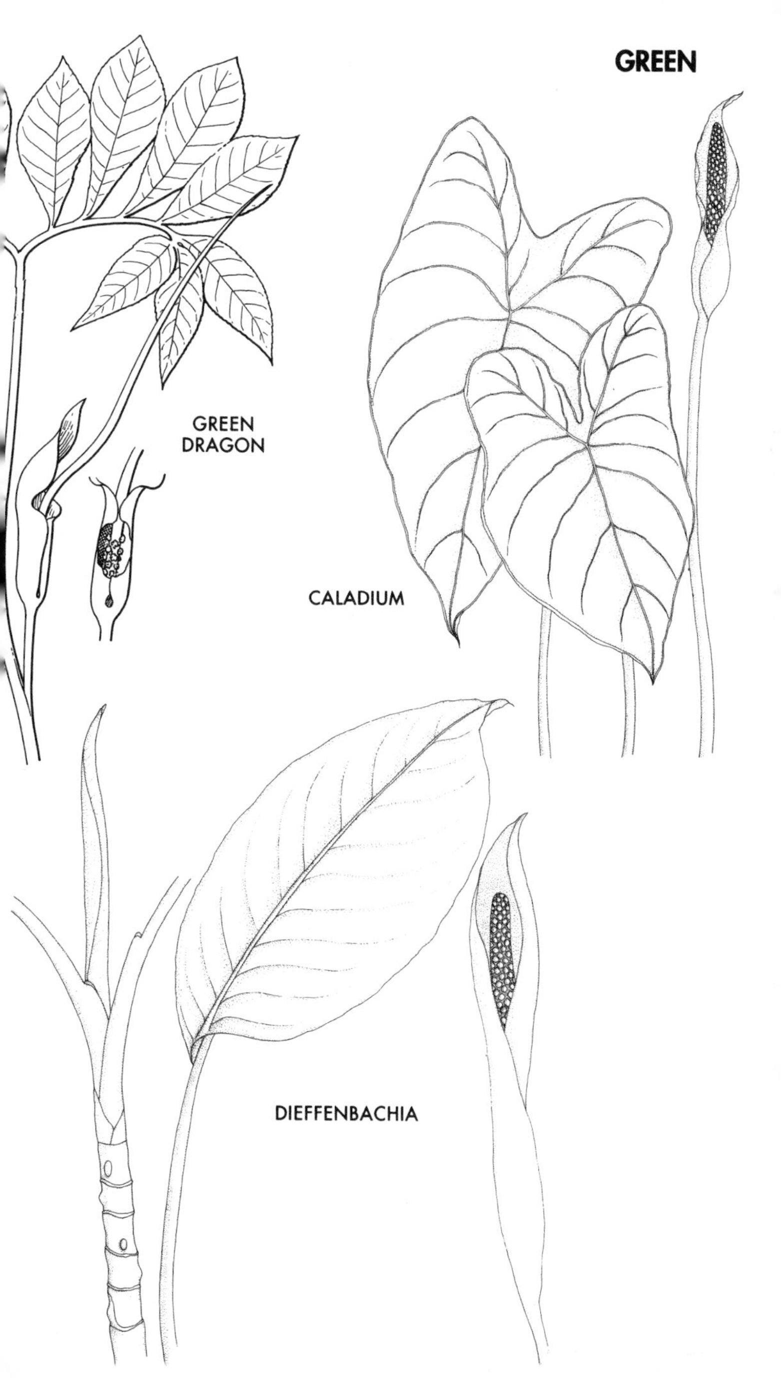
GREEN
GREEN
DRAGON
CALADIUM
DIEFFENBACHIA

PENCIL TREE, PENCIL CACTUS **Whole plant**
Euphorbia tirucalli L. Spurge Family
A *succulent,* spineless tree to 30 ft. Green, *cylindrical,* succulent stems often form a crown. Leaves are minute. Flower structures (cyathia) in clusters at ends of branches.
Where found: Widely grown as a house plant. Sometimes persisting outdoors in Florida. Alien. Tropical Africa.
Comments: See Cypress Spurge below.

CYPRESS SPURGE **Pl. 30** **Whole plant**
Euphorbia cyparissias L. Spurge Family
Smooth, creeping perennial; to 1 ft. With numerous linear leaves to 1½ in. long. Flower structures (cyathia) in umbels, with showy yellowish oval to triangular bracts beneath.
Where found: Roadsides, fence rows. Naturalized Me. to Penn., Ill., to Kans., Colo. Alien. Europe.
Comments: Most of the 1,600 species of spurges *(Euphorbias)* contain a milky sap with a toxic component that can cause dermatitis and, if ingested, severe internal poisoning. Symptoms may include burning and severe irritation of the mouth, throat, and stomach. Avoid contact with the skin, eyes, or mucous membranes. Toxins include the diterpenoids phorbol and ingenol. These substances are capable of inducing nonmalignant tumors.

POINSETTIA **Pl. 32** **Whole plant**
Euphorbia pulcherrima Willd. ex Klotzsch Spurge Family
A shrub to 10 ft. in its natural range; in the U.S. it is usually seen as a house plant. Leaves elliptical to fiddle-shaped, usually lobed; to 7 in. long. Floral parts subtended by *showy red* (or yellowish green) leaflike bracts; winter-flowering.
Where found: Winter house plant. Alien. Native to tropical Mexico, Central America.
Comments: Poinsettia's reputation as a poisonous plant is largely based on one questionable attribution in the death of a child in Hawaii in 1919. Only a small percentage of those reporting Poinsettia ingestion to poison control centers report symptoms, mainly vomiting, diarrhea, and local irritation. Nevertheless, it is best to keep the plant away from children.

CAPER SPURGE, MOLE PLANT **Whole plant**
Euphorbia lathyris L. Spurge Family
Unbranched annual or biennial; to 3 ft. Leaves linear to lance-shaped, arranged in whorls of 4 on a solitary stem. Flower structures (cyathia) in umbels.
Where found: Grown in gardens (especially in the West to deter gophers). Sometimes naturalized. Alien. Europe.
Comments: See Cypress Spurge above.

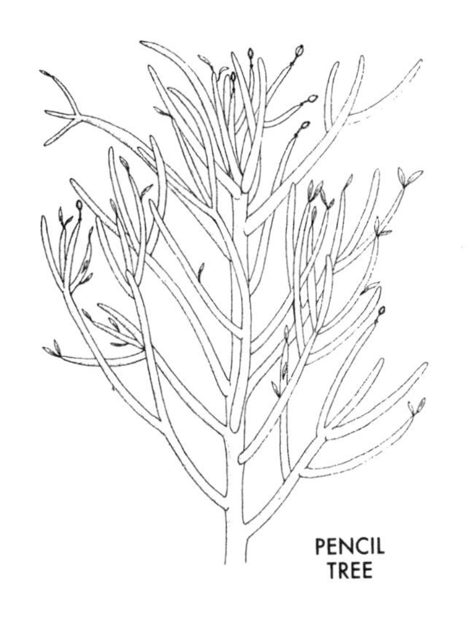

PENCIL TREE

CYPRESS SPURGE

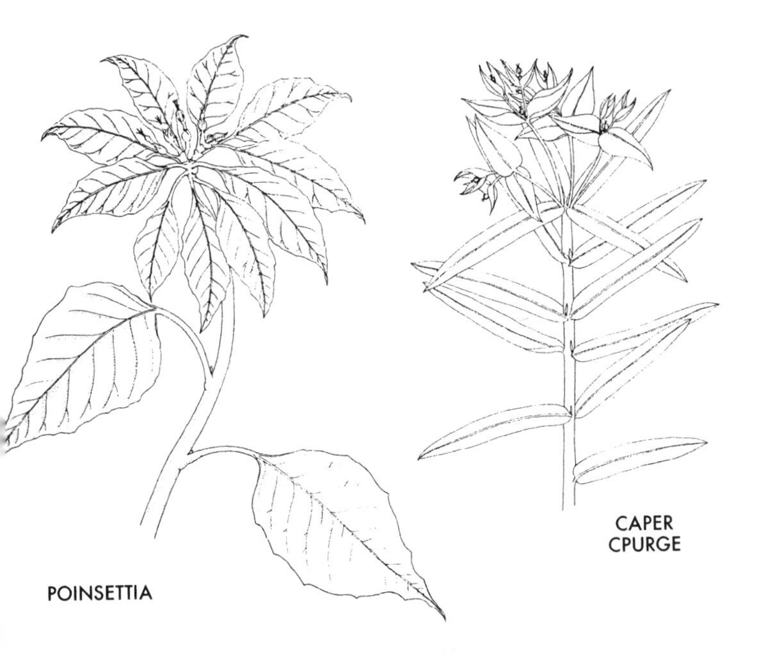

CAPER CPURGE

POINSETTIA

TALL PLANTS WITH LEAVES TO 1 FT. OR MORE

CASTOR BEAN **Pl. 23** **Seeds**
Ricinus communis L. Spurge Family

Bushy annual (or perennial) with green, red, or purple stems; 3–15 ft. Large leaves palmate, with 5–11 oval to lance-shaped, toothed lobes. Female flowers in burlike clusters above, male flowers beneath; July–Sept. Seed capsule a bur with soft spines, encasing 3 oval, brown-white, mottled seeds.

Where found: Widely grown in the American South and West. Tropical Africa.

Comments: The seeds of this plant produce castor oil. However, the cake of the seed contains the highly toxic lectin ricin. A single seed can contain enough ricin (1 mg) to kill an adult. The hard shiny seeds are attractive, inviting ingestion, and are used in making necklaces. Pierced seeds can release toxins that can enter the body through a scratch or through the mouth, should a child suck on a necklace made from the seeds. Ricin can cause burning of the throat and mouth, vomiting, severe abdominal pain, diarrhea, thirst from loss of fluids and electrolytes, convulsions, and death. The leaves are less toxic, but they may cause contact dermatitis.

GREEN HELLEBORE **Pl. 18** **Whole plant**
Veratrum viride Ait. Lily Family

Large, smooth perennial; 2–8 ft. Leaves alternate, large (to 12 in. long, 6 in. wide), oval, pleated (ribbed), strongly sheathing. Flowers yellowish (turning dull green) 6-pointed stars, in a large, pyramidal, terminal cluster; April–July.

Where found: Wet woods, swamps, ditches. Me. to Ga. mountains; Tenn. to Wisc.

Comments: Like the European *V. album*, the eight North American "false hellebores" contain highly toxic alkaloids that affect the heart and nervous system. Formerly used in medicine for heart ailments, resulting in overdoses and cases of poisoning. Contains numerous toxic steroidal alkaloids, including veratrin, veratramine, and verastrosine. Ingestion of the plant may cause severe slowing of heart rate and respiration, decreased arterial resistance, irregular heartbeat, excessive salivation, stomach pain, vomiting, diarrhea, spasms, and paralysis. The root is especially dangerous. Although a relatively large amount of the leaves must be ingested to be lethal in most farm animals, the plant produces lush, succulent growth in early spring that may be attractive to livestock. See also California Corn Lily *V. californicum*, page 78, which is better known for the many cases of deformities it has produced in sheep. Several additional North American *Veratrums* (not shown) may also cause similar poisoning. These include *V. album, V. fimbriatum, V. insolitum, V. parviflorus, V. tenuipetalum, and V. woodii,* some of which are sometimes placed in the related genus *Melanthium*.

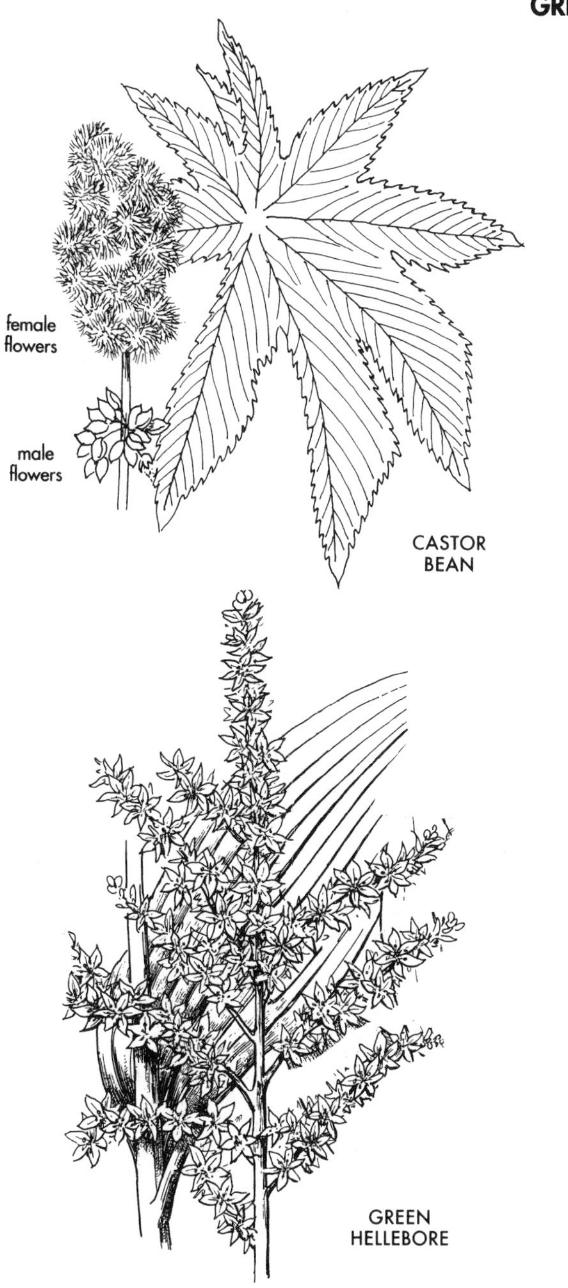
female
flowers
male
flowers
CASTOR
BEAN
GREEN
HELLEBORE

FLOWERS INCONSPICUOUS IN UPPER AXILS OR TERMINAL CLUSTERS

COCKLEBUR **Seeds**

Xanthium strumarium L. Aster Family

Highly variable, weedy annual; to 5 ft. Leaves oval to heart-shaped at base, irregularly lobed and toothed, on long stalks. Flowers greenish, inconspicuous in leaf axils. The burlike oval fruits are covered with *hooked prickles;* July–Nov.

Where found: Fertile damp waste places, often along stream banks. Throughout. Found worldwide.

Comments: The seeds are the most poisonous part of the plant, but they are seldom consumed because of the burs in which they are encased. However, the toxic components, including carboxyatractyloside, and possibly hydroquinone, are apparently present in the young sprouts, which when consumed by livestock produce loss of appetite, nausea, prostration, depression, possibly convulsions, and death. Nervous system symptoms are most pronounced. Pigs are more likely to eat the plant than other livestock. Humans may develop dermatitis from contact with the plant.

AMERICAN WORMSEED **Pl. 34** **Seeds, Whole plant**

Chenopodium ambrosioides L. Goosefoot Family

Stout *aromatic* annual, biennial, or perennial; 3–5 ft. Leaves wavy-toothed, rather like *oak leaves,* smooth or with *yellow glands.* Flowers in spikes, *among the leaves; Aug.–Nov.*

Where found: Weed of waste places and pastures. Much of the U.S. Alien. Tropical America.

Comments: While the leaves are used as a flavoring for bean dishes in the Southwest and Mexico, harvesting the plant may cause dermatitis or allergic reactions. Overdoses of the seed oil, traditionally used to expel intestinal worms, has caused deaths in humans and animals. Excessive handling of plant may cause dizziness from fumes released by the essential oil.

DURANGO ROOT **Root**

Datisca glomerata (Presl.) Baill. Datisca Family

Stout perennial with a *casual resemblance to marijuana;* 3–6 ft. Stems in clusters. Leaves alternate, opposite, or whorled. Upper leaves lance-shaped overall; to 6 in. long, with *prominent lobes at base; strongly toothed;* lower leaves with more divisions. Flowers in dense clusters in leaf axils; May–June. Fruit a capsule.

Where found: Washes and dry stream beds. Calif. Coast Ranges, Sierra Nevada, south to Mexico.

Comments: Flowering and fruiting plants are toxic to cattle, though it is generally unpalatable and eaten only when other feed is unavailable. The root was used to stun fish by California Indians. Symptoms of poisoning in cattle include depression, listlessness, watery diarrhea, rapid respiration, and death. Unlikely to be consumed. The root of the related species *Durango cannabina* L., native to Asia Minor and India, has been used to produce a yellow dye and as a purgative.

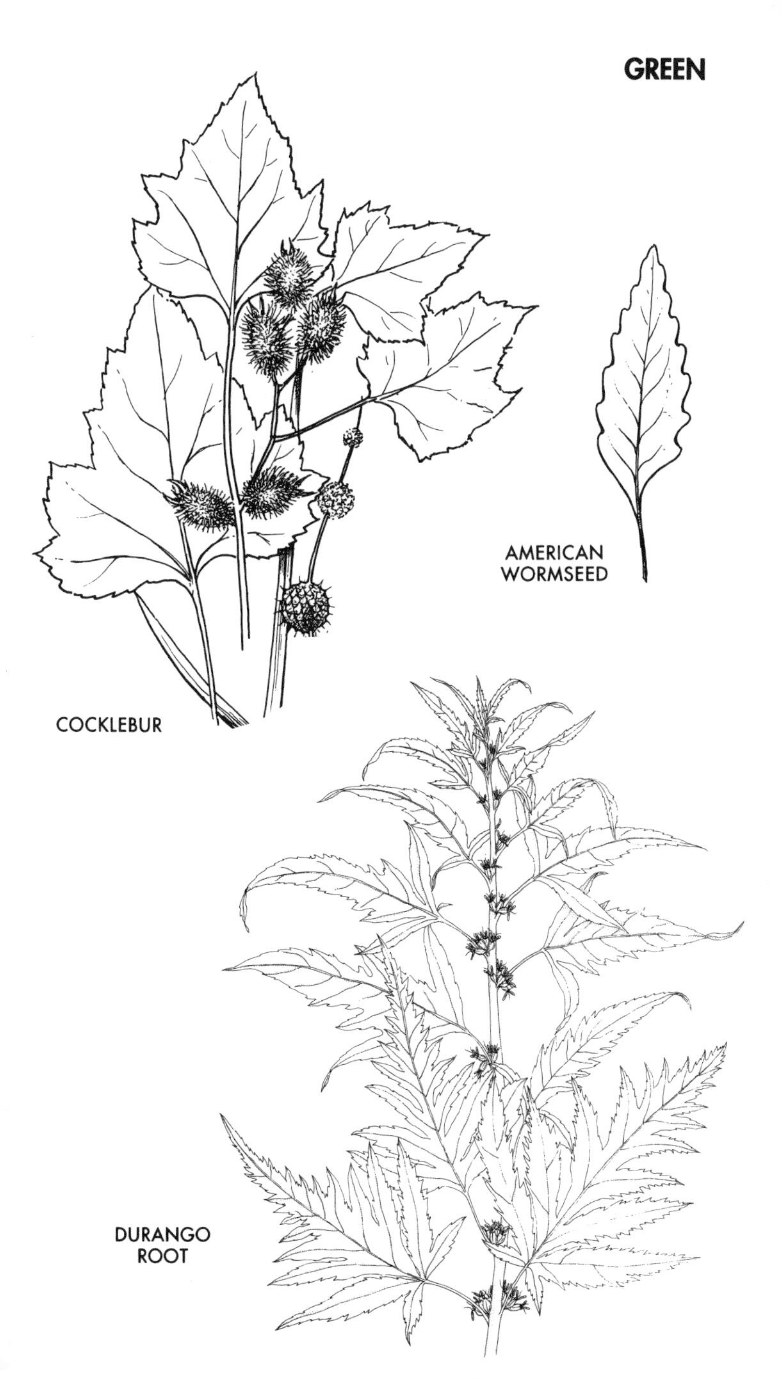
GREEN
AMERICAN
WORMSEED
COCKLEBUR
DURANGO
ROOT

ASTER FAMILY
NODDING, INCONSPICUOUS FLOWERS

COMMON RAGWEED **Pollen**
Ambrosia artemisiifolia L. Aster Family

Annual; 1–3 ft. Leaves opposite on lower stalk, alternate above; much divided into small lobes, the lobes often toothed, with white downy hairs above and. Flowers in spiky racemes, inconspicuous; July–Oct. Many people blame their ragweed allergies on the showy and abundant goldenrods (*Solidago* species), which bloom at the same time.

Where found: Waste ground, disturbed soils. Much of our area, except the desert Southwest.

Comments: While not poisonous, ragweeds are probably responsible for more human suffering than any other plant group in the U.S. Thousands suffer hay fever and allergies because of its pollen. Normally unpalatable to livestock, ragweeds become palatable after application of the herbicide 2,4-D, which can accumulate in ragweeds, causing secondary poisoning. They also cause contact dermatitis. False or Bastard Feverfew *Parthenium hysterophorus* (not shown), a weedy plant from Florida to Texas, has recently been implicated as a generally unrecognized culprit of late-summer allergies in Gulf Coast states, causing ragweedlike symptoms.

GIANT RAGWEED **Pl. 23** **Pollen**
Ambrosia trifida L. Aster Family

Large, coarse annual; 6–15 ft. Whole plant *rough to touch,* with stiff hairs. Leaves opposite, *deeply 3(5)-lobed.* Lobes with pointed tips. Individual flowers small, greenish, upside down on conspicuous erect spikes; July–Oct.

Where found: Alluvial waste places. Often forms vast, pollen-producing stands.

Comments: See Common Ragweed above.

WORMWOOD **Pl. 23** **Whole plant**
Artemisia absinthium L. Aster Family

Fragrant perennial; 1–4 ft. Leaves silver-green, strongly divided into *blunt segments,* with *silky silver hairs on both sides.* Flowers small, drooping, greenish yellow; July–Sept.

Where found: Waste ground. Escaped from cultivation in the n. U.S. Alien. Europe.

Comments: This intensely bitter herb is commonly grown in herb gardens and was formerly used for flavoring the now-outlawed beverage absinthe. Contains a toxic monoterpene, thujone, in its essential oil. In the form of absinthe, thujone may be addictive and cause personality changes, forgetfulness, delirium, and lesions of the brain cortex, possibly leading to convulsions, brain damage, and death. It is found in a number of other common plants, including Yarrow *Achillea millefolium* and some strains of Tansy *Tanacetum vulgare.*

COMMON RAGWEED

GIANT RAGWEED

WORMWOOD

MISCELLANEOUS BROWN/GREEN FLOWERS

JACK-IN-THE-PULPIT **Pl. 18** **Whole plant**
Arisaema triphyllum (L.) Schott Arum Family
Perennial; 1–2 ft. Leaves 1–2, divided into *3 leaflets.* Spathe *cuplike, with curved flap on top,* green to purple-brown with whitish to green stripes; May–July. Fruits *bright red,* in clusters.
Where found: Moist woods. S. Canada, Me. to Fla.; La. to Minn., including eastern Great Plains.
Comments: Ingesting any part of fresh plant may cause intense burning from irritating calcium oxalate crystals. Root considered edible when dried, aged, and properly prepared.

BLUE COHOSH **Pl. 19** **Berries**
Caulophyllum thalictroides (L.) Michx. Barberry Family
Smooth-stemmed, waxy perennial; 1–2 ft. Leaves compound with 3 (to 5) 2–3-*lobed leaflets.* Flowers greenish yellow (or darker) in terminal clusters, with 6 *petallike sepals* and 6 stamens; April–June, before leaves expand. Fruits blue berries.
Where found: Moist rich woods. N.B. to S.C.; Ark., N.D. to Man.
Comments: The raw blue berries have reportedly poisoned children, apparently irritating the gastrointestinal tract, especially the mucous membranes of the intestines. Fresh berries, leaves, and roots may cause contact dermatitis. Contains a number of alkaloids and saponins. Root used medicinally.

SKUNK CABBAGE **Pl. 17** **Whole plant**
Symplocarpus foetidus (L.) Nutt. Arum Family
Large-leaved, *skunk-scented* perennial; 1–2 ft. Leaves broad ovals. Flowers broad, *hooded,* green to purple-brown; spathe sheathed, with clublike spadix; April–early July.
Where found: Wet woods. N.S. to Ga., north to Ill., Iowa.
Comments: Ingesting leaves may cause gastrointestinal burning and inflammation. Roots, formerly used medicinally, considered narcotic and emetic.

HENBANE **Whole plant**
Hyoscyamus niger L. Nightshade Family
Annual or biennial; 8–30 in. Leaves oval-oblong, strongly wavy-toothed; lower leaves short-stalked, upper ones somewhat clasping. Flowers dirty yellow, with *violet veins;* inner base *dark purple;* June–Sept. Globular capsule with flared persistent calyx, many black seeds.
Where found: Infrequent, escaped. Throughout. Alien. Europe.
Comments: Long used in medicine. Contains the toxic alkaloids hyoscyamine, scopoloamine, and atropine. Ingestion may cause salivation, headache, nausea, hallucinations, increased cardiac output, coma, and death. Poisoning now rare.

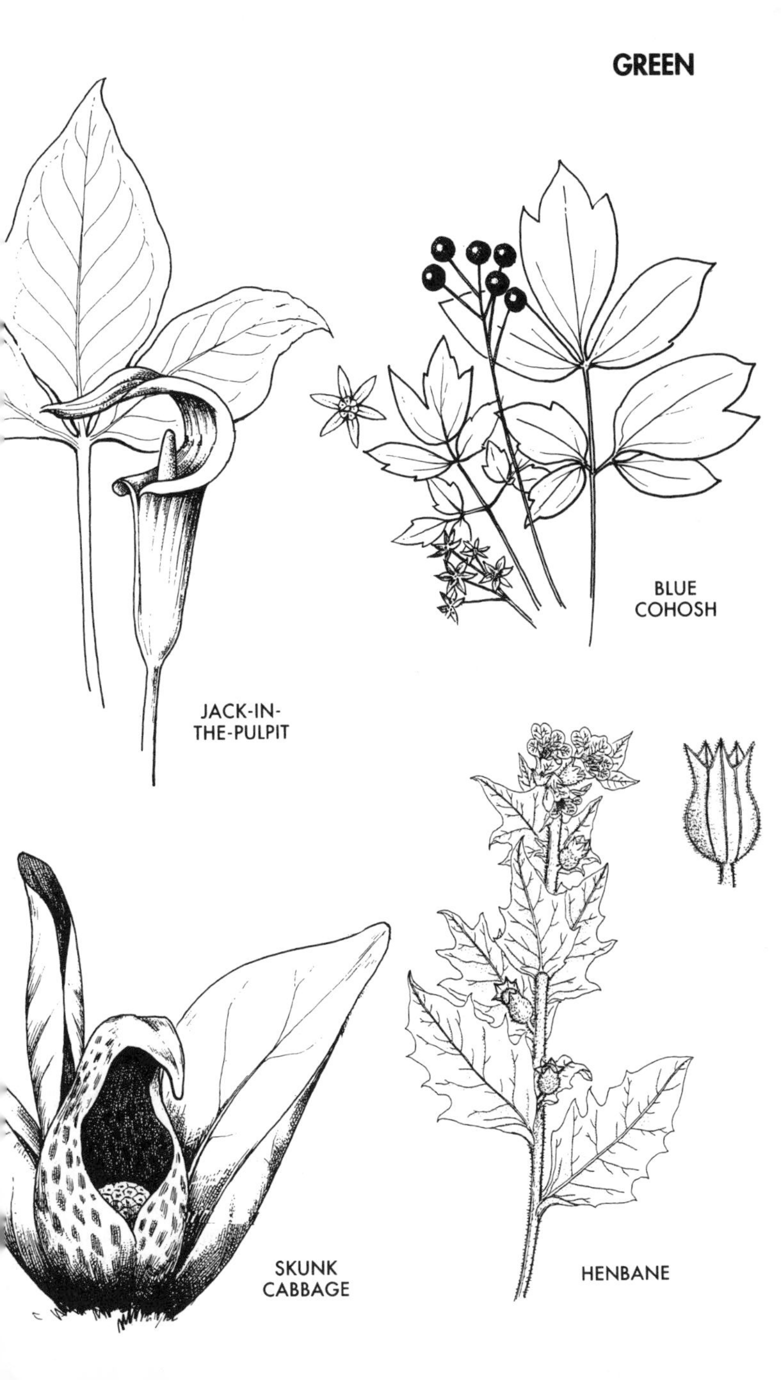
BLUE COHOSH
JACK-IN-THE-PULPIT
SKUNK CABBAGE
HENBANE

PARASITIC SHRUBS; MISTLETOES

EASTERN MISTLETOE **Pl. 40** **Whole plant**

Phoradendron serotinum (Raf.) M.C. Johnston Mistletoe Family

Thick-branched, *parasitic,* semievergreen perennial; forming spheres 1–2 ft. across. Leaves opposite, *shiny,* succulent, oblong to oval, broadest at apex; to 3 in. long. Fruits *translucent white* berries; July–Sept.

Where found: Deciduous trees. N.J. to Fla.; Mo., e. Texas, to Ohio, Minn.

Comments: Mistletoe family members are potentially poisonous. Children may be attracted to the berries of holiday decorations. The plant contains toxic amines and a toxic lectin, phoratoxin. Eating a few berries may result in vomiting, diarrhea, and moderate stomach and intestinal pain. In severe cases there may be labored breathing, dramatically lowered blood pressure, and heart failure. Deaths have been reported from ingesting tea made from berries. While all parts are considered potentially toxic, human fatalities have involved ingestion of the berries. Usually avoided by livestock, but cattle have died from eating parasitic mistletoes.

GREENLEAF MISTLETOE, HAIRY MISTLETOE **Whole plant**

Phoradendron tomentosum (DC.) Englm. ex Gray
Mistletoe Family

Shrubby, parasitic; 1–3 ft. Twigs and leaves usually downy. Leaves *leathery, yellow-green,* in opposite pairs on green woody twigs. Berries white or pinkish; Nov.–Dec.

Where found: Deciduous trees, including hackberries, oaks, walnuts, etc., in lowlands, flood plains. S.-cen. Okla., Texas panhandle to Mexico, Calif.

Comments: See Eastern Mistletoe above.

MESQUITE MISTLETOE **Whole plant**

Phoradendron californicum Nutt. Mistletoe Family

A twiggy shrub with leaves reduced to scales, in large clusters, commonly parasitic on mesquite bushes; 2–24 in. Slender branches have tendency to droop. Berries *red.*

Where found: Mojave and Colorado deserts. Utah (rare), Nev., Ariz., Calif. to Baja Calif.

Comments: See Eastern Mistletoe above. There are at least 17 *Phoradendron* species in North America. We also have about 28 species of Dwarf Mistletoe *Arceuthobium* species (not shown), mostly found on cone-bearing trees in the western U.S. European Mistletoe *Viscum album* (see Plate 40), naturalized in Sonoma Co., Calif., is used medicinally in Europe in prescribed, safe-dosage forms and amounts. Oral preparations are used to treat mild hypertension. Injectable forms are used for treatment of arteriosclerosis. Oral forms of the European species are said to produce reactions only in large amounts. The viscotoxins, toxic proteins, are found in the leaves but not in the berries, which may explain why poison control centers in Europe have not reported symptoms from ingesting berries and consider their toxicity to be slight. Berries of American mistletoes are toxic.

EASTERN
MISTLETOE

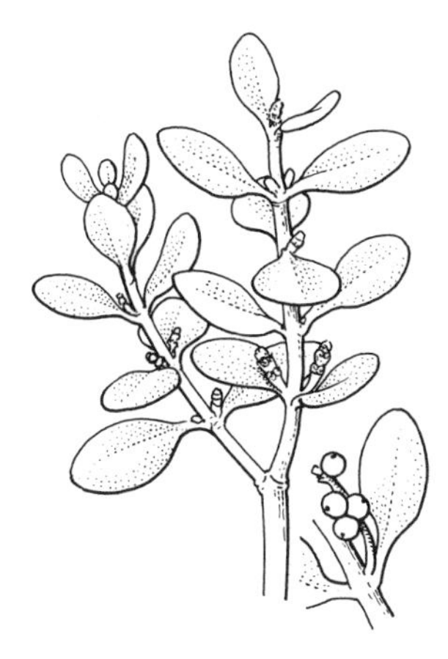

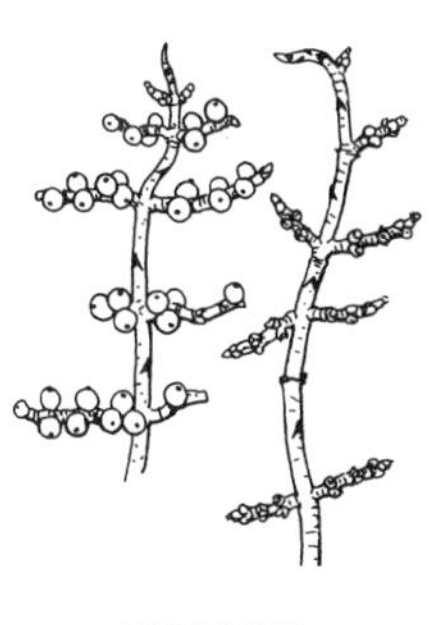

MESQUITE
MISTLETOE

GREENLEAF
MISTLETOE

COMMON JUNIPER **Pl. 38** **Fruits, leaves**

Juniperus communis L. Cypress Family

Low evergreen shrub; to 5 ft. Leaves *needlelike, in whorls of 3,* often with 2 white bands above (or 1 white band divided by a green midrib). Fruits on short stalk; globe-shaped, bluish black with whitish film that rubs off, usually 3-seeded; May–June.

Where found: Sandy or dry sites. Canada to S.C., Ind., Ill., to Minn. Mostly absent from Plains states, Mont. to N.M. and Calif.

Comments: Fruits used to flavor gin and as a diuretic. Eating a few berries usually does not produce symptoms, though large or frequent doses may cause digestive irritation or kidney failure. Not often involved in cases of human poisoning.

CANADA YEW **Seeds, leaves**

Taxus canadensis Marsh. Yew Family

Straggling evergreen; 1–3 ft. (rarely to 7 ft.). Twigs *smooth,* green; reddish brown on older branches. Leaves 2-ranked needles, 3/8–1 in. long, with fine abrupt points. Needles red-tinted in winter. Fruits encased in showy, juicy, *cuplike red arils* (pulp); seeds stony.

Where found: Rich woods. Nfld. to w. Va.; ne. Ky. to Iowa, north to Man.

Comments: See Western Yew. The genus *Taxus* includes seven species from northern temperate regions. The common English Yew *Taxus baccata* (see Plate 36), commonly grown as an ornamental plant, is considered the most poisonous woody plant in Britain. Severe poisoning is reportedly rare and usually involves thorough chewing of the toxic seeds within the nontoxic red aril. The poisonous Japanese Yew *T. cuspidata* (not shown) is also commonly grown in the U.S. The Florida Yew *T. floridiana* (not shown) occurs in nw. Florida.

WESTERN YEW **Seeds, leaves**

Taxus brevifolia Nutt. Yew Family

Small tree; 10–30 ft. Bark thin, reddish brown, somewhat shredded. Leaves linear, 2-ranked, flat, 1/2–2/3 in. long; *midrib raised above and below;* spreading in flat sprays. Fruits berrylike, encased by a *scarlet cuplike aril;* on underside of branches; April–May.

Where found: Understory tree of pine and fir woods. Rare. Mont. to Santa Cruz Mountains of Calif., north through Pacific Northwest to Alaska.

Comments: Yews contain taxine alkaloids. Leaves and seeds (but not the red arils) are toxic. Initial symptoms include dizziness, blurred vision, and dry mouth, followed by salivation, intense abdominal cramps, vomiting, and dilation of the pupils. Severe cases result in coma, cardiac or respiratory failure, and death. Chewing the leaves may also cause allergic reactions and a rash. Ingesting as few as 50 leaves has resulted in death. The Western Yew has been the subject of intense interest as a source of taxol, a promising drug for the treatment of ovarian cancer.

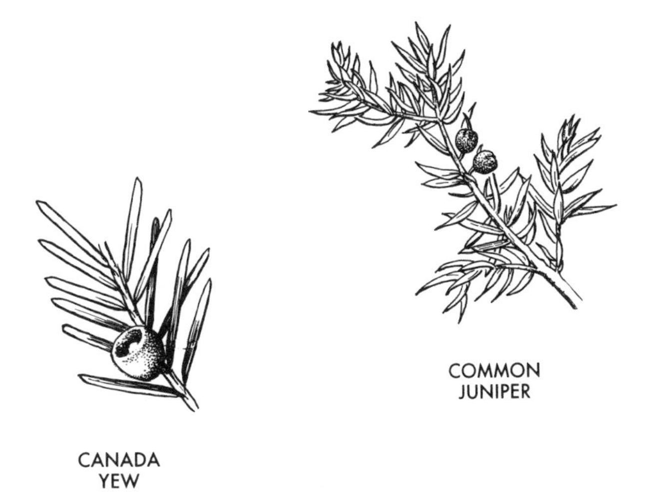
COMMON
JUNIPER
CANADA
YEW

WESTERN
YEW

DAPHNE **Whole plant**

Daphne mezerum L. Mezerum Family

Rounded, deciduous shrub; 1–4 ft. Leaves alternate, simple; 2–3 in. long. Flowers fragrant, lilac to white, with 4 showy calyx lobes, in clusters of 2–5, among branches of previous year's growth; early spring, before leaves. Fruits red (or yellow).

Where found: Ornamental. Naturalized N.S. to N.Y.; west to Ohio. Alien. Europe.

Comments: Children may be attracted to fruits; even a few can be fatal. Contains poisonous glycosides. May cause burning and ulceration of the mouth, throat, and stomach, with vomiting, abdominal pain, and bloody diarrhea. In severe cases, death may be preceded by delirium, convulsions, and coma. Leaves are a skin irritant.

DOUGLAS BUTTERWEED **Leaves**

Senecio douglasii DC. Daisy Family

Bushy, woolly-hairy shrub; 2–6 ft. Leaves 1–3 in. long, pinnately divided, 3–9 linear lobes. Flowers yellow; July–Oct.

Where found: Dry soils, chaparral. Common; Texas to Nev., Calif.

Comments: Not generally eaten, but suspected of livestock poisoning.

YAUPON HOLLY **Berries, leaves**

Ilex vomitoria Ait. Holly Family

Evergreen shrub; 6–15 ft. Leaves elliptical, leathery, *round-toothed; 1–2 in. long.* Berries red (or yellow) in clusters hugging the branches; Sept.–Nov.

Where found: Sandy woods. Se. Va. to Fla.; Texas to Ark.

Comments: Berries may cause vomiting or diarrhea if eaten to excess by small children. Leaves used by Indians to make a ceremonial beverage, "black tea" or "yaupon tea." If drunk in large amounts, induces vomiting. Contains caffeine.

BOXWOOD **Pl. 36** **Leaves**

Buxus sempervirens L. Box Family

Evergreen shrub; *stems winged or angular.* Leaves opposite, leathery, oval, to 3/4 in. long; light beneath, with whitish midrib. Flowers small, in axillary clusters. Fruit a 3-celled capsule with black seeds.

Where found: Commonly grown as a hedge. Sometimes naturalized. Throughout.

Comments: Ingesting any plant part may cause abdominal pain, vomiting, and diarrhea. Large doses can result in convulsions, respiratory failure, and death. May cause dermatitis.

STAGGERBUSH **Pl. 37** **Leaves**

Lyonia mariana (L.) D. Don. Heath Family

Slender, deciduous shrub; to 7 ft. Leaves *thin, oblong to oval, untoothed;* to 4 in. Flowers white to pinkish, bell-shaped, in *umbel-like racemes,* in clusters on old leafless branches; April–June.

Where found: Sandy pine thickets. S. R.I., N.Y. to Fla.; e. Texas to Ark.

Comments: Leaves and nectar (and honey made from the nectar) contain lyoniatoxin (lyoniol A). May cause burning in mouth, then salivation, vomiting, diarrhea, and a prickling sensation of skin. Severe poisoning can result in coma and convulsions. Poisonous to grazing animals.

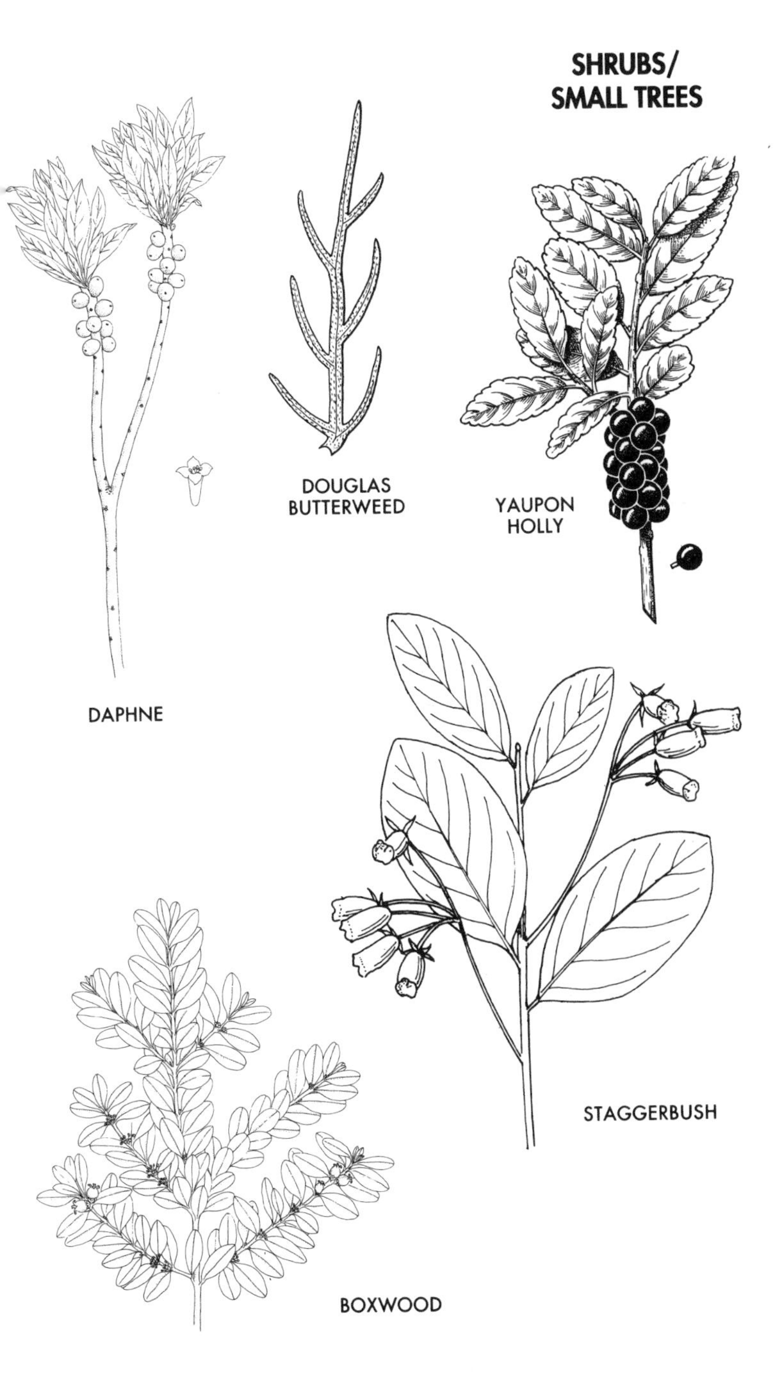
SHRUBS/
SMALL TREES
DAPHNE
DOUGLAS
BUTTERWEED
YAUPON
HOLLY
STAGGERBUSH
BOXWOOD

EVERGREEN SHRUBS WITH OPPOSITE LEAVES AND 5–PARTED FLOWERS

MOUNTAIN LAUREL **Pl. 37** **Leaves**
Kalmia latifolia L. Heath Family

Shrub or small tree; 5–30 ft. Leaves mostly alternate, leathery, oval, hairless, without teeth, *stalked;* to 5 in. long. Flowers *terminal,* numerous, white to pink-rose, with purple markings, to 1 in. wide; May–July.

Where found: Sandy or rocky woods. N.B. to Fla.; La. to s. Ind., and Ont. Sometimes cultivated elsewhere.

Comments: See Sheep Laurel below.

SHEEP LAUREL, LAMBKILL **Pl. 37** **Leaves, flowers**
Kalmia angustifolia L. Heath Family

Slender shrub; 3–5 ft. Leaves opposite, leathery, elliptical to lance-shaped, to 2 in. wide. Flowers pink to red-purple, angular cup–shaped, to 1/2 in., in clusters from *axils of previous year's leaves;* May–July.

Where found: Dry acid soils. Nfld. to Va.; Ga. to Mich. Sometimes cultivated elsewhere.

Comments: Seven species of *Kalmia* are known from North America (and Cuba), two of which have become naturalized in Europe. Toxic diterpenes known as acetylandromedols (andromedotoxins) are considered the poisonous constituents of *Kalmia* species and other toxic heath family members. While the plants are seldom involved in cases of poisoning, children have been poisoned from sucking on the flowers or playing with the leaves. Symptoms may include salivation, burning of the mouth, watery eyes and nose, listlessness, dizziness, vomiting, diarrhea, abdominal cramps, and itching and burning of mucous membranes and skin. Cardiac disturbances including low blood pressure and slow pulse are also reported. Convulsions, paralysis, and death have been reported. Honey from the flowers of poisonous heath family members is considered potentially toxic.

SWAMP LAUREL **Leaves**
Kalmia polifolia Wang. Heath Family

Somewhat branched shrub; to 3 ft. Twigs *sharply 2–edged.* Leaves opposite, linear to lance-shaped, papery, stalkless, to 1 3/4 in. long. Flower clusters *terminal,* with folded bract beneath, rose-purple, to 5/8 in., across; May–June.

Where found: Bogs. Lab. to Conn.; Pa., Minn.; Calif. to Alaska.

Comments: See Sheep Laurel above.

WESTERN LAMBKILL **Leaves**
Kalmia microphylla (Hook.) Heller Heath Family

Sometimes treated as a variety of *K. polifolia* (see above). Low, straggling alpine shrub; 8 (rarely to 24) in. Leaves opposite, oval to lance-shaped, mostly stalkless; to 1 1/2 in. long, typically less than 3/4 in. Flowers in *terminal* clusters.

Where found: Mountains. Colo., Calif. to B.C.

Comments: See Sheep Laurel above.

MOUNTAIN
LAUREL

SHEEP
LAUREL

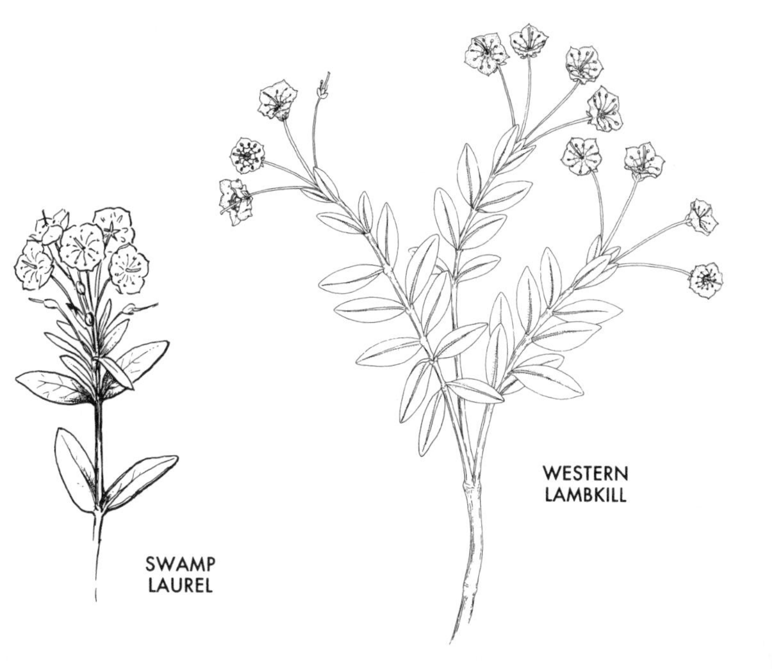
SWAMP
LAUREL
WESTERN
LAMBKILL

MISCELLANEOUS EVERGREEN HEATHS

GREAT RHODODENDRON **Pl. 37** **Leaves**

Rhododendron maximum L. Heath Family

Evergreen shrub to small tree; 10–30 ft. Forms thickets. Leaves oblong, untoothed, *leathery*, with edges *rolled under*; to 10 in. long. Flowers rose-pink (white), spotted, in showy clusters; June–July.

Where found: Damp woods. S. Me. to Ga.; Ala. to Ohio.

Comments: Toxins and symptoms same as Sheep Laurel *Kalmia angustifolia* (see page 158).

CALIFORNIA ROSE-BAY **Leaves**

Rhododendron macrophyllum D. Don ex G. Don Heath Family

Evergreen shrub to small tree; 3–15 ft. Leaves elliptical, smooth, untoothed; 4–10 in. long. Flowers in dense, showy, terminal corymbs; rose to white, $1\frac{1}{2}$ in. across.

Where found: Woods and thickets. Calif. to Wash.

Comments: Toxins and symptoms same as Sheep Laurel *Kalmia angustifolia* (see page 158).

MOUNTAIN FETTERBUSH **Leaves**

Pieris floribunda (Pursh ex Sims) Benth. & Hook. Heath Family

Attractive evergreen shrub; 3–5 ft. Leaves alternate, leathery, usually with *hairs along margins*. Small white flowers in dense terminal raceme clusters; April–May.

Where found: Bald knobs and moist slopes. Mountains. Va. to N.C., Tenn. to W. Va.

Comments: Toxins and symptoms same as Sheep Laurel *Kalmia angustifolia* (see page 158). A Japanese native, Lily of the Valley Bush *P. japonica*, a shrub or small tree to 10 ft., is often grown as an ornamental.

SIERRA LAUREL **Leaves**

Leucothoe davisiae Torr. ex Gray Heath Family

Evergreen shrub; 2–5 ft. Leaves alternate, stiff, and somewhat leathery, oblong; without teeth or slightly toothed; 1–3 in. long. Flowers whitish, in erect terminal racemes; 2–4 in. long; June–July.

Where found: Wet, boggy mountain meadows, 6,000–8,500 ft. Calif. to s. Ore.

Comments: One of the most poisonous members of the heath family. Only 1–2 ounces of leaves is a lethal dose for sheep. Poisoning symptoms similar to those of Sheep Laurel *Kalmia angustifolia*, page 158.

DROOPING LEUCOTHOE, FETTERBUSH **Leaves**

Leucothoe axillaris (Lam.) D. Don. Heath Family

Evergreen shrub; 3–6 ft. Leaves alternate, leathery, shiny, sharply pointed, broadly oval; to 6 in. long; on stalks to $\frac{1}{2}$ in. long. White flowers in dense racemes; April–May.

Where found: Wet woods, swamps, bogs; coastal plain. Se. Va. to Fla.; Miss. to La.

Comments: See Sierra Laurel and Sheep Laurel *Kalmia angustifolia* on page 158.

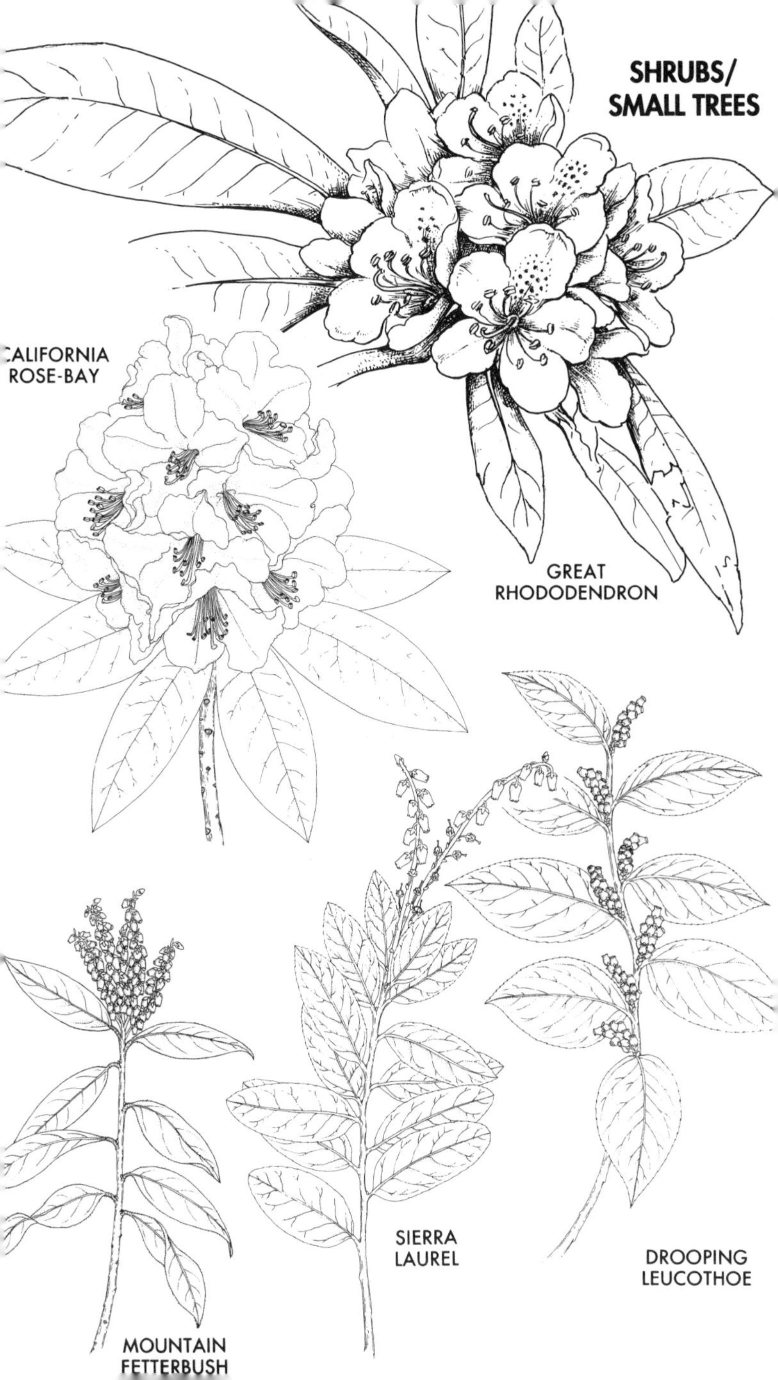
SHRUBS/
SMALL TREES
CALIFORNIA
ROSE-BAY
GREAT
RHODODENDRON
SIERRA
LAUREL
DROOPING
LEUCOTHOE
MOUNTAIN
FETTERBUSH

PRIVET **Fruits**

Ligustrum vulgare L. Olive Family

Much-branched shrub; 6–25 ft. Leaves persistent but deciduous, *opposite,* smooth, elliptical to lance-shaped, untoothed, dark green above, lighter beneath; to 2 in. Flowers fragrant, whitish, in small panicles; June–July. Fruits glossy black berries with *2 violet seeds.*

Where found: Cultivated. Commonly escaped. Alien. Europe.

Comments: Fruits are said to have poisoned children in Europe, but documentation is weak. May cause abdominal pain, vomiting, and diarrhea. In most documented cases reported to European poison control centers, no symptoms occurred. Of cases that did produce symptoms, digestive disturbances were most commonly reported. At least five additional *Ligustrum* species are commonly grown in the U.S.

WILD HYDRANGEA **Pl. 20** **Leaves**

Hydrangea arborescens L. Saxifrage Family

Shrub; 3–6 ft. Leaves opposite, oval, toothed, to 5 in. long. Flowers in flat-topped clusters, to 6 in. wide, with *papery, white, sterile, calyx lobes along perimeter of flower clusters;* June–Aug.

Where found: Rich, shady slopes. N.Y. to n. Fla., La.; Okla. to Ind., Ohio.

Comments: A cyanogenic glycoside, hydrangin, is found in the leaves. Ingestion may cause nausea, vomiting, and bloody diarrhea. Root has traditionally been used as a diuretic. Unlikely to be consumed. Several *Hydrangea* species are commonly grown as ornamentals, especially the Japanese native *H. macrophylla.*

WAHOO **Pl. 38** **Leaves, fruits**

Euonymus atropurpureus Jacq. Staff Tree Family

Shrub or small tree; 6–25 ft. Leaves opposite, oblong-oval, *stalked, hairy beneath.* Flowers purplish; June–July. Fruits purplish, *smooth,* irregularly 4-lobed; seeds covered with *scarlet aril* (pulp); May–Oct.

Where found: Rich woods. Ont. to Tenn., Ala.; Texas to N.D.

Comments: See Strawberry Bush. Historically, Wahoo has been valued for its emetic (inducing vomiting) and laxative effects.

STRAWBERRY BUSH **Leaves, fruits**

Euonymus americanus L. Staff Tree Family

Erect deciduous shrub; 3–6 ft. Stems green, 4-angled. Leaves opposite, lance-shaped to oval, lustrous, sharp-pointed, stalkless. Flowers 1–3, greenish purple; *petals stalked;* May–June. Fruits *scarlet, warty;* June–Oct.

Where found: Rich woods. Se. N.Y. to Fla.; Texas, Okla. to Ill.

Comments: *Euonymus* species contain cardiotonic glycosides and possibly alkaloids in the leaves, stems, and fruits. Cases of poisoning are reported for the European Spindle Tree *E. europeaus* L., historically producing symptoms such as colic, diarrhea, fever, circulatory problems, and collapse 8–15 hours after ingestion. Modern cases of poisoning have produced only mild symptoms. Human poisoning is not reported for the American species, but they should not be ingested.

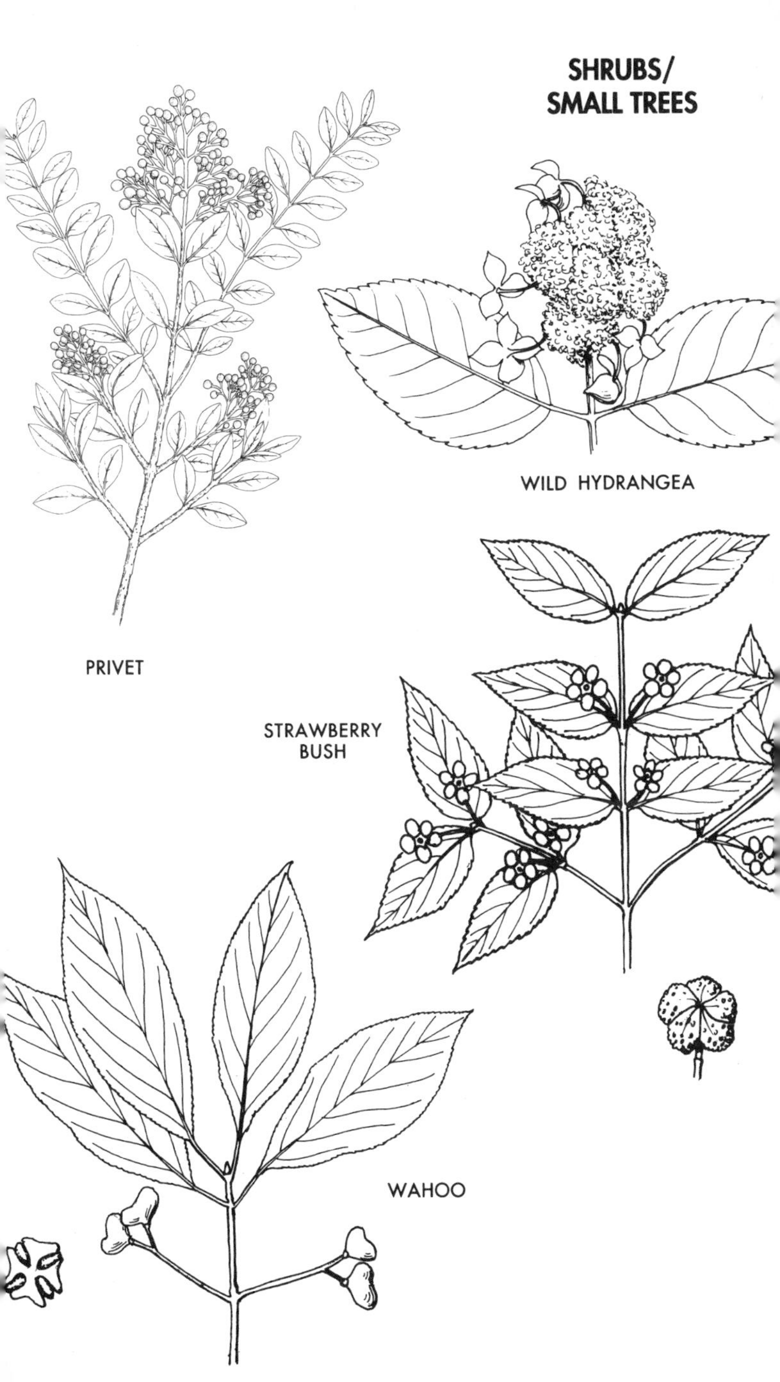
SHRUBS/
SMALL TREES
WILD HYDRANGEA
PRIVET
STRAWBERRY
BUSH
WAHOO

ELDERBERRIES
COMPOUND LEAVES

AMERICAN RED ELDERBERRY **Whole plant**
Sambucus racemosa L. ssp. *pubens* (Michx.) House
Honeysuckle Family

Deciduous shrub; to 15 ft. Pith *reddish brown.* Leaves compound; 5–9 elliptical to lance-shaped leaflets; to 4 in. long. Flowers creamy white, in loose *pyramidal clusters* to 4 in. long; June–July. Fruit *scarlet,* inedible.

Where found: Rocky soils, steam banks, moist woods. Nfld. to Ga.; Tenn. to N.D., Alta.

Comments: Also known as *S. pubens* Michx. This is the American subspecies of the European Red Elderberry *S. racemosa.* Only blue-fruited elderberries are considered edible. Fruits of red-berried elders contain a bitter yellow oil. Elderberries also accumulate nitrates from the soil. The leaves, stems, and roots contain cyanogenic glycosides. Ingestion of these plant parts can cause severe diarrhea.

COMMON ELDERBERRY **Pl. 38** **Whole plant**
Sambucus canadensis L. Honeysuckle Family

Deciduous shrub; 8–12 ft. Stem with *large white pith.* Leaves opposite (in pairs), with 5–11 elliptical to lance-shaped, sharply toothed leaflets. Flowers white-yellow, fragrant, in flat, umbrellalike clusters; to 10 in. across; May–July. Fruits purplish black; July–Sept.

Where found: Rich soils, field edges, roadsides. N.S. to Ga.; Texas to Man.

Comments: Elders have often been used in folk medicine to induce vomiting and purging. Avoid consumption of *raw* elderberries. The berries are considered edible *after cooking.* All other plant parts, especially fresh, are potentially poisonous. The dried flowers, made into tea, have been used in folk medicine and are probably nontoxic in relatively small amounts. Children have developed symptoms from placing toys such as pea shooters or whistles made from the stems in their mouths. All seven species of elder found in North America are potentially toxic. Mexican Elderberry *S. mexicana* (not shown), native to the Southwest, has caused cases of nausea, vomiting, abdominal pain, and dizziness.

BLUE ELDERBERRY **Whole plant**
Sambucus cerulea Raf. Honeysuckle Family

Sometimes spelled *S. caerulea.* Shrub or small tree; to 50 ft. Leaves compound, 5–7 oblong leaflets; to 6 in. long. Flowers yellowish white in umbrellalike clusters; to 6 in. across; June–July. Fruits *black-blue,* with whitish bloom that rubs off.

Where found: Open coniferous forest in mountains. Idaho to Utah; Calif., to B.C., Alta.

Comments: See Common Elderberry.

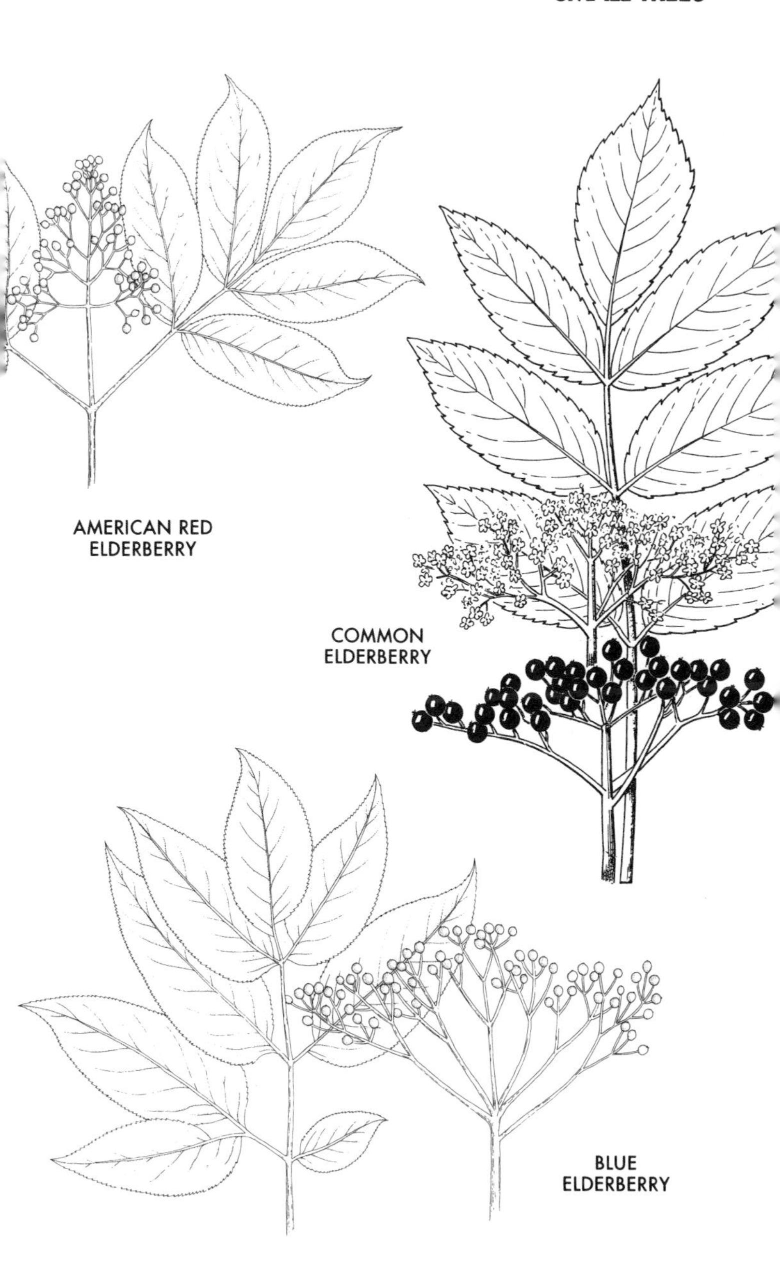
AMERICAN RED
ELDERBERRY
COMMON
ELDERBERRY
BLUE
ELDERBERRY

MISCELLANEOUS LEGUMES WITH PINNATE LEAVES

CATCLAW **Leaves, twigs**

Acacia greggii Gray Pea Family

Rounded, much-branched shrub or small tree, usually 3–6 (20) ft. Armed with *clawlike curved spines.* Leaves 2-pinnate; leaflets 3–7 pairs, oval, broadest at apex, 1/8–1/4 in. long. Flowers creamy white to pale yellow, in spikes to 2 1/4 in. long. Pod often *twisted, constricted between seeds,* to 4 in. long, seeds flat, *nearly round in outline.*

Where found: Dry brush. Texas to Ariz.; se. Calif. to Nev.; n. Mexico.

Comments: The leaves and twigs contain cyanogenic glycosides, which have poisoned cattle. Poisoning is due to hydrocyanic acid, causing rapid breathing, convulsions, paralysis, coma, and death. Poisoning usually occurs around the first frost. Unlikely to be consumed by humans.

PURPLE SESBANIA **Seeds**

Sesbania punicea (Cav.) Benth. Pea Family

Shrub; to 6 ft. Leaves pinnate, with 6–20 even-numbered pairs of leaflets to 1 in. long. Flowers *showy vermilion,* to 3/4 in. long, in hanging 4-in.-long *racemes.* Fruits 4-winged pods, to 4 in. long.

Where found: Ornamental. Naturalized from Fla. to La., S. Calif. Alien. South America.

Comments: This and other *Sesbania* species are toxic. Ingestion of the seeds has caused death in fowl and other livestock. May cause depression, diarrhea, rapid and irregular pulse, difficulty in breathing, coma, and death. Contains saponins and the alkaloid sesbanine. Children may be attracted to the showy flowers or unusual pods.

GUAJILLO **Leaves, pods**

Acacia berlandieri Benth. Pea Family

Hairy, sparingly branched shrub with short straight spines; 3–6 (12) ft. tall. Leaves 2-pinnate, with numerous oblong-linear leaflets less than 1/2 in. long (20–50 pairs). *Fragrant creamy white flowers* in dense clusters from leaf axils. Fruit a heavy, thick-walled, flat, velvety legume to 6 in. long.

Where found: Limestone ridges. Texas to ne. Mexico. Dominant in some areas.

Comments: Affects range animals in sw. Texas. If animals consume large amounts of the leaves and pods over several months, they may develop "Guajillo wobbles," a condition involving loss of muscular coordination, primarily of the hind legs. Animals may be incapacitated, then die. A toxic amine is believed to be responsible for poisoning. Unlikely to affect humans.

SHRUBS/ SMALL TREES

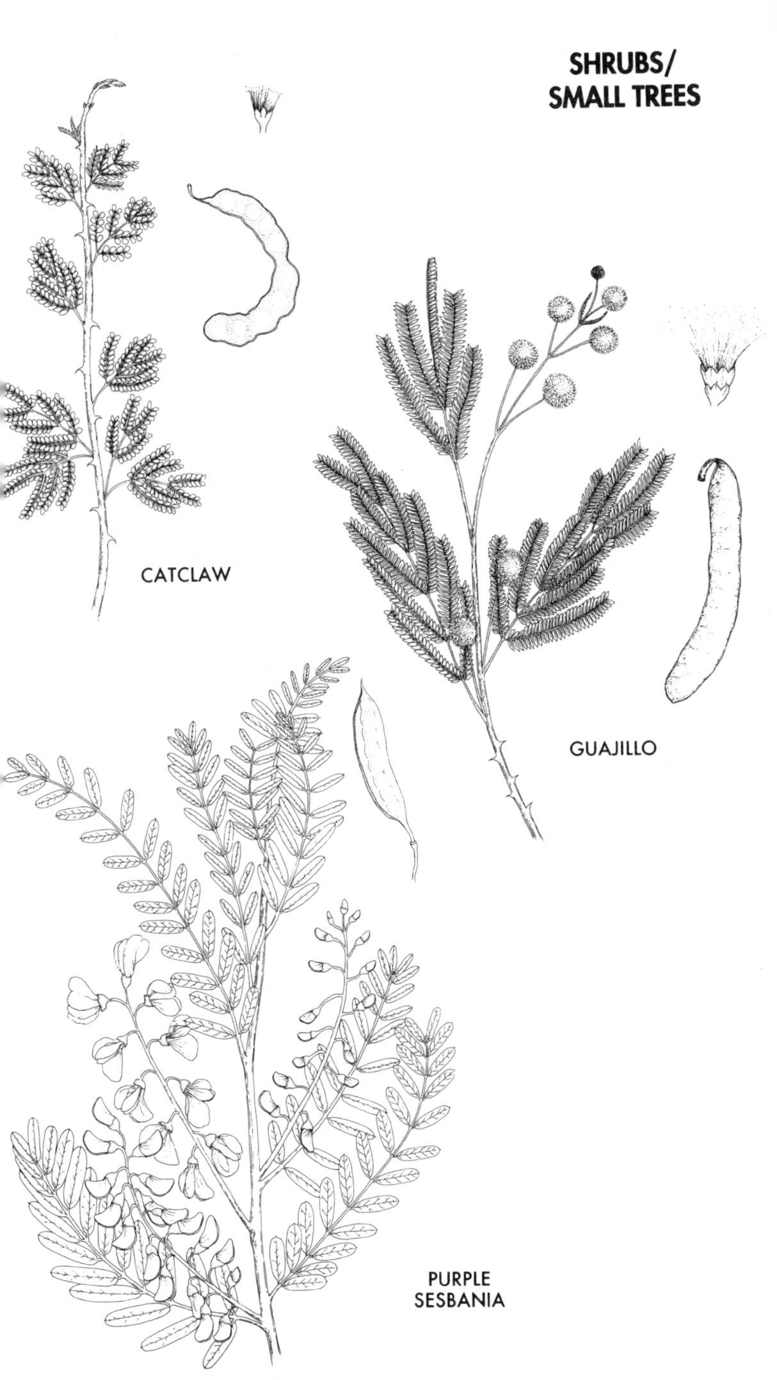

SCOTCH BROOM **Whole plant**
Cytisus scoparius (L.) Link Pea Family
Evergreen shrub, *stems angled;* 3–10 ft. Leaves divided into 3 tiny leaflets, 1/4–1/2 in. long. Flowers yellow, to 1 in., single or paired in axils, *style strongly curved;* May–June. Fruit a flat pod to 2 in. long.
Where found: Pine barrens, sandy soil, roadsides. Eastern and western North America. Alien. Europe. Invasive weed in Calif.
Comments: Contains cytisine and sparteine (alkaloids). Sparteine can delay the development and conduction of cardiac impulse. Cytisine is rapidly absorbed in the digestive tract, including the mouth, causing abdominal pain and diarrhea. Reports of human poisoning are rare and suspect.

GOLDEN CHAIN **Seeds, flowers**
Laburnum anagyroides Medic. Pea Family
Showy, hardy shrub or small tree; 20–30 ft., with wide-spreading or drooping branches. Alternate leaves divided into 3 leaflets, on long stalks; underside of elliptical leaflets has light gray down. Flowers golden yellow, to 3/4 in. long; *10–20 in showy hanging racemes to 18 in. long;* May–June. Pods silky, persistent, to 2 in. long, with flat dark brown seeds.
Where found: Cultivated, sometimes naturalized. Alien. Europe.
Comments: Golden Chain has been responsible for more poisonings in Britain than any other shrub except the Yews. Ingestion of the highly toxic seeds can cause vomiting, diarrhea, colicky pain, lack of coordination, convulsions, coma, and death. Contains the toxic alkaloid cytisine. Poisoning is usually the result of children playing with the pealike seed pods, then chewing the seeds, or even sucking on the showy flowers.

BIRD OF PARADISE BUSH **Pods**
Caesalpinia gilliesii (Wallich ex Hook.) A. Dietr. Pea Family
Unbranched, showy shrub or small tree; to 15 ft. Leaves twice-pinnate, 7–15 main divisions with 7–10 pairs of smooth, oblong leaflets to 3/8 in. long. Flowers yellow in terminal clusters; stalk covered with *yellow glands;* stamens *bright red;* May–Sept.
Where found: Widely cultivated as a container plant. Naturalized in deep South, Southwest. Alien. South America.
Comments: The unripe seed pods may severely irritate the digestive tract, causing nausea, vomiting, and diarrhea. A case of poisoning of two boys was recorded in Arizona. Symptoms appeared within 30 minutes and lasted for about 24 hours.

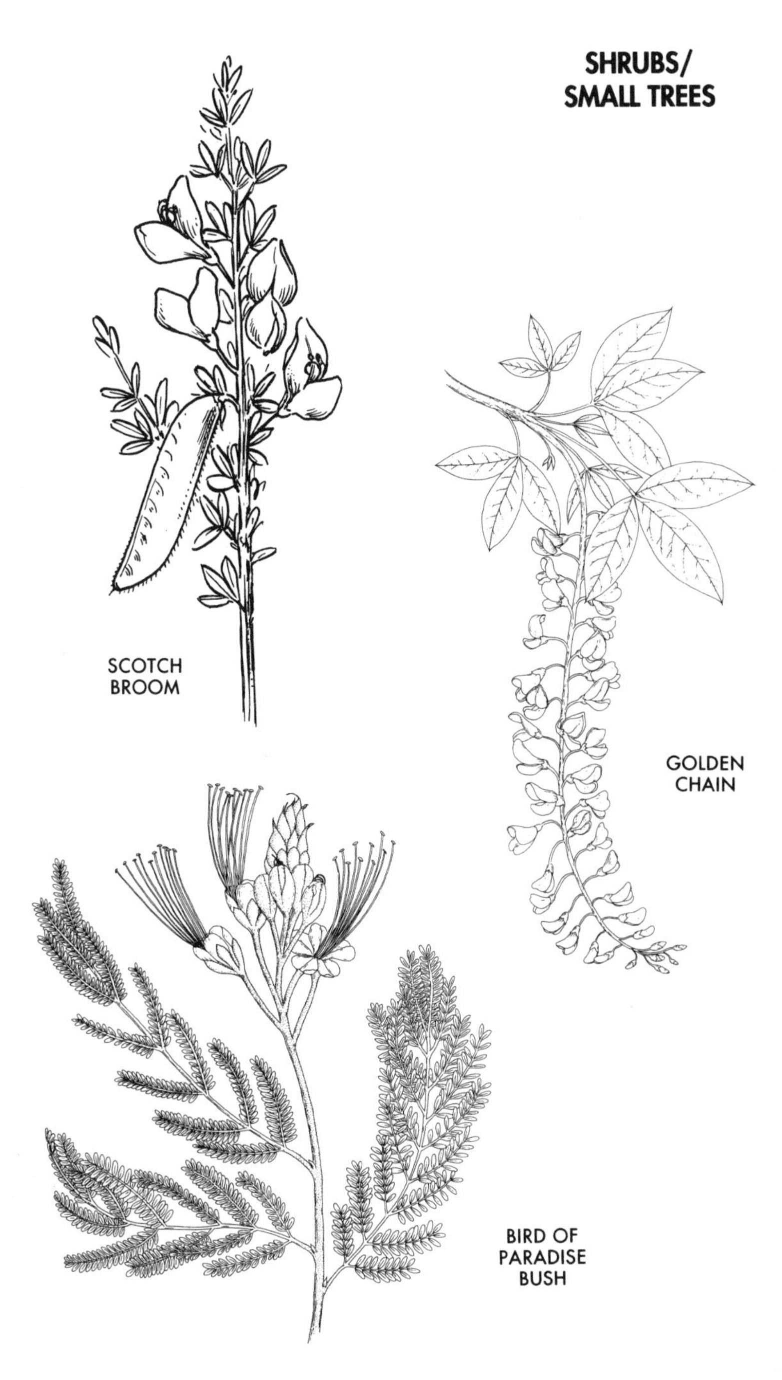
SHRUBS/
SMALL TREES
SCOTCH
BROOM
GOLDEN
CHAIN
BIRD OF
PARADISE
BUSH

MISCELLANEOUS SUBTROPICAL SHRUBS

JESSAMINE **Fruits**

Cestrum spp. Nightshade Family

Jessamines are trees and shrubs of tropical America with simple, untoothed, mostly narrow or elliptical leaves. Flowers are 5-parted, showy, trumpetlike, often greenish white, terminal or in axils, about 1 in. long. Fruits are small berries.

Where found: Willow-leaved Jessamine *C. parqui* L'Her., Day Jessamine *C. diurnum* L., and Night-blooming Jessamine *C. nocturnum* L. are grown in the southern U.S. and naturalized from s. Fla. to s. Texas.

Comments: Ingestion of any part may produce symptoms in humans or livestock, including fever, excitability, gastroenteritis, hallucinations, salivation, and paralysis.

OLEANDER **Pl. 36** **Whole plant**

Nerium oleander L. Dogbane Family

Shrub, 3–20 ft. Dull, dark green leaves in whorls of 3 or opposite; lance-shaped, without teeth; to 10 in. Flowers white, yellow, pink, or purple, 1–2 in., trumpetlike, 5-parted; in showy clusters.

Where found: Widely cultivated as a container plant or in the ground in milder areas, especially Calif. Occasionally naturalized. Alien. Europe.

Comments: If ingested, the leaves can cause vomiting, dizziness, and heart dysfunction due to a cardiac glycoside, oleandrin. Toxic to both livestock and humans. Ingesting sufficient quantities of the leaves can cause death, though few human fatalities are recorded. Sucking the nectar from the flowers has caused poisoning in children. Smoke from the burning branches can also cause poisoning. Severe poisoning is rare because vomiting expels the toxins before they are absorbed into the system.

PHYSIC NUT, CORAL PLANT **Fruits**

Jatropha multifida L. Spurge Family

Shrub, 3 –7 ft.; or small tree up to 20 ft. Leaves rounded, 4–8 (12) in., with 7–11 lobes and untoothed or incised margins. Flowers small, red, in branched clusters. Fruits yellow, 3-angled.

Where found: S. Fla. to tropical America. Cultivated elsewhere.

Comments: All *Jatropha* species are potentially toxic. Some are used medicinally. Attractive fruits have poisoned children. May produce burning in throat, violent vomiting, stomach pain, muscle cramps, bloody diarrhea, coma, and death.

GOLDEN DEWDROP **Fruits**

Duranta repens L. Verbena Family

Highly variable large shrub or small tree, often spiny, with trailing branches; to 18 ft. Leaves opposite, oval, untoothed or with coarse teeth on upper half; to 4 in. long. Lilac flowers up to 1/2 in. across in racemes. Fruits yellow.

Where found: Native to Key West, Fla. Naturalized in s. Texas. Grown in Hawaii.

Comments: Ingestion of fruit may cause gastrointestinal irritation, vomiting, drowsiness, fever, convulsions, and death.

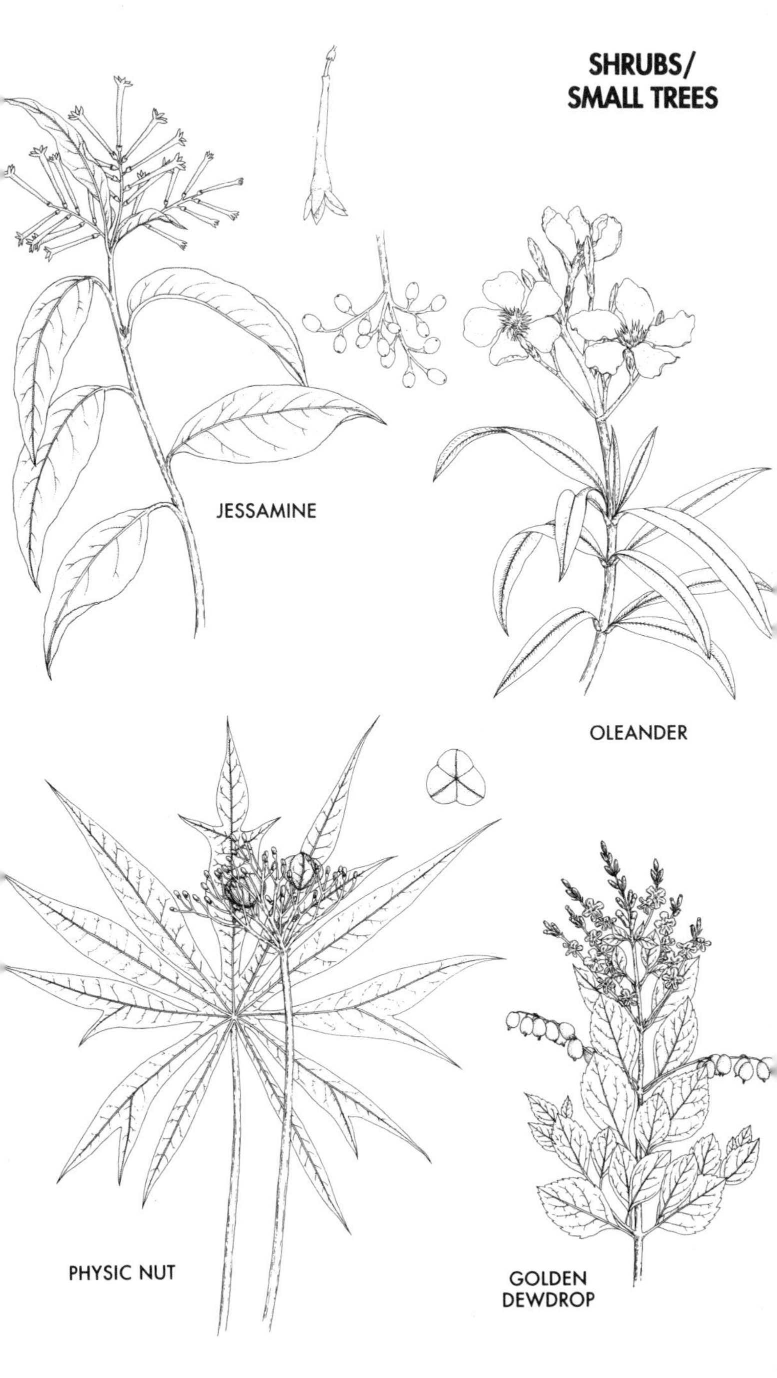
SHRUBS/
SMALL TREES
JESSAMINE
OLEANDER
PHYSIC NUT
GOLDEN
DEWDROP

POISON IVY, POISON SUMAC, AND POISON OAK

Poison ivys, oaks, and sumac are the best known and least liked of all poisonous plants in North America. Fifty percent of the population is allergic to some degree to this plant group, which causes painful, irritating dermatitis in about two million people in the U.S. each year. All plant parts, especially the sap, contain irritant, nonvolatile, phenolic substances called urushiol, or toxicodendrol. The oily mixture, found in resin canals, is released when the plant is bruised, even slightly. Contact, even indirect contact such as patting a dog that has brushed against the plant, causes severe dermatitis including redness, itching, swelling, and blistering. Sometimes requires hospitalization. Eating the fruits or inhaling smoke from burning plants causes similar internal irritation. Droplets of the toxin can become airborne in smoke particles and cause internal irritation. The plant is most dangerous in spring and summer when sap is abundant. Members of this highly variable and confusing plant group have sometimes been placed in the genus *Rhus* (now reserved for the "good sumacs," which have red berries). *Toxicodendrons* have hairy or smooth whitish fruit in pendulous clusters from leaf axils.

EASTERN POISON OAK — **Whole plant**

Toxicodendron toxicarium (Salisb.) Gillis — Cashew Family

Differs from Poison Ivy in that it is a shrub, never a climbing vine. Fruits hairy.

Where found: Sandy acid woods, fields. N.J., southward.

POISON SUMAC — **Whole plant**

Toxicodendron vernix (L.) Kuntze — Cashew Family

Shrub to small tree; 6–20 ft., with hairless buds and twigs. Leaves pinnate; 7–13 untoothed leaflets. Fruits white, smooth; Aug.–Nov.

Where found: Swampy pine woods. Sw. Que., Me. to Fla.; Texas to Minn.

POISON IVY — **Pl. 20** — **Whole plant**

Toxicodendron radicans (L.) Kuntze — Cashew Family

Trailing vine, or climbing by aerial roots. Leaves alternate, with 3 leaflets, irregularly toothed, smooth above, hairy beneath. Flowers pendent, in small axillary clusters, male and female on separate plants; June–July. Fruits smooth or *hairy;* Aug.–Nov.

Where found: Woods, margins. Throughout eastern North America.

WESTERN POISON OAK — **Whole plant**

Toxicodendron diversilobum Greene (Torr. & Gray) — Cashew Family

Erect, bushy shrub; 3–6 ft., with stiff *smooth* branches. Compound leaves with 3 leaflets, round or even lobed, shiny, paler beneath; suggestive of oak leaves. Fruit white, smooth.

Where found: Thickets, woods. B.C., Wash. to Calif., Mexico.

RYDBERG'S POISON IVY — **Whole plant**

Toxicodendron rydbergii Greene (Small ex Rydb.) — Cashew Family

Erect, usually unbranched shrub; 4–12 in., *stems hairy.* Leaves *crowded at end of branches,* alternate, 3 leaflets, with unequal lobes. Fruits white, waxy; Aug–Nov.

Where found: Wooded slopes. N.S. to Va.; Texas to Ariz., north to Alta.

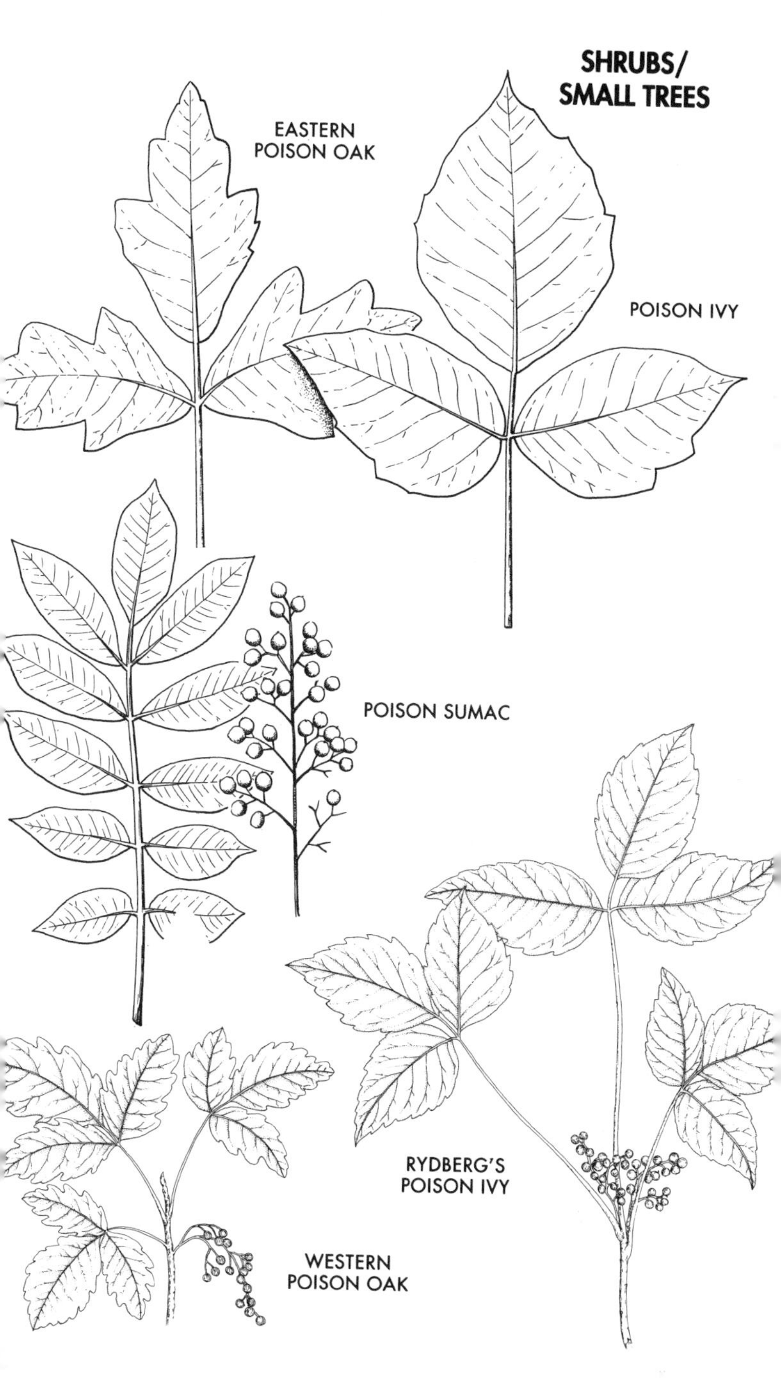
SHRUBS/
SMALL TREES
EASTERN
POISON OAK
POISON IVY
POISON SUMAC
RYDBERG'S
POISON IVY
WESTERN
POISON OAK

MISCELLANEOUS TREES

RED CEDAR **Pl. 38** **Leaves, fruits**

Juniperus virginiana L. Cypress Family

Spire-shaped tree; 10–50 ft. Leaves reduced to *overlapping scales, twigs 4-sided.* Fruits hard, roundish, blue-green berries with pitted surface.

Where found: Infertile soils, pastures, roadsides. Canada, Me. to Ga.; Texas to Minn., Mich.

Comments: If taken in large amounts, the berries can irritate the kidneys. Potential poisoning may result from excessive use of the berries in herbal teas. Children may be attracted to the fruits, but they are bitter and unlikely to be ingested in any quantity. The unpalatable branches have been questionably implicated in livestock poisoning. Oil distilled from branches, used to induce abortion, has resulted in fatalities.

AMERICAN HOLLY **Fruits**

Ilex opaca Ait. Holly Family

Evergreen tree, to 90 ft. Leaves shiny or waxy, wavy-margined, *with few to many spine-tipped teeth.* Male and female flowers on separate trees; greenish white, relatively small and inconspicuous. Fruits reddish (rarely yellow); Sept.–Oct.

Where found: Mixed woods. E. Mass. to Fla.; Texas to Ill.

Comments: Fruit-bearing sprigs are a popular Christmas decoration. Ingestion of the fruits can induce violent vomiting, diarrhea, and stupor. However, the fruits are extremely bitter and are unlikely to be consumed in quantity. Considered most dangerous to small children; should be kept out of their reach.

GINKGO, MAIDENHAIR TREE **Fruits**

Ginkgo biloba L. Ginkgo Family

Deciduous tree; 30–120 ft. Leaves *fan-shaped,* 2-lobed, 2–3 in. Male and female flowers on separate trees. Fruits *foul-smelling,* globe-shaped, similar in appearance to persimmons; Sept–Oct.

Where found: Ornamental. Often grown in cities.

Comments: The juice of the fruits can cause contact dermatitis, severely irritating the skin. If eaten, fruit pulp will cause gastrointestinal irritation. After cooking or boiling, the seeds are eaten in moderation in Asia (no more than 10 seeds for an adult). Westerners have been poisoned from eating the processed seeds like peanuts, instead of in careful moderation. Pharmaceuticals made from Ginkgo leaves are among the best-selling drugs in Europe, used for treating a wide variety of ailments and diseases. Clinical use focuses on the treatment of poor circulation, heart disease, eye disease, ringing in the ears (tinnitus), chronic cerebral circulation insufficiency, brain trauma, dementia, and senility. Leaves have not been reported to be toxic.

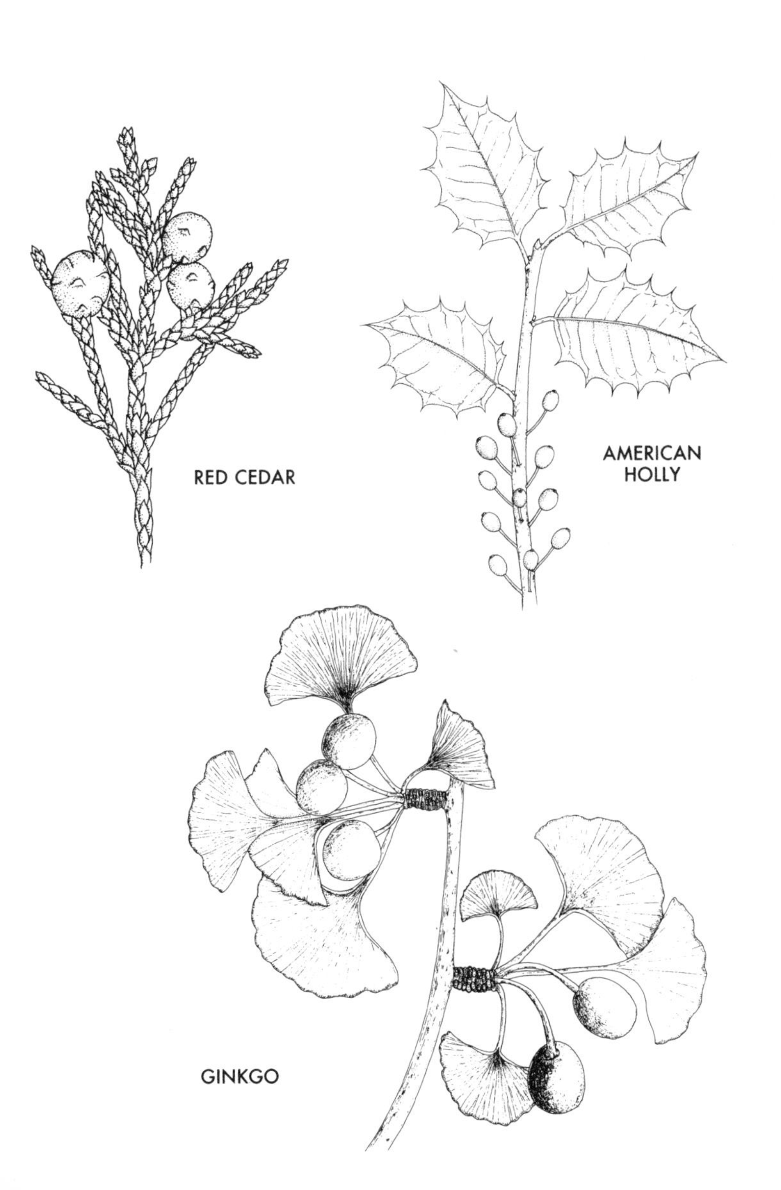
RED CEDAR
AMERICAN HOLLY
GINKGO

MISCELLANEOUS TREES

TUNG TREE **Seeds**

Aleurites fordii Hemsl. Spurge Family

Small tree with milky sap; 15–40 ft. Leaves oval or heart-shaped; 5–10 in. long. Flowers white, pinkish striped, to 1 in.; May–June. Fruit roundish, pendent, brown, smooth; with 3–7 white seeds.

Where found: Cultivated. N.C. to Fla.; Texas, Calif. Alien. Asia.

Comments: Nut is the source of tung oil. Seeds have been mistaken for Brazil nuts or chestnuts. Eating one seed can result in vomiting, diarrhea, abdominal pain, weakness, dizziness, and in severe cases dehydration and respiratory and cardiac irregularities. Fatalities reported. May cause dermatitis. Cattle have been poisoned by eating fruits.

WESTERN SOAPBERRY **Whole plant**

Sapindus drummondii Hook & Arn. Soapberry Family

[*S. saponaria* L. var. *drummondii* (Hook. & Arn.) L. Benson]

Deciduous tree to 50 ft. Leaves pinnate, to 8 in.; leaflets 8–18, lance-shaped, to 3 in. long. Flowers small, in panicles to 10 in. long. Fruits yellowish, smooth but minutely pitted; to 1/2 in.

Where found: Wood margins, field edges. Mo. to La.; Texas, Kans. to Ariz.

Comments: All parts toxic, especially fruits and seeds; used as fish poison. May produce gastroenteritis or dermatitis. Poisoning rare.

BRAZILIAN PEPPER, FLORIDA HOLLY **Pl. 40** **Berries**

Schinus terebinthifolius Raddi Cashew Family

Shrub or tree, to 20 ft. Crushed leaves *with peppery scent;* 5–13 oblong leaflets; to 2 1/2 in. Fruits bright red, in clusters. **California Pepperbush** *S. molle* (also shown) has rose-red berries.

Where found: Very weedy in S. Fla. and Hawaii. Alien. Brazil. *S. molle* from Peru.

Comments: Trimming blooming plant can cause dermatitis, itching, and possible eye irritation. Flowers or crushed fruits may cause respiratory irritation. Berries cause gastroenteritis, nausea, vomiting, burning throat, and diarrhea.

YELLOW OLEANDER **Whole plant**

Thevetia peruviana (Pers.) K. Schum. Dogbane Family

Evergreen, milky-sapped tree; 10–20 ft. Leaves like those of oleander. Flowers yellow, to 2 in. across. Fruit hard, to 1 in., red, then black.

Where found: S. Fla., sw. U.S. Alien. Tropical America.

Comments: Contains cardiac glycosides. See also page 170.

CHINABERRY TREE **Fruits**

Melia azedarach L. Mahogany Family

Low, branched tree; 40 ft. Leaves 2-pinnate, 1–3 ft., with numerous lance-shaped leaflets 1–2 in. long. Flowers in large panicles, lilac, fragrant, *stamen tubes purple.* Fruits ovoid, smooth, yellowish; to 1/2 in.

Where found: Cultivated. Escaped. Se. U.S. Alien. Asia.

Comments: Ingesting fruits has killed livestock and humans. Other parts also toxic. Fruits severely irritating, causing nausea, vomiting, bloody diarrhea, or constipation. May also cause respiratory failure, excitement or depression, and death. Six to eight fruits have caused a fatality in a child. Used in folk medicine. Toxicity varies widely in different strains and locations.

CALIFORNIA PEPPERBUSH

× 1/3

BRAZILIAN PEPPER

× 1/3

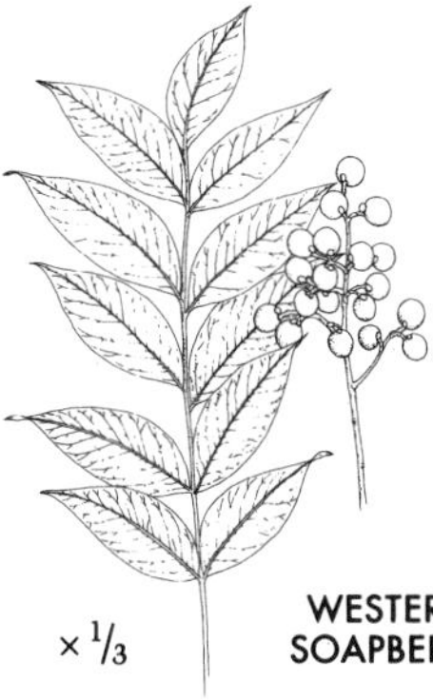

WESTERN SOAPBERRY

× 1/3

TUNG TREE

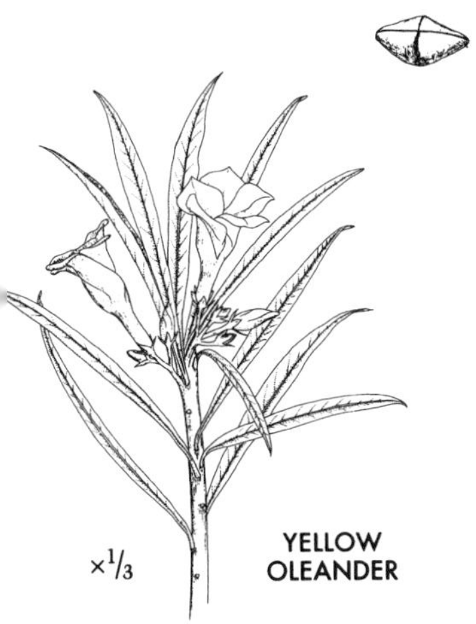

× 1/3

YELLOW OLEANDER

× 1/3

CHINABERRY TREE

× 1/3

DECIDUOUS TREES WITH OPPOSITE, PALMATE COMPOUND LEAVES

HORSECHESTNUT **Pl. 39** **Fruits**
Aesculus hippocastanum L. Horsechestnut Family
Deciduous tree to 100 ft., with *large sticky buds.* Leaflets 5–7, *stalkless,* toothed; to 12 in. Flowers whitish, mottled with red and yellow spots; May. Fruits *warty* or *spiny;* Sept.–Oct.
Where found: Street or shade tree. Naturalized. Alien. Europe.
Comments: The seed husks and seeds of Horsechestnut are bitter, inedible, and potentially poisonous, but they are seldom eaten. A mixture of saponins known as aescin is found in the seeds. Various extracts of Horsechestnut including aescin are widely used in many parts of the world (but not in the U.S.) in topical products to reduce swelling and seal capillaries. Applied externally, aescin is not systemically absorbed. Poorly absorbed if ingested, but may irritate mucous membranes, causing upset stomach. No human fatalities reported from North America. Various *Aesculus* species have been implicated in livestock poisoning.

SWEET BUCKEYE, YELLOW BUCKEYE **Fruits**
Aesculus flava Soland. Horsechestnut Family
[*A. octandra* Marsh.]
Large tree; to 90 ft. Leaflets 5–7, oval (broadest at apex) to oblong, to 9 in. long, toothed. Flowers with *gland dots,* yellow (rarely purple-red).
Where found: Rich woods. Sw. Pa. to Ga.; Ala., to s. Ill.
Comments: See Horsechestnut above.

CALIFORNIA BUCKEYE **Pl. 39** **Fruits**
Aesculus californica (Spach) Nutt. Horsechestnut Family
Broad tree; 10–40 ft. Leaflets 5, oval-oblong, stalked, to 7 in. Flowers pale pink to white. Fruits smooth, pear-shaped.
Where found: Low elevations. Calif.
Comments: See Horsechestnut above.

OHIO BUCKEYE **Pl. 39** **Fruits**
Aesculus glabra Willd. Horsechestnut Family
Small tree, 20–40 ft. Leaflets 4 (rarely to 7); toothed; 4–15 in. long. Buds not sticky, scales of buds with *strong ridges.* Twigs *foul-smelling* when broken. Flowers yellow, April–May. Fruits with *weak prickles.*
Where found: Rich, moist woods. W. Pa., W. Va., e. Tenn., cen. Ala.; to cen. Okla., Neb., to Iowa.
Comments: See Horsechestnut above.

RED BUCKEYE **Fruits**
Aesculus pavia L. Horsechestnut Family
Shrub or small tree; 12–15 ft. Leaflets 5–7, oval to lance-shaped (broadest at apex). Flowers *red* (rarely yellow) *in oblong panicles.*
Where found: Rich woods. N.C. to Fla.; Texas to Mo., s. Ill.
Comments: See Horsechestnut above.

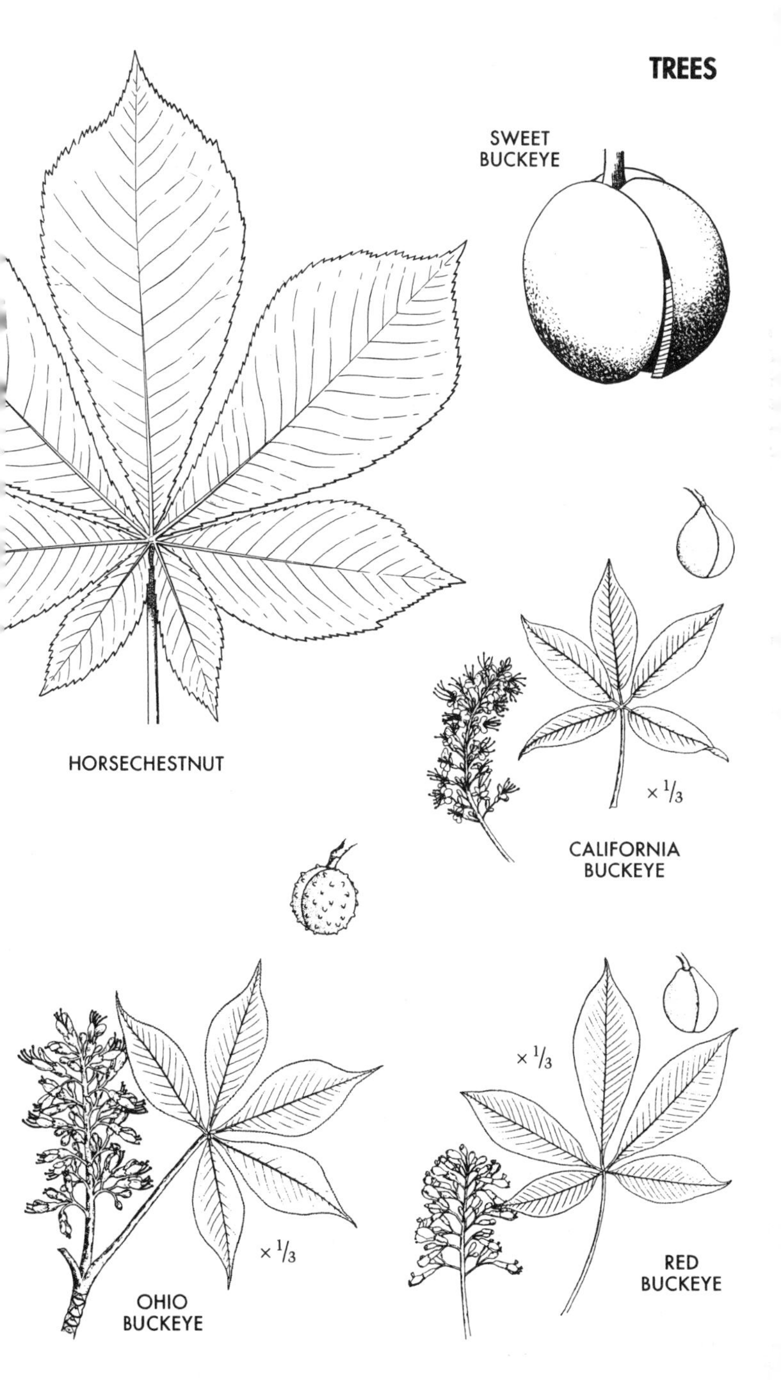
SWEET
BUCKEYE
HORSECHESTNUT
× 1/3
CALIFORNIA
BUCKEYE
× 1/3
× 1/3
OHIO
BUCKEYE
RED
BUCKEYE

DECIDUOUS TREES WITH ALTERNATE LEAVES, BLACK OR REDDISH FRUITS

CAROLINA BUCKTHORN **Pl. 39** **Fruits**

Rhamnus caroliniana Walt. Buckthorn Family

Small tree; 10–30 ft. Leaves elliptical to oval; without teeth or barely fine-toothed, smooth beneath when mature except in var. *mollis,* which has leaves hairy beneath. Veins hug leaf edges. Flowers small. Fruits black; 3-seeded, *not grooved on back.*

Where found: Woods. Va. to Fla.; Texas to Neb.

Comments: Fruits and bark of all buckthorns are potentially toxic, containing purgative anthracene glycosides that act on the large intestine. We have 15 native or naturalized *Rhamnus* species. The European Alder Buckthorn *R. frangula* and Common Buckthorn *R. cathartica* (not shown), planted and naturalized in the United States, are used in laxative preparations. The fresh bark or berries of buckthorns can be violently laxative and may irritate the skin or mucous membranes. Deaths in children from eating berries are reported from Europe. Such poisoning is rare.

CASCARA SAGRADA **Pl. 39** **Fruits**

Rhamnus purshiana DC. Buckthorn Family

Small tree; 8–20 ft. Leaves alternate, *in tufts at end of branchlets,* elliptical-oblong, slightly toothed, 2½–5 in. long, leaf stalks soft-hairy. Flowers 5-petaled. Fruits black with 3 (rarely 2) seeds.

Where found: Woods, mountain slopes. Calif. to Wash.

Comments: The aged bark is widely used in laxative products. Fresh bark and fruits are potentially toxic.

BLACK CHERRY, WILD CHERRY **Pl. 34** **Seeds, bark**

Prunus serotina Ehrh. Rose Family

Deciduous tree; 30–90 ft. Leaves oval to lance-shaped, *more than twice as long as wide, blunt-toothed;* with *whitish brown hairs on prominent midrib.* Flowers in drooping slender racemes; April–June. Fruits black berries.

Where found: Dry woods. N.S. to Fla.; Texas to N.D.

Comments: Seeds, fresh and dried leaves, twigs, and bark contain potentially deadly cyanogenic glycosides such as amygdalin and prunasin. When ingested, these substances are transformed into highly toxic hydrocyanic acid, which has an almond odor, in essence producing cyanide poisoning. Symptoms, which can develop without warning, include respiratory failure, loss of voice, muscle twitching or spasms, coma, and death. Fruit is considered harmless with seeds removed. Children have been poisoned by chewing on the twigs. There are about 48 *Prunus* species in the wild in North America. Others are cultivated, including plums, apricots, peach, and almonds. All produce poisonous amygdalin in the seeds.

CHOKECHERRY **Fruits, bark**

Prunus virginiana L. Rose Family

A shrub or small tree; 12–60 ft. Leaves oval, *twice as long as wide, sharp-toothed, midrib hairless.* White flowers in a thick raceme; April–July. Fruit round, to ⅓ in., lustrous, scarlet to blackish red.

Where found: Thickets. Nfld. to N.C.; Mo., Kans., west to N.M., Ariz., Wyo., Mont.; Calif. to B.C.

Comments: See Black Cherry above.

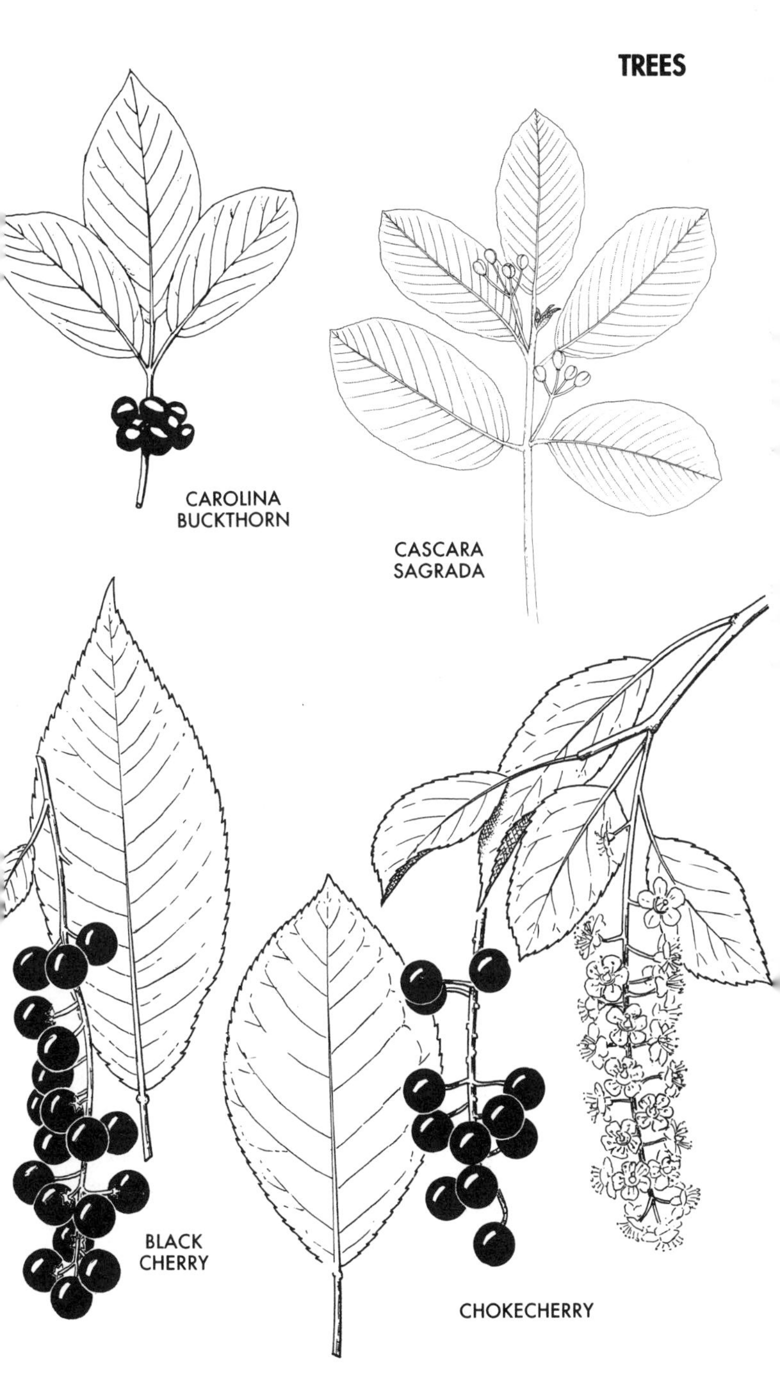
CAROLINA
BUCKTHORN
CASCARA
SAGRADA
BLACK
CHERRY
CHOKECHERRY

KENTUCKY COFFEE TREE **Fruits**

Gymnocladus dioica (L.) K. Koch Pea Family

Tree; 50–60 ft. Leaves 2-pinnate; 7–13 leaflets *opposite, untoothed,* about 1 in. wide, tips acute. Whitish flowers in racemes in leaf axils; April–June. Fruits leathery, flat pods to 10 in. long, 2½ in. wide, with large, hard, black seeds.

Where found: Rich woods. Cen. N.Y. to Tenn.; Ark. to S.D.

Comments: Seeds contain the alkaloid cytisine, which may cause vomiting, diarrhea, cardiac irregularities, and coma. Poisoning rare.

CORALBEAN **Fruits**

Erythrina flabelliformis Kearney Pea Family

Small tree; to 15 ft. Stems and leafstalks *prickly.* Leaves 3-divided; leaflets broadly oval or *fan-shaped,* the outermost on a long stalk. Flowers showy, bright red, elongated; appear in spring before leaves. Pods large, with bright red seeds.

Where found: Sw. N.M., s. Ariz., to Mexico.

Comments: The genus has 100 tropical shrubs and trees (rarely herbs). Eight species are native or naturalized in the U.S. About 30 species are grown as ornamentals. All are considered toxic. In Mexico, the seeds are used to make necklaces. Contains various alkaloids and cyanogenic glycosides. May cause depression of the central nervous system, vomiting, and diarrhea.

MESCAL BEAN **Fruits**

Sophora secundiflora (Ortega) Lag. ex DC. Pea Family

Evergreen shrub or small tree, 8–40 ft. Leaves alternate, pinnate, 4–6 in.; leaflets oblong, 5–13, yellow-green and shiny above, silky below. Flowers *sweetly fragrant, violet-blue;* to 1 in. long, in dense 1-sided raceme. Pod woody, *constricted,* with white hairs; seeds *bright red,* to ½ in.

Where found: Hills, canyons. Sw. Texas, N.M. to Mexico. Widely planted as an ornamental in the deep South.

Comments: Seeds highly toxic, containing cytisine. One thoroughly chewed seed may be fatal to a child. Unbroken seeds may pass through the digestive tract without poisoning. Symptoms include vomiting, diarrhea, headache, delirium, hallucinations, excitement, and coma. Nonfatal doses may cause sleep lasting up to 3 days.

BLACK LOCUST **Pl. 40** **Fruits**

Robinia pseudoacacia L. Pea Family

Tree armed with stout *paired thorns to 1 in. long, twigs smooth;* 70–90 ft. Leaves *pinnate;* 7–21 elliptical to oval leaflets. Fragrant *white* flowers in racemes; May–June. Legumes *smooth,* flat; 2–6 in. long.

Where found: Dry woods. Pa. to Ga.; La., Okla. to Iowa.

Comments: Children have been poisoned by eating the seeds and inner bark or chewing on fresh twigs. Contains lectins known as toxalbumins (robin and phasin), which may interfere with protein synthesis in the small intestine. Also contains a glycoside (robitin). If ingested, may cause loss of appetite, vomiting, bloody diarrhea, depression, weak pulse, and coldness of arms and legs. Poisoning rare.

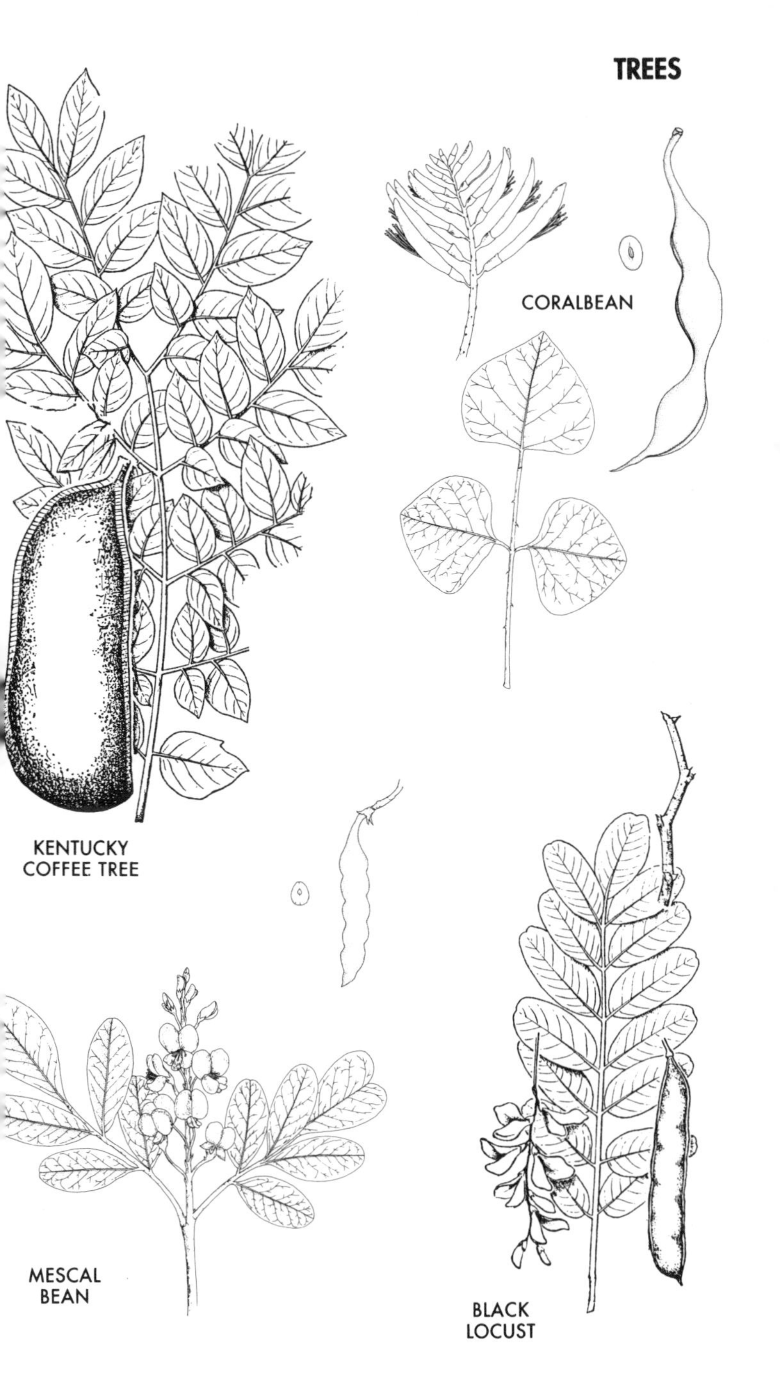
CORALBEAN
KENTUCKY
COFFEE TREE
MESCAL
BEAN
BLACK
LOCUST

CLEMATIS

PITCHER'S CLEMATIS **Pl. 35** **Leaves**

Clematis pitcheri Torr. & Gray Buttercup Family

Semiwoody vine. Leaves compound; leaflets usually 5–9, leathery, variable in shape, without teeth or unlobed. Flowers solitary, purplish, urn-shaped, petallike sepals; May–Aug.

Where found: Near steams. Ind. to s. Mo.; Texas to Neb.

ROCKY MOUNTAIN CLEMATIS **Leaves**

Clematis occidentalis (Horem.) DC. Buttercup Family

[*Clematis pseudoalpina* (Kuntze) A. Nels.]

Woody, climbing and spreading vine to about 10 ft. Leaves compound; leaflets 3, lance-shaped, oval (broadest at apex) or round, without teeth or blunt-toothed. Flowers solitary, somewhat bell-shaped, the 4 petallike sepals blue-violet (rarely white). Seeds with feathery plumes.

Where found: Moist woods, mountain brush. Mont., Utah to Colo.; Ore. to B.C.

Comments: See Virgin's-bower below.

VIRGIN'S-BOWER **Pl. 35** **Leaves**

Clematis virginiana L. Buttercup Family

Somewhat woody, sprawling vine. Leaves divided into *3 sharp-toothed oval leaflets.* Flowers numerous, white, with *4 narrow, petallike sepals;* July–Sept. Feathery plumes attached to seeds.

Where found: Thickets, wood edges. N.S. to Ga.; La., e. Kans., north to Man.

Comments: Like many buttercup family members, the fresh sap of *Clematis* species contains protoanemonin, usually in small and variable amounts. Fresh sap can be highly irritating to skin or mucous membranes, causing blistering or ulceration, especially in *Ranunculus* (buttercup) species. Contact with the fresh plant may cause skin irritation through absorption of the protoanemonins. Symptoms of fresh plant poisoning may include gastrointestinal distress, colic, diarrhea, and nephritis. In severe cases, causes paralysis of the central nervous system. *Clematis* species are generally avoided by livestock. Though *Clematis* should be regarded as potentially toxic, cases of poisoning are very rare in the United States. Handling leaves may cause dermatitis in susceptible individuals.

WESTERN VIRGIN'S-BOWER **Leaves**

Clematis ligusticifolia Nutt. Buttercup Family

Semiwoody vine. Leaves compound; leaflets 5–7, lance-shaped to oval-lance-shaped. Flowers numerous, with thin, spreading, cream-colored, petallike sepals; July–Aug.

Where found: N.D. to N.M.; Calif. to B.C.

LEATHER-FLOWER **Leaves**

Clematis viorna L. Buttercup Family

Climbing vine; to 10 ft. Leaves compound; leaflets 3–7, oval to lance-shaped, smooth. Flowers solitary, purplish, urn-shaped, to 1 in. long.

Where found: Woods. Penn. to Ga.; Texas to Iowa, Ill.

Comments: See Virgin's-bower above.

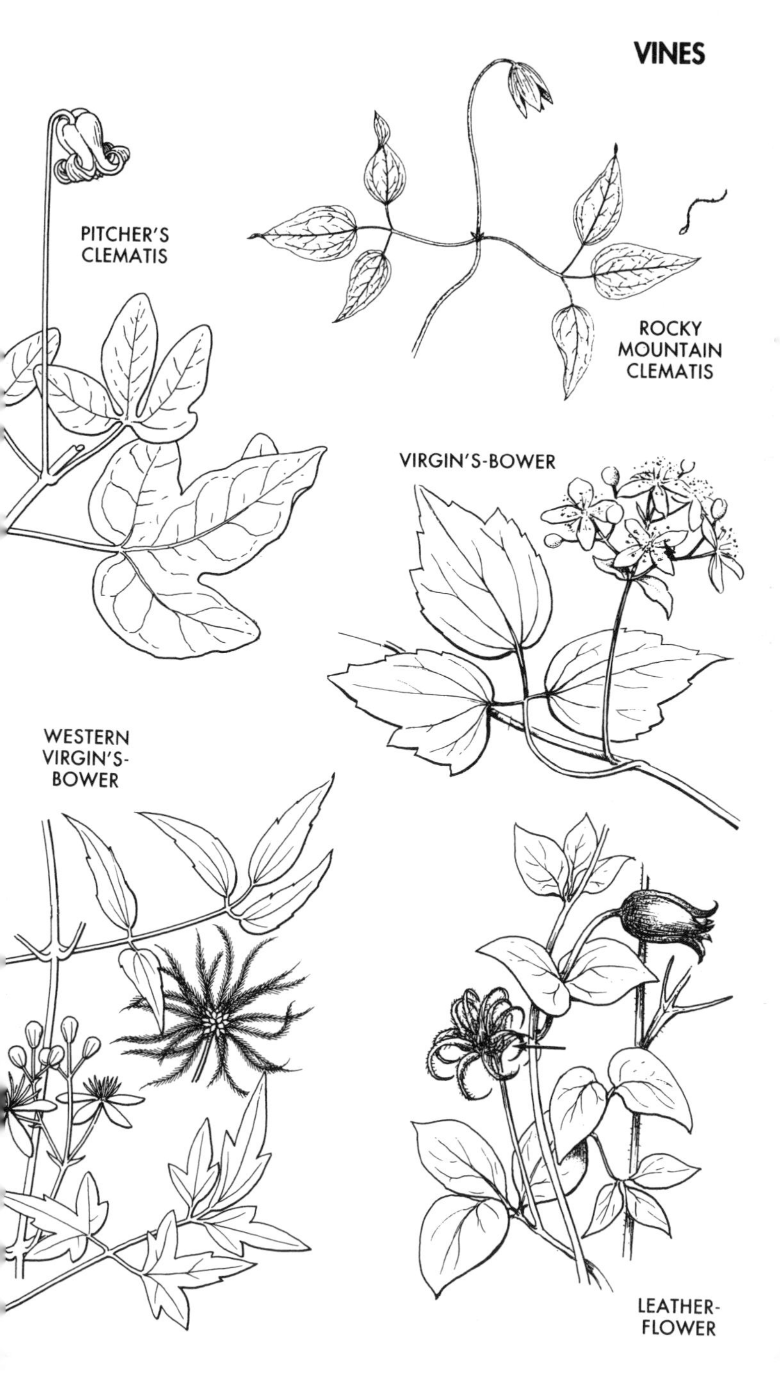
PITCHER'S CLEMATIS
ROCKY MOUNTAIN CLEMATIS
VIRGIN'S-BOWER
WESTERN VIRGIN'S-BOWER
LEATHER-FLOWER

JAPANESE WISTERIA **Whole plant**

Wisteria floribunda (Willd.) DC Pea Family

High-climbing deciduous vine. Leaves odd-pinnate, large; leaflets 13–19, oval to elliptical; to 3 in. long. Flowers showy, purplish (pink or white) in slender pendent racemes, *gradually opening from base; to 3/4 in. long.* Legumes and ovary velvet-hairy; to 6 in. long. See related species on Plate 35.

Where found: Widely cultivated in South. Sometimes escaped.

Comments: All parts of Wisteria are considered potentially toxic. Contains a glycoside, wistarine, that may cause nausea, abdominal pain, repeated vomiting, irritation of gastrointestinal tract and mucous membranes. As few as 2–3 seeds may be poisonous to children. Chinese Wisteria *W. sinensis* (not shown), also widely cultivated, has somewhat larger flowers, 7–13 leaflets, with most flowers opening at same time.

WISTERIA **Whole plant**

Wisteria frutescens (L.) Poir. Pea Family

Low-climbing deciduous vine. Leaves odd-pinnate; paired oval leaflets usually 5–7. Flowers lavender, pealike, in dense clusters; *to 1/2 in. long;* April–Sept. Legumes and ovary *smooth.* See related species on Plate 35.

Where found: Md. to Fla., west to Texas.

Comments: See Japanese Wisteria above.

YELLOW JESSAMINE **Whole plant**

Gelsemium sempervirens L. St. Hil. Logania Family

Twining, climbing, evergreen vine. Leaves opposite, lance-shaped to oval, *shiny.* Flowers yellow, open funnel-shaped with *5 rounded petal lobes at end,* sweetly fragrant; Feb.–June.

Where found: Woods, thickets. Se. Va. to Fla.; Texas to Ark.

Comments: Several cases of poisoning of children have been reported from chewing or sucking nectar from the flowers. Symptoms may include sweating, pain in the eyes, double vision, headache, dry mouth, respiratory failure, nausea, prostration, and death. Contains numerous alkaloids including gelsemine, gelseminine, gelsemoidine, and sempervirine. Alkaloids are highly concentrated in the flower nectar. Honey derived from the flowers is regarded as toxic.

TRUMPET VINE **Pl. 33** **Whole plant**

Campsis radicans (L.) Seem. ex Bureau Bignonia Family

High vine, climbing by means of aerial rootlets; often weedy. Leaves opposite, pinnate, with 5–13 oval-lance-shaped, sharp-toothed leaflets. Flowers in clusters, bright orange, leathery, funnel-shaped; June–Sept. Fruits tapering, cylindrical pods to 8 in; splitting, with numerous winged seeds within.

Where found: Fence rows, thickets. N.J. to Fla., Texas to S.D.

Comments: Flowers or leaves may cause dermatitis, producing inflammation and blisters. Potentially poisonous if ingested. A common plant in its range that is poorly researched.

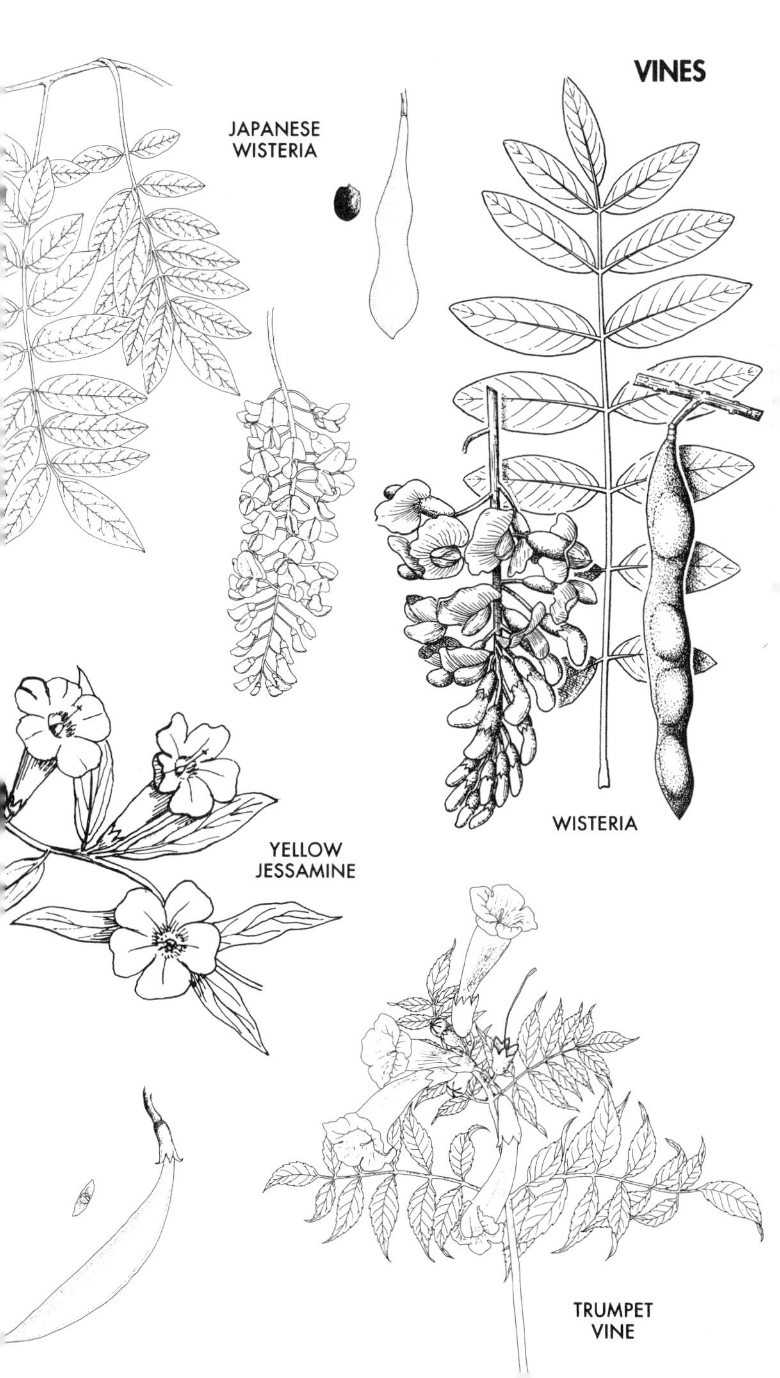
JAPANESE
WISTERIA
WISTERIA
YELLOW
JESSAMINE
TRUMPET
VINE

FRUITS OR SEEDS BRIGHT RED/ORANGE

WOODY NIGHTSHADE **Pl. 22** **Whole plant**
Solanum dulcamara L. Nightshade Family
Woody climbing vine. Leaves oval, usually with 1 or 2 *prominent lobes at base.* Flowers violet (or white) stars, *petals back-curved,* 2 green spots at base; May–Sept. Fruits ovoid, red.
Where found: Waste places. Much of North America. Alien. Europe.
Comments: Different genetic races of the plants produce different alkaloids. Berries and leaves contain toxic steroidal alkaloids. Alkaloid content is highest when seeds are mature but unripened. Ingestion of fewer than 10 fruits usually results in gastric disturbances. Historically, the ingestion of 10 fruits has been implicated in deaths in children, but recent German experience suggests that relatively large amounts of fruits must be ingested to produce serious symptoms (200 is a lethal dose).

BALSAM PEAR **Pl. 34** **Seeds, fruits**
Momordica charantia L. Gourd Family
Annual vine. Leaves deeply lobed, blunt to sharp-pointed. Flowers solitary, yellow, to 1 in. across. Fruits large, oblong to oval, yellow-orange, *warty,* splitting when mature, *covered by red arils.*
Where found: Tropics, cultivated. Naturalized in se. U.S.
Comments: Properly prepared, the bitter unripe fruits are eaten as a vegetable, especially in Asian communities. Seeds or fresh fruit pulp may cause vomiting and severe diarrhea.

ROSARY PEA **Seeds**
Abrus precatorius L. Pea Family
Twining vine to 10 ft. Leaves pinnately compound, with 16–30 small leaflets. Flowers rose to violet, small, in racemes in leaf axils. Legume oblong; to 2 in. long, seeds *glossy, 3/4 scarlet, 1/4 black.*
Where found: Tropical vine, sometimes escaped. Common in s. Fla.
Comments: A phytotoxin, abrin, is found in the seed. Eating one seed may cause nausea, vomiting, severe abdominal pain and diarrhea, burning in throat, cold sweat, prostration, rapid weak pulse, circulatory collapse, coma, and death. One to three hours may pass before onset of symptoms. Unbroken seeds may pass harmlessly through the digestive system. One thoroughly chewed seed may be fatal to a child. The attractive seeds are used to make necklaces.

BITTERSWEET **Pl. 35** **Whole plant**
Celastrus scandens L. Staff Tree Family
Climbing shrub to 50 ft. Leaves *oval-oblong,* sharp-pointed, fine-toothed, to 4 in. long. Fruits 3-valved, scarlet-orange, *splitting;* seeds *scarlet.*
Where found: Rich thickets. Que. to Ga., Ala; Okla. to N.D.
Comments: Roots, leaves, and fruits are potentially poisonous. May produce vomiting, diarrhea, chills, and convulsions. Fruits in dried arrangements may be attractive to children. Cases of human poisoning are rare and poorly documented. Oriental Bittersweet *C. orbiculatus,* with 2–3 flowers in axillary clusters, is occasionally naturalized; locally weedy.

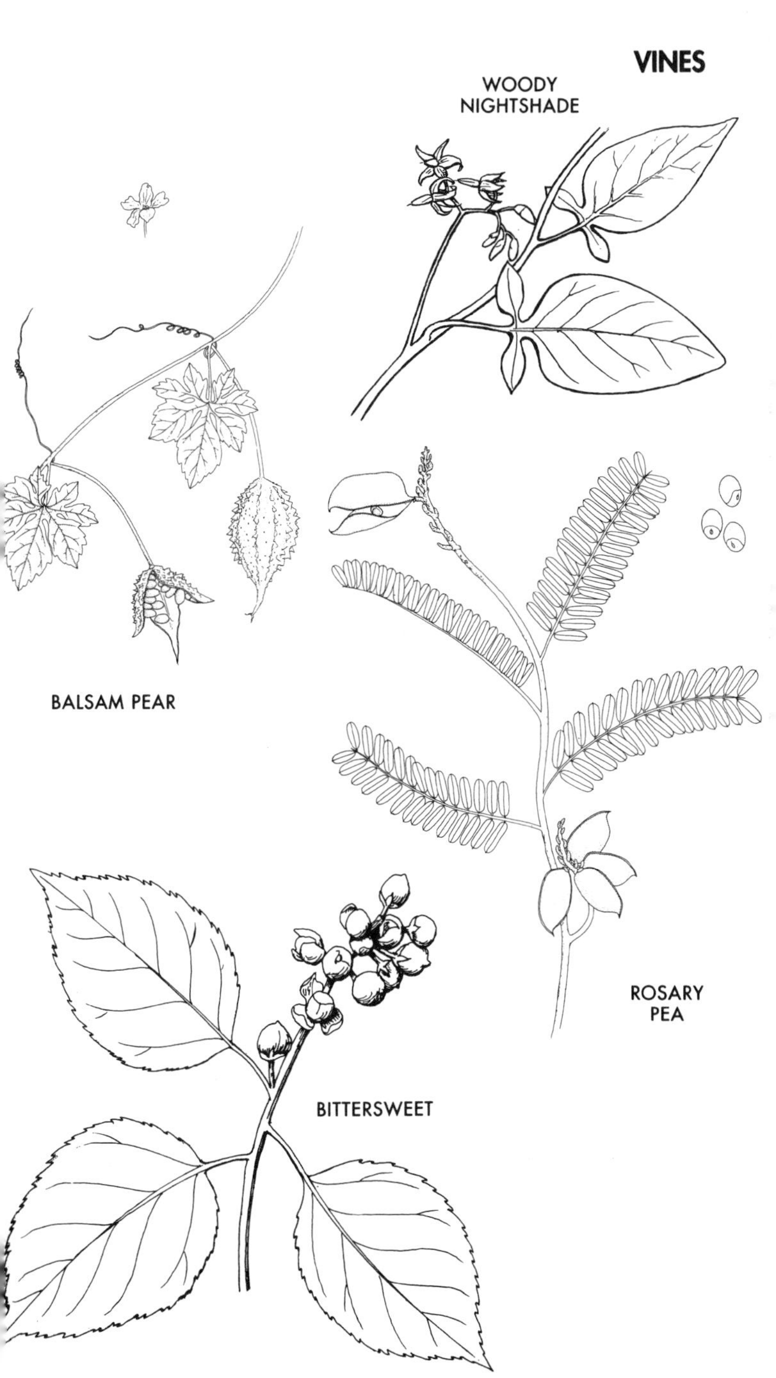
WOODY
NIGHTSHADE
BALSAM PEAR
ROSARY
PEA
BITTERSWEET

CANADA MOONSEED **Fruits**
Menispermum canadense L. Moonseed Family
Stout, climbing woody vine; 8–12 ft. Root bright yellow within. Leaves with *3–7 angles or lobes,* smooth, *stalk attached just above heart-shaped base.* Flowers small, whitish, in loose clusters; June–Aug. Fruits bluish black or purplish, with a bluish film that rubs off, wrinkled. Seed is flat, grooved, with rough patterns, moon-shaped; Aug–Oct.
Where found: Rich, moist thickets. Que., w. New England, south to Ga.; Ark. to Okla.
Comments: Root and fruits contain a number of alkaloids including berberine, menispine, menispermine, and dauricine. A number of other alkaloids have been identified from the roots. The small purplish fruits have been mistaken for grapes by children, resulting in convulsions or seizures. Fatalities from eating the fruits have been reported.

VIRGINIA CREEPER **Pl. 20** **Berries**
Parthenocissus quinquefolia (L.) Planch. Grape or Vine Family
Climbing or creeping vine with *adhesive disks on much-branched tendrils.* Leaves palmate; *leaflets* 5 (3–7), elliptical to oval, sharply toothed, smooth. Flowers small, in terminal clusters; May–Aug. Fruits blue, inedible berries.
Where found: Thickets, woods, waste places. Weedy. Me. to Fla.; Texas to Minn.
Comments: Berries, which may be attractive to children, have been reported to cause poisoning in a number of cases. Results ranging from gastric upset to death have been reported. However, evidence is circumstantial. The fruits and the leaves contain small amounts of calcium oxalate. Their ingestion should be avoided.

COMMON IVY **Pl. 35** **Fruits**
Hedera helix L. Ginseng Family
Trailing and climbing evergreen vine with aerial roots. Leaves alternate, leathery, oval, with 3–5 lobes (or without teeth). Flowers inconspicuous, in racemelike umbels. Fruit a black berry, superficially resembling a blueberry.
Where found: Widely cultivated around homes, fences, roadsides. Naturalized Va. to Fla., Miss., northward. Dozens of cultivated varieties have been developed. Alien. Europe.
Comments: Reports of deaths of children cited in older European literature are suspicious. Contains saponins. Eaten in sufficient quantity, the berries presumably can cause a burning sensation in the throat and gastrointestinal upset with vomiting and diarrhea. The berries are dry and insipid and are unlikely to be eaten in any quantity. Also suspected of causing contact dermatitis.

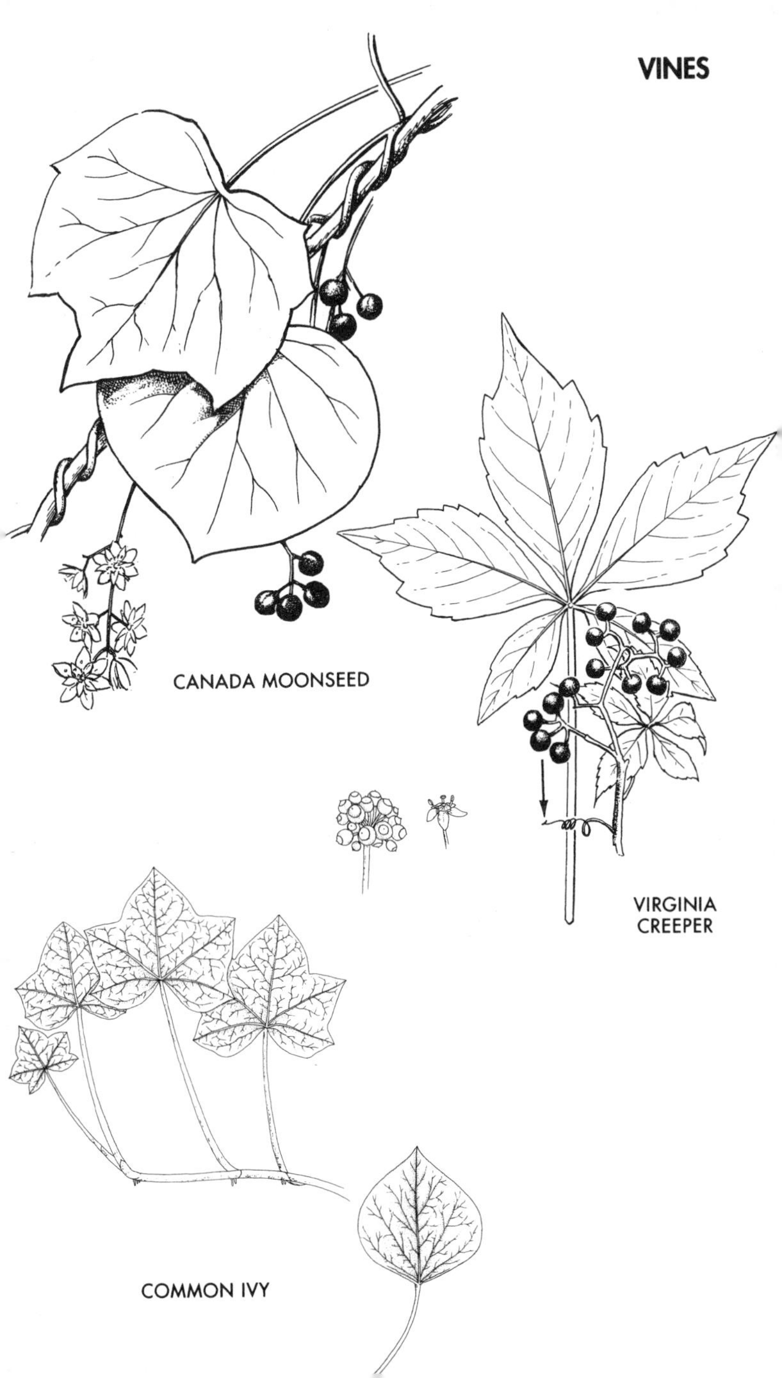
CANADA MOONSEED
VIRGINIA CREEPER
COMMON IVY

POISON IVY AND POISON OAK

Members of this highly variable and confusing plant group have sometimes been placed in the genus *Rhus* (now reserved for the "good sumacs," which have red berries). *Toxicodendrons* have hairy or smooth whitish fruit in pendulous clusters from leaf axils. Poison ivys, poison oaks, and poison sumac are probably the best known and least liked of all poisonous plants in North America. Fifty percent of the population is allergic to some degree to this plant group, which causes painful, irritating dermatitis in about two million people in the United States each year. All plant parts, including and especially the sap, contain irritant, nonvolatile, phenolic substances referred to as urushiol or toxicodendrol. The oily mixture, found in resin canals, is released when the plant is bruised. Contact with this substance causes severe contact dermatitis, including redness, itching, swelling, and blistering. Sometimes requires hospitalization. Eating the fruits causes similar internal irritation. It is most dangerous in spring and summer when sap is abundant but is toxic all year long. Droplets of the toxin can become airborne in smoke particles or ashes. The oily toxin may also be carried on the fur of pets that brush against the plant, or on equipment such as fishing poles and garden tools, then rubbed onto the skin of a human. Thus it is possible to contract poison ivy without actually coming into contact with the plant.

POISON IVY **Pl. 20** **Whole plant**
Toxicodendron radicans (L.) Kuntze Cashew Family
Highly variable trailing vine, or climbing by aerial rootlets. Its hairy trunk, resembling "hairy rope," hugs tree trunks as it climbs. Leaves are alternate, with 3 leaflets, outermost leaf on longer stalk; irregularly toothed, smooth above, hairy beneath. Flowers pendent, in small clusters in leaf axils, male and female on separate plants; June–July. Fruits smooth or *hairy;* Aug.–Nov.
Where found: Woods, margins. Throughout eastern North America.

WESTERN POISON OAK **Whole plant**
Toxicodendron diversilobum (Torr. & Gray) Green Cashew Family
An erect, bushy shrub, but sometimes sprawling, producing a vine-like appearance; 3–6 ft., with stiff *smooth* branches. Compound leaves with 3 leaflets, round or even-lobed, shiny, paler beneath; suggestive of oak leaves. Fruits white, smooth.
Where found: Thickets, woods. B.C., Wash. to Calif., Mexico.
Comments: See also on page 172.

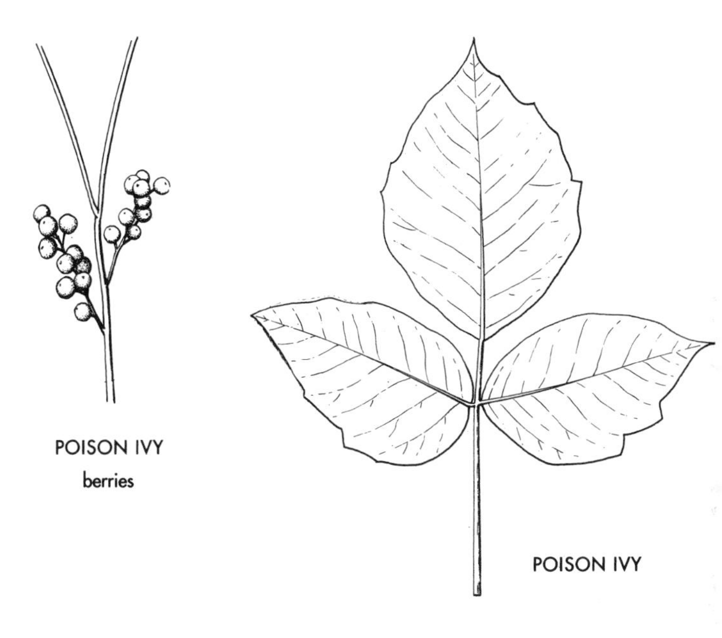

POISON IVY
berries

POISON IVY

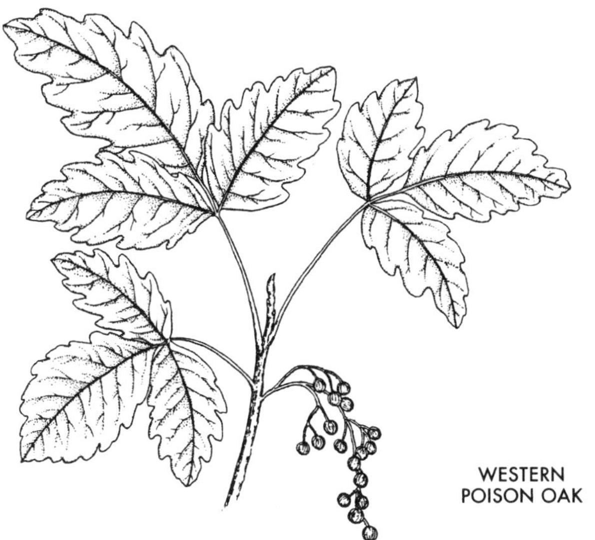

WESTERN
POISON OAK

COONTIE **Whole plant**

Zamia spp. Cycad Family

Low, evergreen, crown-forming plants with fernlike or palmlike leaves. They have terminal oblong male and female cones. Seeds fleshy.

Where found: Dry sand barrens. S. Ga. to Fla. Cultivated elsewhere in warmer climates. There are six species in the United States.

Comments: The seeds and roots contain cycasin, a glycoside. May produce persistent vomiting, diarrhea, colic, depression, and muscular paralysis. After special processing, an edible starch is prepared from the root. A member of DeSoto's army is said to have died after eating the raw root.

BRACKEN FERN **Pl. 20** **Whole plant**

Pterdium aquilinum (L.) Kuhn Fern Family

Our most common fern, 3–6 ft. tall; forms large colonies. Highly variable. Leaves (fronds) triangular in outline, divided into 3 parts; leaflets blunt-tipped, upper ones not cut to midrib, underside mostly woolly, smooth or nearly smooth above.

Where found: Barren soils, many plant communities. Much of North America; mostly absent from plains.

Comments: Known to have poisoned livestock, especially in the Pacific Northwest. An enzyme in the fern, thiaminase, disturbs thiamin metabolism. Massive doses of thiamin have successfully treated Bracken poisoning. Poisoning occurs when animals are fed hay with a high Bracken content. Livestock will quickly develop a rough coat, listlessness, discharges from the mouth and nostrils, and high temperature. They will die after eating their weight in Bracken over a period of one to three months. Bracken also contains cancer-causing compounds that can be directly imparted to cow's milk.

MALE FERN **Roots, whole plant**

Dryopteris filix-mas (L.) Schott. Fern Family

Yellow-green, leathery, scaly-stemmed fern, 7–20 in., with blackish, thick-wiry rhizomes with one to few crowns. Leaves (fronds) elliptical-oblong, divided into about 20 lance-shaped, pointed leaflets; leaflets narrow-oblong, cut nearly to midrib, with rounded alternating subleaflets, or remotely toothed.

Where found: Rocky woods, Me. to N.Y., Vt. to Mich; n. Calif. (rare) to B.C. Common in Europe.

Comments: Since ancient times, the root was used in medicinal preparations against tapeworm. Historically, deaths are primarily attributed to overdoses or hypersensitivity to medicinal preparations. These have largely been replaced by synthetic drugs, so there are now only rare accidental poisonings. A mixture of compounds known as filicin is the toxin, primarily found in the hairlike scales at the base of the fronds and rhizomes. May cause gastrointestinal irritation, cramps, blurred vision, and even blindness.

FERNS
COONTIE
BRACKEN FERN
MALE FERN

MISCELLANEOUS DEADLY MUSHROOMS

DEATHCAP *Amanita phalloides* (Vaillant ex Fr.) Sec.

Solitary or clustered, medium to large mushroom; variable in all respects. Cap *greenish,* smooth, somewhat sticky, *roundish to flat,* darker in center, often with radiating streaks, 2½–6 in. Stalk smooth, then developing bands of variable scales, white to greenish yellow, tapering to bulbous base; 3–5 in. Gills white, close, free. Volva cuplike, white to green, usually with projection on one side. Ring persistent, membranous, pendent, near top of stalk. Summer–fall. See *Peterson Field Guide to Mushrooms*, page 231.

Where found: Forests, clearings. E. North America, Pacific Coast.

Comments: Contains cyclopeptides known as amatoxins and phallotoxins, among the most lethal of known poisons. Can cause destructive cellular changes of the kidneys and liver. Deathcap poisoning is often fatal, accounting for 90 percent of mushroom fatalities in Europe. Onset of symptoms may not occur for 10 or more hours.

FOOL'S MUSHROOM *Amanita verna* (Bull.) Pers.

Solitary or group-forming, medium to large fungus. Cap *pure white,* smooth (sticky when moist), without warts, egg-shaped, becoming convex to bell-shaped, 1½–4 in. across, on *slender bulbous stalk,* 3–8 in. long. Gills crowded, nearly touching but free from stalk. Volva cuplike, persistent, usually closely hugging stalk, splitting above, with prominent free limb. Ring skirtlike on upper stalk. Spring–fall. See *Peterson Field Guide to Mushrooms,* page 238.

Where found: Deciduous and coniferous woods. East, Midwest to Pacific Northwest.

Comments: See *Amanita phalloides* above. Close relatives, which are nearly impossible to distinguish from field characteristics, include: Destroying Angel *A. virosa,* Two-spored Death Angel *A. bisporigera,* and Slender Death Angel *A. tenuifolia* (not shown).

AUTUMN SKULLCAP *Galerina autumnalis* (Pk.) A.H.S. & Sing.

[Pholiota autumnalis]

Single, group- or cluster-forming, small to medium, brown mushroom. Cap smooth, with tacky layer, brown or yellow-brown to rust; 1½–2 in. across. Gills close, broadly attached to stem. Stalk cylindrical, hollow, streaked with white fibrils; 1½–3½ in. Ring thin, high on stalk, sometimes not evident when mature. See *Peterson Field Guide to Mushrooms,* page 296.

Where found: Decaying logs of hardwoods or conifers. Widespread.

Comments: Contains amatoxins; see *Amanita phalloides* above. *G. venenata* and *G. marginata* (not shown) may be confused with *G. autumnalis,* as may other "little brown mushrooms."

DEADLY CONOCYBE *Conocybe filaris* Fr.

[C. rugosa, Pholiota rugosa, Pholiotina filaris]

Scattered to many in groups, small, fragile, "little brown mushroom." Caps bell-shaped, smooth, wet or dry, ¼–1 in. across. Stalk just ⅜ to 2½ in., orange-brown. Gills close, broad, notched, dull white to rust. Veil membranous, forming a delicate ring. Late summer–fall. See *Peterson Field Guide to Mushrooms,* page 308.

Where found: Well-watered lawns, moss, wood chips. Throughout.

Comments: Contains amatoxins; see *A. phalloides* above.

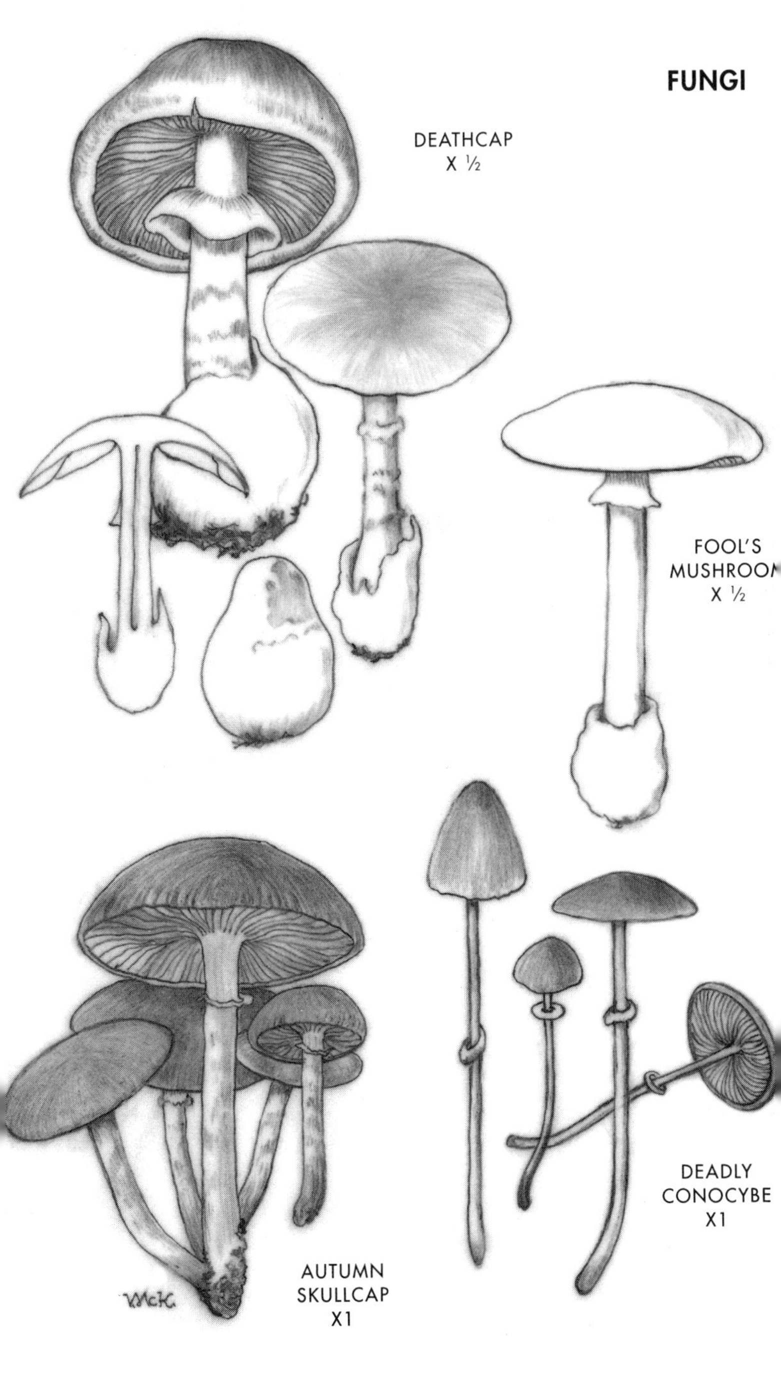
DEATHCAP
X ½
FOOL'S
MUSHROOM
X ½
DEADLY
CONOCYBE
X1
AUTUMN
SKULLCAP
X1

DEADLY MUSHROOMS RESEMBLING MORELS

FALSE MOREL *Gyromitra esculenta* (Pers.) Fr.
[Helvella esculenta]
Solitary or group-forming, *reddish brown* (or yellowish), *wrinkled*, brainlike mushroom. Cap is *irregularly rounded*, club-shaped, or flattish; *wrinkled, folded, or lobed*, suggesting a morel, but without the distinct pits of morels; chambered within; 1½–4 in. across. Stalk thickish, lighter in color than cap, hollow or with cottony filaments, grooved. Spring. See *Peterson Field Guide to Mushrooms*, page 51.
Where found: Woods beneath conifers, sometimes under deciduous woods. Much of North America, especially in North; southward in mountains.
Comments: In Europe, this mushroom is reported to cause 2–4 percent of mushroom fatalities; while comparatively few poisonings have been reported in the U.S., a number of fatalities have been attributed to this mushroom in this century. Contains gyromitrin, which forms the deadly compound monomethylhydrazine upon hydrolysis (though the compound is highly volatile and may be destroyed by cooking). If eaten raw, symptoms usually appear within 6–8 hours, including nausea, vomiting, watery or bloody diarrhea, cramps, and abdominal pain. Severe cases include jaundice, liver damage, high fever, dizziness, convulsions, coma, and possibly death 2–4 days after ingestion. While mushroom connoisseurs savor this mushroom when it is thoroughly cooked and properly prepared, it is best left alone.

HOODED FALSE MOREL *Gyromitra infula* (Schaeff ex Fr.) Quel.
[Helvella infula]
Solitary or group-forming, medium to large, *saddle-shaped* and *irregularly lobed* mushroom. Cap reddish brown to dark brown, wrinkled; 1½–4 in. across. Stalk hollow, thick, *without ribs;* 1–3 in. long. Spring. See *Peterson Field Guide to Mushrooms*, page 53.
Where found: Mostly on rotting deciduous wood or in ground litter. Throughout; frequent in Pacific Northwest.
Comments: See *Gyromitra esculenta* above.

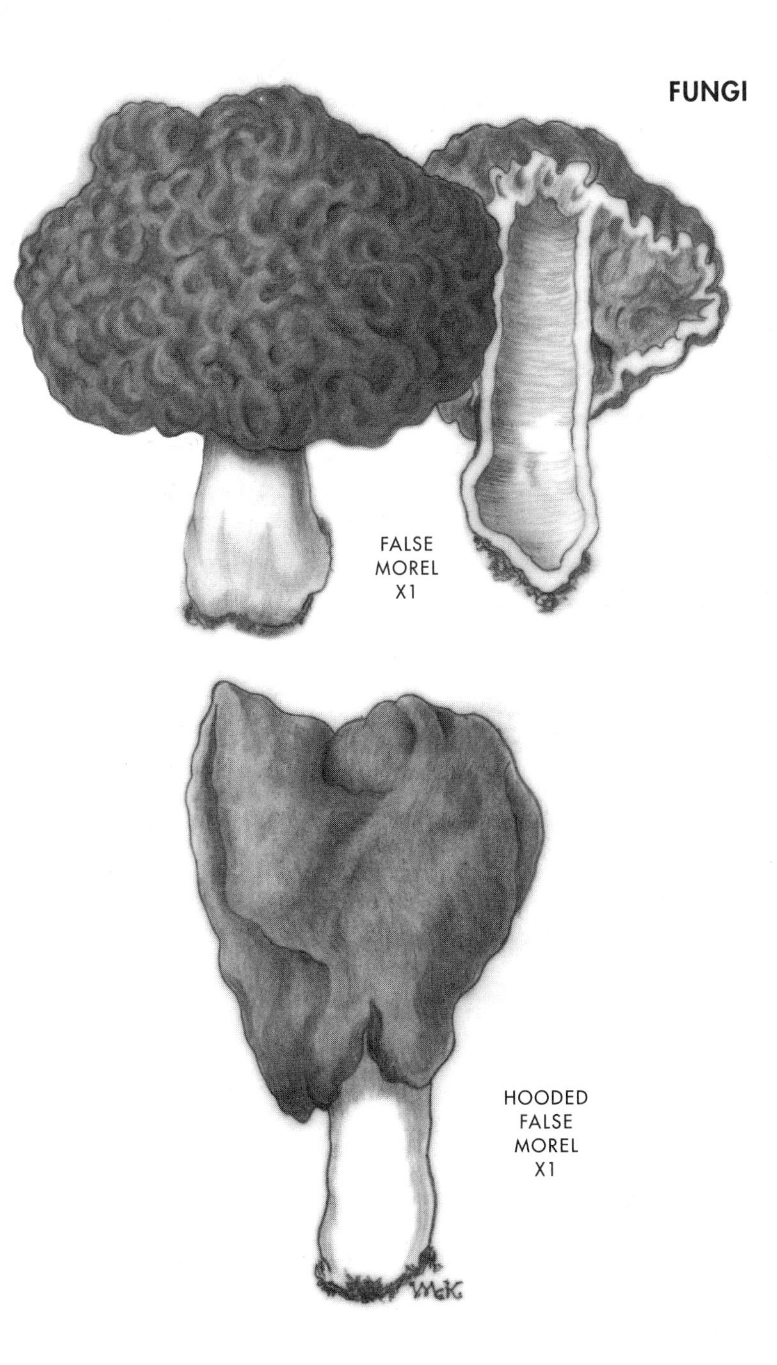
FALSE
MOREL
X1
HOODED
FALSE
MOREL
X1
McK

NEUTOTOXIC MUSHROOMS VEILED, WITH FREE GILLS

FLY AGARIC *Amanita muscaria* (L. ex Fr.) Pers. ex Hook.

Scattered or group-forming, medium to large mushroom. Cap *shiny, reddish* (varying to orange, yellow, or whitish), with prominent *white to yellow warts;* 2–10 in. across. Stalk quite thick, easily separates from cap, tapering upward from bulbous base; 2–7 in. Gills broad, close, free or slightly attached to stalk, whitish. Universal veil (covering emerging "button") white to yellow, at maturity forming 2–3 concentric rings at and above upper bulb. Partial veil (ring) at upper stalk thin, fragile, drooping, often streaked with "gill lines," or smooth. Summer–fall (winter in Pacific states). See *Peterson Field Guide to Mushrooms,* page 227.

Where found: Coniferous or hardwood forest floors. Much of North America.

Comments: This well-known poisonous mushroom, depicted in fairy tales and legend, contains ibotenic acid and muscimol. Muscimol is a toxic breakdown product of ibotenic acid. The toxins affect the central nervous system. The toxins of *Amanita muscaria* and *A. pantherina* are chemically unrelated to the toxins in other *Amanita* species. The Fly Agaric is a well-known intoxicant, reported to produce sensual derangement, erratic behavior, delirium, deep "deathlike" sleep, and in some cases, death.

PANTHERCAP *Amanita pantherina* (D.C. ex Fr.) Sec.

Solitary or several, medium to large, brownish, buff or dull yellow mushroom. Cap brownish with whitish, often pyramidal patches, margins with fine radial lines; 1–6 in. across. Stalk thick, tapering upward from a rounded bulb, with collar at upper edge of bulb; 2–7 in. tall. Gills free or barely touching stalk, close or crowded, white with delicate scalloped edges. Veil ring above is thin, whitish, pendent, persistent. Spring, late summer–fall, winter in Calif. See *Peterson Field Guide to Mushrooms,* page 229.

Where found: Coniferous or mixed hardwood forest floors. Northern U.S., especially Rocky Mountains to West Coast.

Comments: See *A. muscaria* above.

FLY AGARIC
X ½

PANTHERCAP
X ½

INKY CAP *Coprinus atramentarius* (Bull.) Fr.

Clump-forming, medium-sized, fleshy, gray-brown mushroom. Cap *egg- to bell-shaped,* shiny, gray, with a *lobed or pleated margin;* upturned with age, flesh turning *inky black* as it decays; 2–3 in. across. Gills free, wide, crowded, straight-sided. Stalk cylindrical, white above, grayish beneath, with *conspicuous false ring;* 2–8 in. Spring–fall. See *Peterson Field Guide to Mushrooms*, page 276.

Where found: Grassy or bare areas, often near tree stumps, from buried wood. Throughout North America.

Comments: Considered edible, but the mushroom deactivates an enzyme that activates alcohol in the system. If alcohol is consumed within 24–48 hours of eating Inky Cap, a toxic reaction may occur, including flushing of face and neck, tingling of extremities, headache, and nausea. Symptoms appear about 30 minutes after consuming alcohol. Effects are usually overcome within a few hours.

NAKED BRIMCAP *Paxillus involutus* (Bat. ex Fr.) Fr.

Solitary or group-forming, medium to large, brown-capped mushroom. Cap convex to flat or depressed, surface with matted fibers or slimy when wet, margin *strongly inrolled;* 1¾–4¾ in. Gills crowded, forked, tapering down stem, yellow to olive, becoming brown when touched; *separate readily from flesh of cap.* Stalk sometimes off-center, firm, solid, smooth; 1¾–4 in. Summer–fall. See *Peterson Field Guide to Mushrooms,* page 313.

Where found: In mixed woods; on ground. Throughout North America.

Comments: Can develop an acid-sour flavor. In some individuals it may produce allergic sensitivity resulting in destruction of red blood cells and kidney failure. Fatalities reported.

GOLDBAND WEBCAP *Cortinarius gentilis* (Fr.) Fr.

Few to colony-forming, medium-sized, smooth, orange-brown mushroom. Cap conical, knob-shaped, becoming flattish, smooth, orange-brown at first, then yellow-orange; 1–2 in. Gills attached, broad, quite widely spaced, brown. Stalk yellowish to orange-brown, cylindrical, somewhat striated; 1–3 in. Veil reduced to webby yellow fibrils halfway up the stalk. Summer–fall. See *Peterson Field Guide to Mushrooms,* page 290.

Where found: Coniferous forests, especially in the West.

Comments: While some members of the large *Cortinarius* genus are considered edible, a number of species contain a toxin, orellanine, that does not produce symptoms until 3 days to 3 weeks after ingestion. Symptoms include excessive thirst, nausea, headache, chills, and kidney failure. Fatalities recorded. Ingestion of *Cortinarius* species should be avoided.

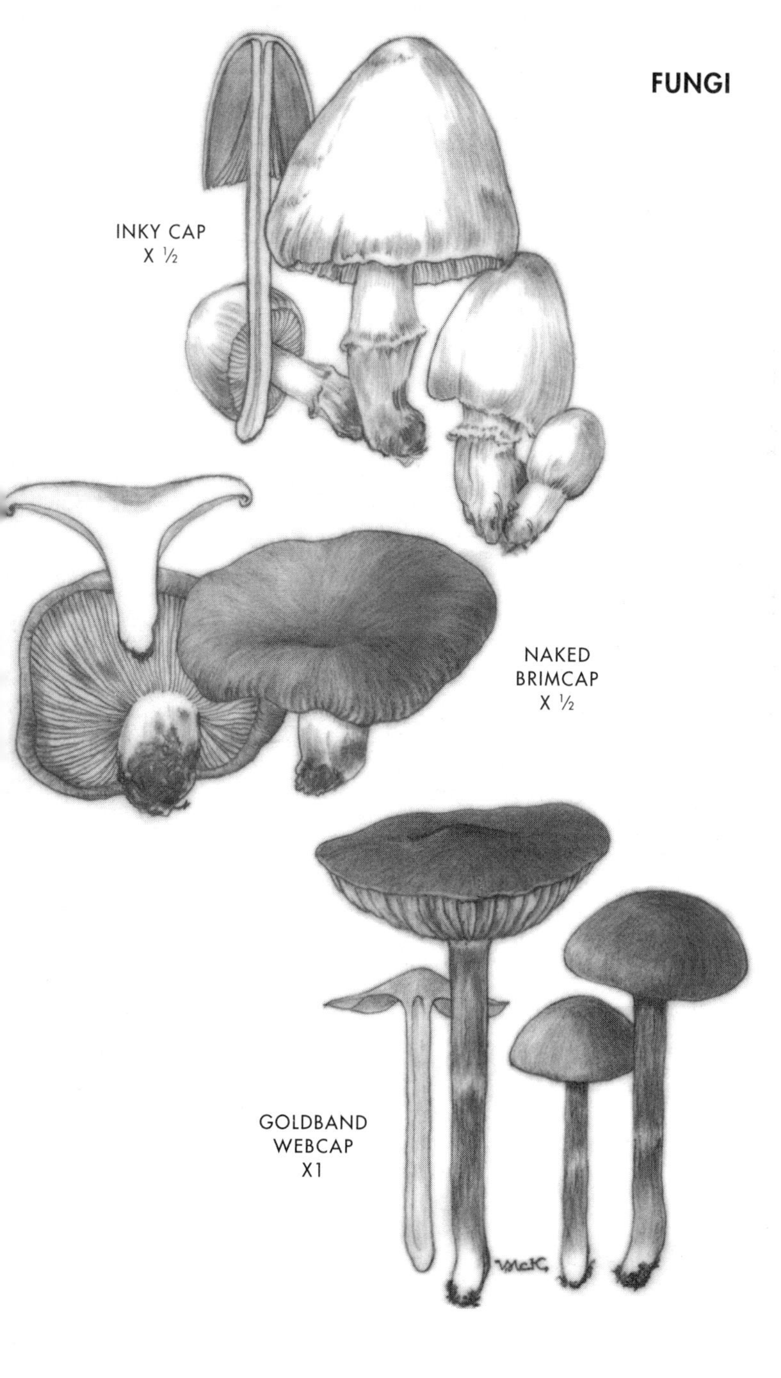
INKY CAP
X ½
NAKED
BRIMCAP
X ½
GOLDBAND
WEBCAP
X1

HALLUCINOGENIC MUSHROOMS

BELL MOTTLEGILL *Panaeolus campanulatus* (Bull.) Quel.

Solitary or clustered, small to medium, thin, brown mushroom. Cap *bell-shaped,* with rounded knob at center, reddish brown to grayish, margin slightly incurved when young, toothed; $^3/_4$–2 in. Gills broad, medium-spaced, notched, grayish then blackening, edges remain white. Stalk brownish, hollow, *very thin;* $2^1/_2$–$5^1/_2$ in. All year. See *Peterson Field Guide to Mushrooms,* page 286.

Where found: Widespread on horse or cow dung.

Comments: Considered hallucinogenic.

BLUESTAIN SMOOTHCAP *Psilocybe cubensis* (Ear.) Sing.

Cluster-forming, small to medium mushroom that stains blue when handled. Cap rounded, conical, or bell-shaped, or flattish with central knob, somewhat sticky, yellowish white to yellow-brown, often split at margin; $^5/_8$–$3\,^1/_2$ in. Gills attached, close, dull yellow at first, turning purplish when mature. Stalk cylindrical with a high ring, smooth, grooved above, staining blue when bruised; $1^1/_2$–6 in. Much of the year. See *Peterson Field Guide to Mushrooms,* page 275.

Where found: Cow and horse dung in pastures in Southeast.

Comments: Hallucinogenic; may cause high fevers and convulsion in children. Possession is illegal.

SHOWY FLAMECAP, BIG LAUGHING MUSHROOM

Gymnopilus spectabilis (Fr.) A.H. Sm.

Usually clustered, medium to large, orange-yellow mushroom. Caps convex, becoming flat, with knob in center, smooth, or fibrous-surfaced to minute scaly, thick-fleshed, tapering to thin margin; $3^1/_4$–7 in. wide. Gills attached, crowded, narrow, pale yellow to yellowish red. Stalk cylindrical or tapering, solid, firm; 2–8 in. Veil thin, on upper stalk. Summer–fall. See *Peterson Field Guide to Mushrooms,* page 298.

Where found: Logs, stumps, and buried wood. Throughout.

Comments: Reported to be hallucinogenic in some people and in others causing painful and serious illness. Ingestion of all *Gymnopilus* species is best avoided. May cause blurred vision, lack of coordination, and in some persons uncontrolled giggling or laughter. Some *Gymnopilus* species, but apparently not *G. spectabilis,* reported to contain psilocybin.

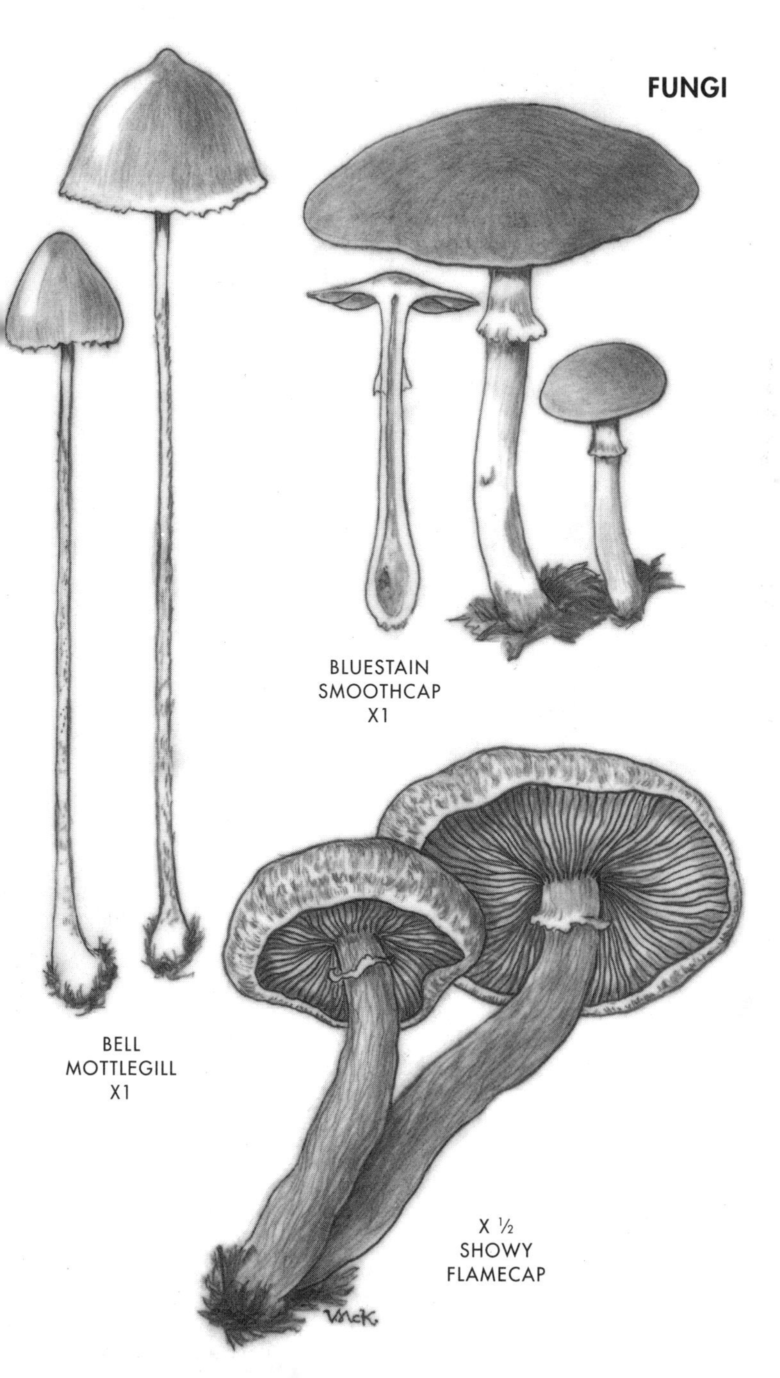
BLUESTAIN
SMOOTHCAP
X1
BELL
MOTTLEGILL
X1
X ½
SHOWY
FLAMECAP
VMcK.

GILLED MUSHROOMS CAUSING SWEATING, TEARS, AND SALIVATION

SWEAT MUSHROOM *Clitocybe dealbata* subsp. *sudorifica* Pk.

Single, group- or cluster-forming small white mushroom. Cap convex to flattish, finally sunken; margin at first incurved, then becoming wavy or flaring-upturned at maturity, smooth; 3/8–1 5/8 in. Gills attached and slightly descending down stalk, close, narrow. Stalk cylindrical or slightly downward-tapered, thin and tough; 1–2 in. Summer–fall, or winter in Calif. See *Peterson Field Guide to Mushrooms,* page 140.

Where found: In grass or lawns, old fields, or open woods.

Comments: May be collected in error during harvest of edible white mushrooms. If ingested, it may cause mild or severe poisoning, with chills, blurred or tunnel vision, and sweating. Tha alkaloid muscarine has been detected in this mushroom. Can be confused with other *Clitocybe* species of Fairy Ring Mushroom *(Marasmius oreades).*

CONIC FIBERHEAD *Inocybe fastigiata* (Schaeff.) Quel.

Solitary or grouped, small to medium-sized mushroom. Cap light yellow-brown, *conical* or bell-shaped, then becoming flat and upturned; margin incurved at first, then upturning or splitting. Gills attached, close, narrow, whitish, becoming olive to brown. Stalks whitish to yellowish, twisted and streaked; 1 5/8–3 1/4 in. Ring absent. Summer–fall. See *Peterson Field Guide to Mushrooms,* page 301.

Where found: Mossy or grassy soil of open woods and pastures. Widespread.

Comments: Contains the alkaloid muscarine, but in amounts that would require ingestion of 7–11 pounds of mushrooms to produce a lethal dose. However, ingestion may cause symptoms of muscarine poisoning.

EARTHBLADE FIBERHEAD *Inocybe geophylla* (Sow. ex Fr.) Kum.

Scattered, grouped, or cluster-forming, small, dry, silky, white mushroom. Caps conical to bell-shaped, then flattening with a central *pointed knob,* with earthy to nauseating odor; 5/8–1 1/4 in. Gills close, attached, broad, whitish, becoming yellow-brown with age. Stalk firm, white, with veil fibrils above; 1–2 in. Summer to late fall. See *Peterson Field Guide to Mushrooms,* page 301.

Where found: Under coniferous or deciduous forests.

Comments: Contains muscarine. Like all *Inocybe* species, may cause symptoms of muscarine poisoning.

FLUFF FIBERHEAD *Inocybe lanuginosa* (Bull. ex Fr.) Kum.

Solitary or clump-forming, small to medium, *dark brown* mushroom. Cap convex to hemispherical, dry, with small, raised, woolly scales; 3/4–1 1/4 in. broad. Gills attached, close, broad, whitish yellow, then becoming brownish. Stalk brownish, scaly; 1–2 in.; no ring left by veil on stalk. Summer–fall. See *Peterson Field Guide to Mushrooms,* page 302.

Where found: On decaying or buried wood. Widespread, not abundant.

Comments: Contains muscarine. Like all *Inocybe* species, may cause symptoms of muscarine poisoning.

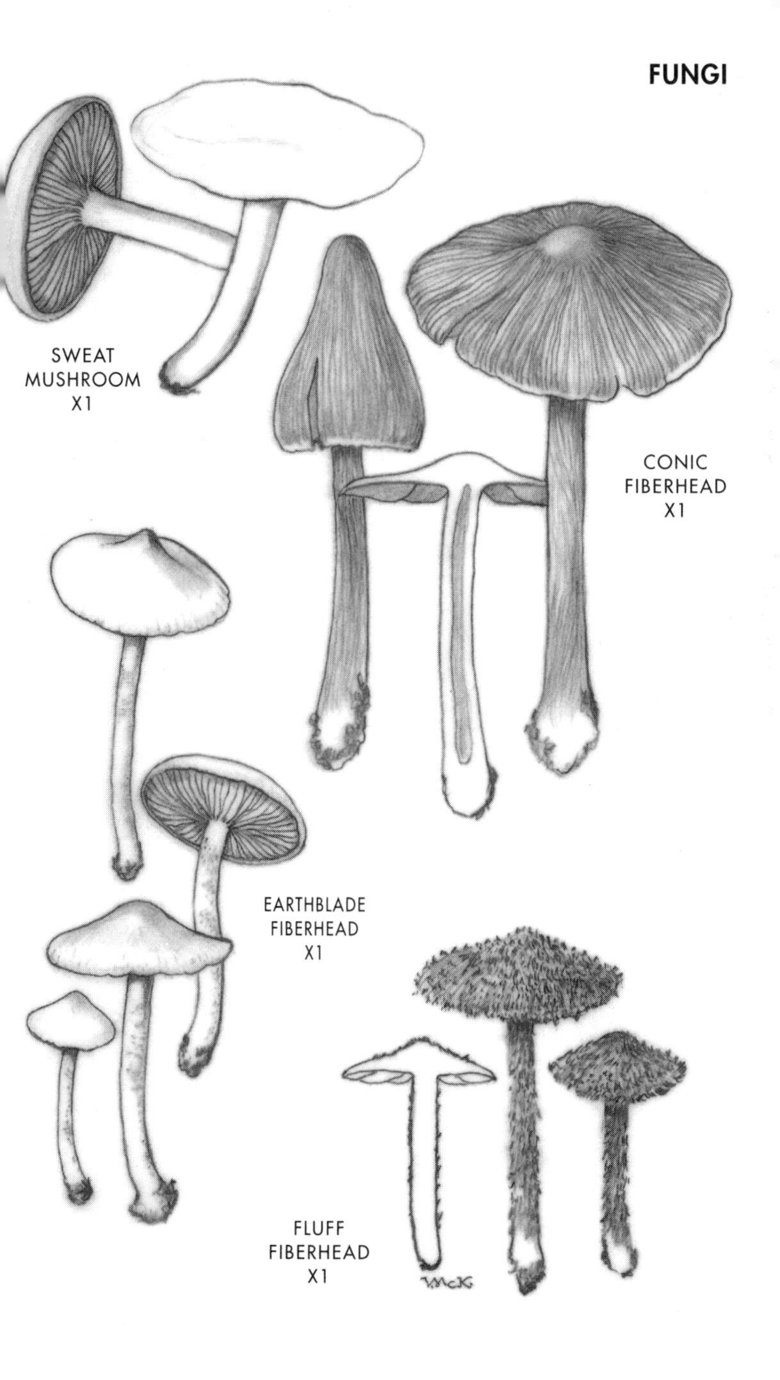
SWEAT
MUSHROOM
X1
CONIC
FIBERHEAD
X1
EARTHBLADE
FIBERHEAD
X1
FLUFF
FIBERHEAD
X1

GRAYSCALE *Agaricus meleagris* J. Schaeff.

Single to grouped, large, whitish mushroom. Cap rounded, then convex and flattish, dry, with *grayish* scales on white background. When cut, stains yellow at first, then becomes purplish brown; bruising emits foul odor; 2–6 in. Gills free, broad, close, grayish at first, becoming pinkish then brown. Stalk off-white to pink-brown, base yellow if cut, tapering to swollen base; 2–7 in. Veil white, producing large ring on upper stalk. Summer–fall. See *Peterson Field Guide to Mushrooms,* page 258.

Where found: Forest floors, lawns, mostly in West.

Comments: Many *Agaricus* species are delicious, such as the common white mushrom *Agaricus bisporus*; others are poisonous. Amateurs have mistaken deadly *Amanita* species for edible *Agaricus* species. Avoid *Agaricus* species with foul-smelling gills, or those in which the base of the stalk becomes yellow if cut.

GREEN GILL *Chlorophyllum molybdites* Mass.

[Lepiota molybdites (Meyer ex Fr.) Sacc., *L. morgani]*

Large white mushroom. Caps bell-shaped to convex, becoming broadly convex; margin thin and incurved, with pink-buff to brownish scles, 2–12 in. Gills free, close, broad, *pale yellowish,* obviously greenish as spores develop. Stalks *slender,* somewhat flaring at base, with smooth or powdered surface, becoming grayish brown when cut or bruised; 4–10 in. Two-layered ring with jagged edge. Late summer–fall. See *Peterson Field Guide to Mushrooms,* page 240.

Where found: Grassy areas. Mid-Atlantic states, south, west to Calif.

Comments: Commonly involved in poisonings. Ingestion of this mushroom may produce vomiting, diarrhea lasting up to two days, faintness, chills, weakness, or mental confusion.

RED-MOUTH BOLETE *Boletus subvelutipes* Pk.

Single to group-forming, medium-sized mushroom, *all parts bruising bluish black.* Cap convex to flattish, brownish red to cinnamon red to yellowish, surface with tiny fibers or somewhat felty; 2½–5¼ in. wide. Tubes touching or descending stalk, pores orange to bright red. Stalk yellowish, dotted; base with *reddish hairs;* 1¼–4 in. long. Late spring–fall.

Where found: Mixed woods. Northeastern U.S.

Comments: Ingestion of all orange- to red-pored boletes should be avoided. They are found throughout North America.

PEPPER BOLETE *Boletus piperatus* Bull. ex. Fr.

Solitary or group-forming, small to medium-sized, *reddish orange to brownish* mushroom. Cap rounded, convex to flattish, darker in center, surface dry, margins minutely hairy; ¾–3½ in. wide. Tubes attached or descending stalk, pore mouths reddish, tubes becoming larger toward margin of cap. Stalk cylindrical, yellowish to reddish brown, *solid* within; base bright yellow; ¾–4¾ in. Summer–fall.

Where found: Coniferous and hardwood forests. Most of southern Canada and northern U.S., south to middle U.S.

Comments: Peppery flavored; may cause variable reactions in sensitive individuals, including severe stomach pains. Some mushroom fanciers eat Pepper Bolete if it is well-cooked, but it is best avoided.

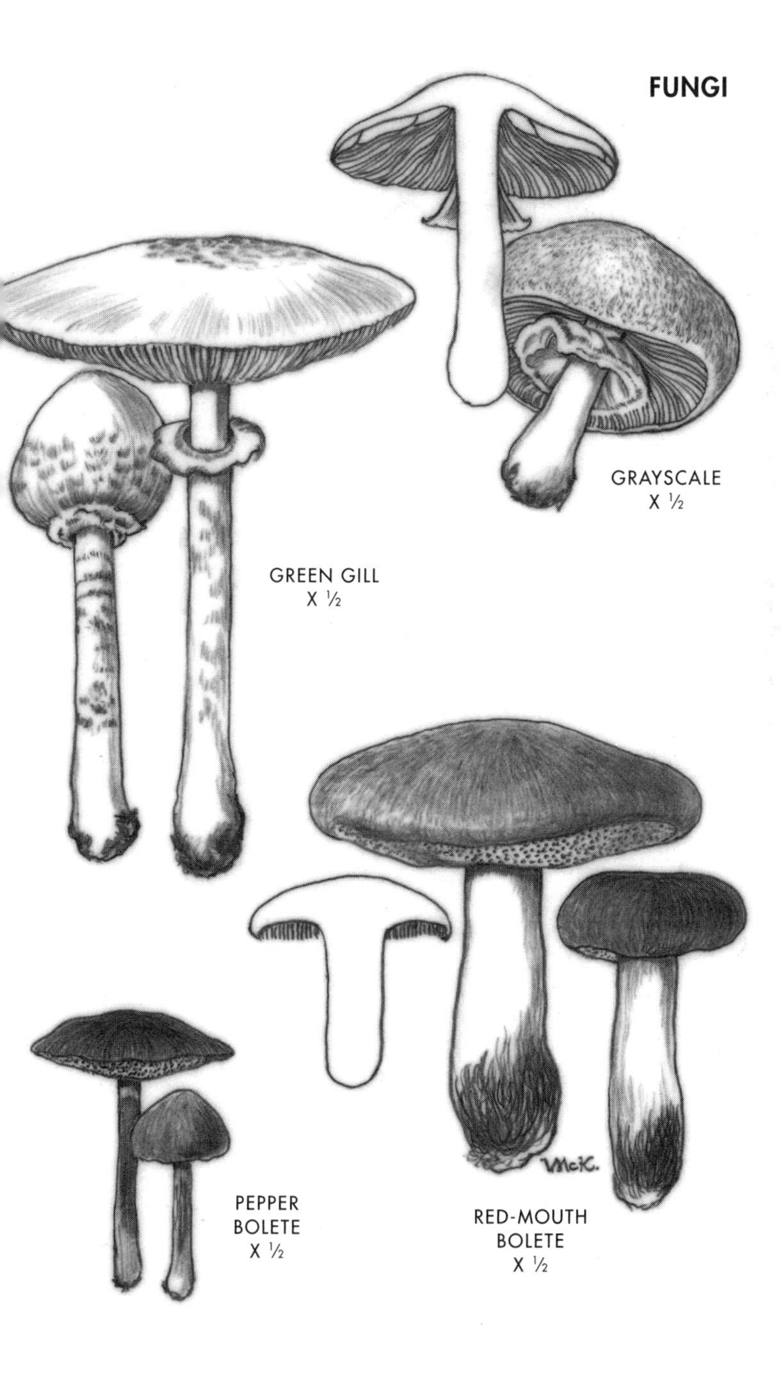
GRAYSCALE
X ½
GREEN GILL
X ½
PEPPER
BOLETE
X ½
RED-MOUTH
BOLETE
X ½
VMcK.

MUSHROOMS WITH ATTACHED GILLS CAUSING GASTROENTERITIS

SICKENER *Russula emetica* (Schaff. ex Fr.) S.F.G.

Scattered or group-forming, medium to large, *fragile, reddish-capped mushroom.* Cap somewhat convex or flattish, center often sunken or depressed, margin somewhat incurved with warts; surface tacky, red to pinkish. Gills close, broad, attached, yellowish gray or white; 1–3 in. Stalk cylindrical, hollow, or with white filaments within, fragile; breaking easily; 2–4 in. long. Summer–fall. See *Peterson Field Guide to Mushrooms,* page 320.

Where found: Coniferous or mixed woods, often on sphagnum moss in boggy areas, and on well-rotted wood. Widespread.

Comments: The many red species of *Russula* may be confused with or misidentified as *R. emetica.* Some species are identified only with technical characteristics, including microscopic details. As the species name *emetica* implies, ingestion of this mushroom will induce vomiting. Some have claimed that the species is edible, but cases of poisoning are documented. While properly prepared mushrooms may not cause reactions in some persons, others may be adversely affected. Recent investigations have detected the alkaloid muscarine in *R. emetica,* while other reports have failed to confirm its presence. This may explain why some people have reportedly eaten the mushroom without ill effects, while others experience gastrointestinal irritation. Best to avoid ingestion.

FRAGILE BRITTLEGILL *Russula fragilis* (Pers. ex Fr.) Fr.

Solitary or scattered, small to medium mushroom with multicolored cap. Cap convex, with a small hump when young, then becoming flat and depressed in center; margin incurved, reddish but with multicolors, dull olive in center, *colors variable;* 1–3 in. wide. Gills attached to nearly free, broad, close to almost distant, yellowish white. Stalk cylindrical, solid or with cottony fibers within, whitish, moist or dry, with vertical wrinkles, smooth or with minute hairs; 1–3 in. long. Summer–fall. See *Peterson Field Guide to Mushrooms,* page 321.

Where found: Coniferous, hardwood, or mixed forests on soil and rotted wood. Widespread.

Comments: Considered poisonous. Ingestion of this and other widespread and difficult-to-identify red *Russula* species, common and frequent in North America, should be avoided.

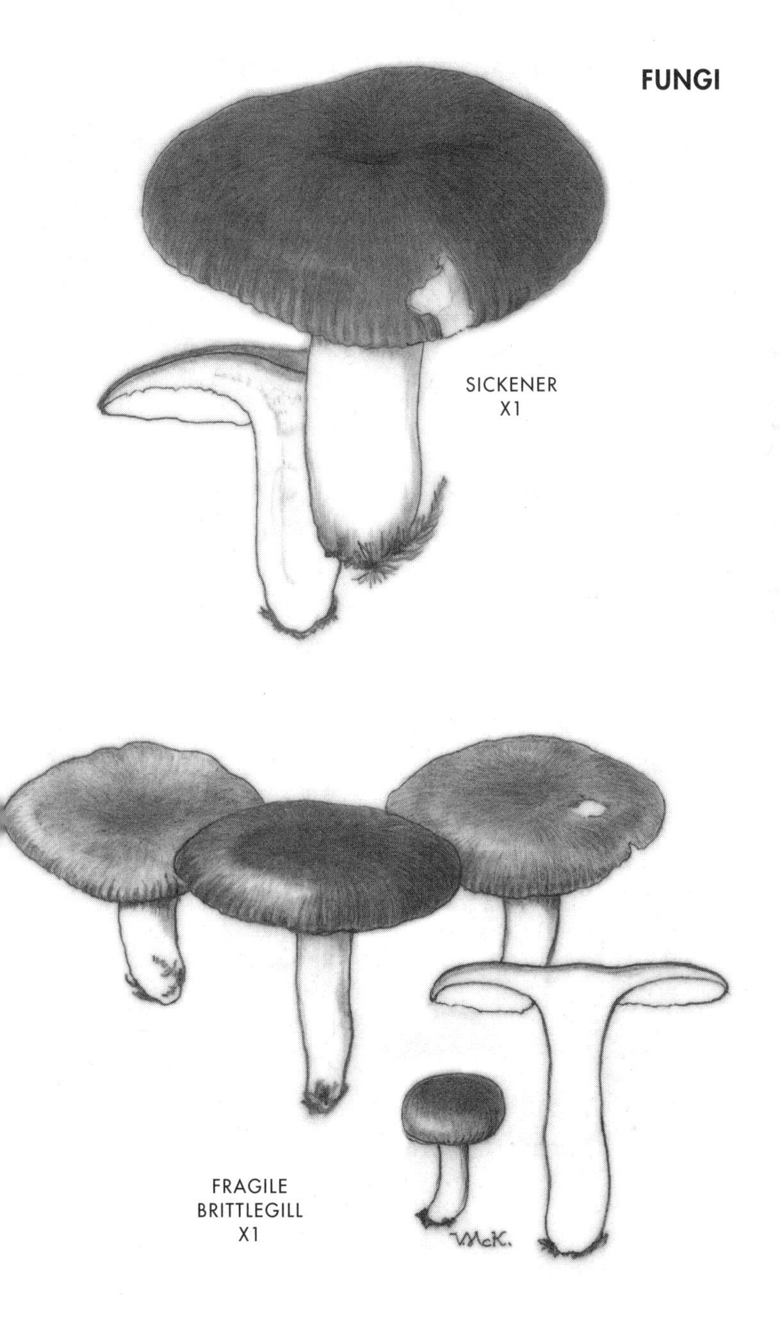
SICKENER
X1
FRAGILE
BRITTLEGILL
X1
McK.

MUSHROOMS WITH ATTACHED GILLS CAUSING GASTROENTERITIS

GRAY PINKGILL, LEAD POISONER *Entoloma lividum* (Bull.) Quel. [*Entoloma sinuatum* (Bull. ex Fr.) Kum.]

Scattered to group-forming, medium to large mushroom with *grayish* cap and faint mealy odor. Cap convex, with a hump when young, then becoming flat or wavy, with downturned or inrolled margin. Surface smooth, *slippery,* dirty grayish or brownish; 2¾–6 in. wide. Gills attached, *close to distant,* broad. Stalks cylindrical, solid, lighter than cap; 1¾–4¾ in. Fall. See *Peterson Field Guide to Mushrooms,* page 311.

Where found: Oak woods, in the eastern U.S.

Comments: Ingestion of small amounts may cause reactions 20 minutes to 4 hours after ingestion, including vomiting, diarrhea, headache, thirst, prostration, and unconsciousness. May cause liver damage. In Europe, *E. lividum* is responsible for 10 percent of mushroom poisoning cases.

STRAIGHT-STALK PINKGILL *Entoloma strictior* (Pk.) Sacc.

Solitary or scattered, *dark brown to grayish brown* mushroom. Cap conical or bell-shaped early, then becoming nearly flat with pointed knob when mature, with smooth surface, dark brownish gray; 1–2 in. Gills attached, close, broad, white, then becoming pinkish. Stalk slender, fragile; twisted, spiraled, or streaked; whitish; 2–4 in. long. Spring-fall. See *Peterson Field Guide to Mushrooms,* page 312.

Where found: Moist ground, bogs, or rotten wood. Much of the eastern U.S.

Comments: See *E. lividum* above. Often confused with *Melanoleuca* species when young. Many *Entolomas* can be identified only through microscopic examination.

POISON PIE *Hebeloma crustuliniforme* (Bull. ex St. Amans) Quel.

Single or group-forming, medium to large, round mushroom with strong *radishlike odor.* Cap convex, margin inrolled when young, then becoming round to flattish with a low knob; surface sticky, cream to brown; 1½–4 in. Gills attached, close, narrow, with minute beaded droplets on edges at first, yellowish white to brown-yellow. Stalk solid, bulbous at base, dirty yellow-white; 1⅝–2¾ in. Late summer–fall. See *Peterson Field Guide to Mushrooms,* page 300.

Where found: Under hardwoods or conifers. Widespread.

Comments: All *Hebelomas* are considered poisonous and should be avoided. Symptoms of reported poisoning include deep sleep followed by severe abdominal cramps, vomiting, and severe diarrhea.

DARK DISK *Hebeloma mesophaeum* (Pers.) Quel.

Group-forming or scattered, small to medium, brown, *shiny* mushroom with *radishlike odor.* Cap rounded, often with a knob, smooth, brown; ¾–2⅜ in. across. Gills attached, close, broad, dirty yellow-white to brownish, edges with minute hairs. Stalk cylindrical, *slender,* surface striated; 1¼–3½ in. long. Ring thin, fibrillose. Fall (or spring). See *Peterson Field Guide to Mushrooms,* page 300.

Where found: Bare spots, mossy ground in coniferous forests. Northern U.S. and Canada.

Comments: See *H. crustuliniforme.*

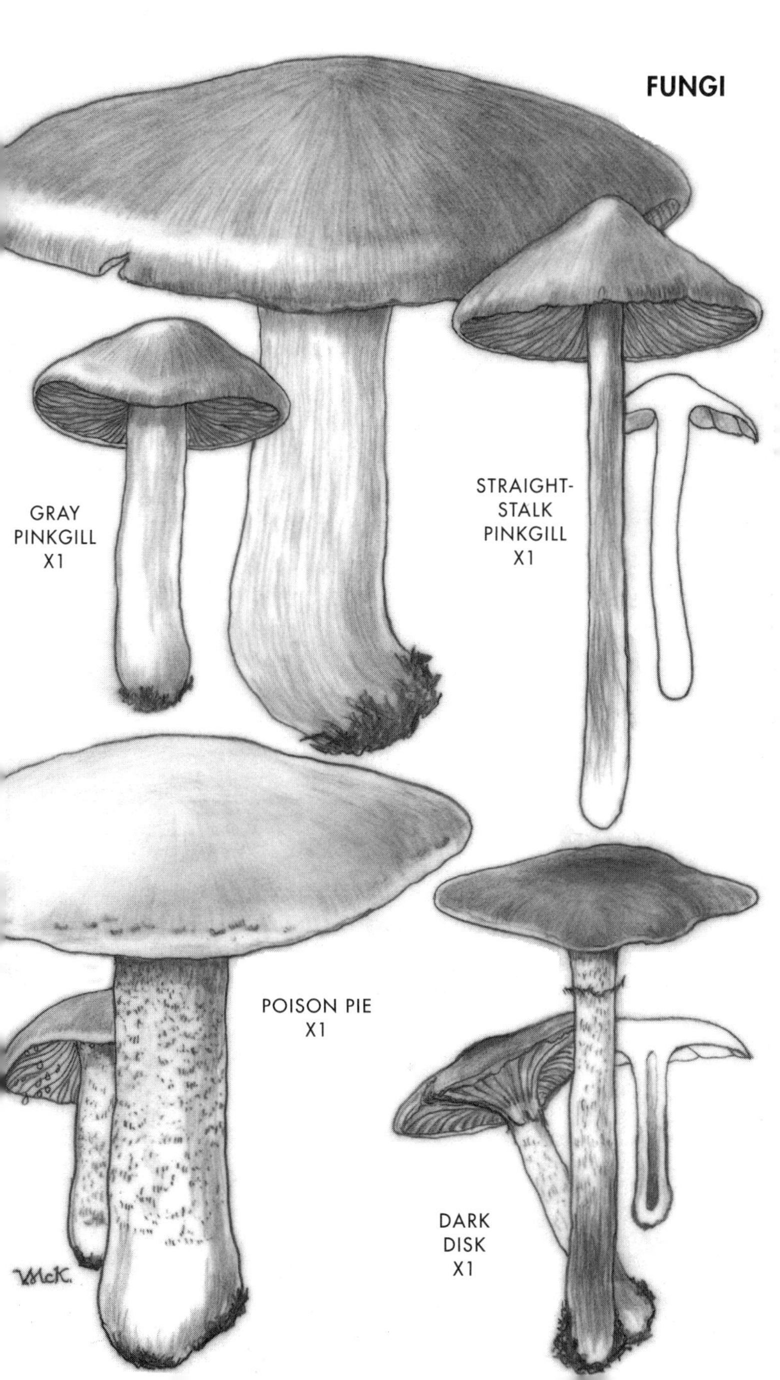
GRAY
PINKGILL
X1
STRAIGHT-
STALK
PINKGILL
X1
POISON PIE
X1
DARK
DISK
X1
VMcK.

MILKCAPS CAUSING GASTROENTERITIS

Over 200 *Lactarius* species are known in North America, many of which may cause gastrointestinal upset or irritation, including vomiting and diarrhea. As the genus name implies, they exude milky juice or latex when fresh. Only milkcaps identified by an expert should be eaten. The following milkcaps should *not* be eaten.

PEPPER MILKCAP *Lactarius piperatus* (Fr.) S.F.G.

Scattered or group-forming, medium to large, whitish mushroom with *white latex.* Cap convex, flat or depressed, becoming vase-shaped, smooth or wrinkled, spotted if damaged; 2–6 in. across. Gills attached, very crowded, forked one or more times, white, then pale yellow. Stalk white, dry; 3/8–1 in. long. Summer. See *Peterson Field Guide to Mushrooms,* page 333.

Where found: Deciduous woods, eastern North America.

SHAGGY BEAR *Lactarius representaneus* Britz. sensu Neuh.

Scattered to cluster-forming, large, *sticky,* yellowish mushroom. Cap convex, flattish, to shallow funnel-shaped, yellowish to orange-yellow, bruising purple; surface hairy; margins stiff, shaggy-hairy; 2 3/8–8 in. Gills attached, close, yellowish; purple when bruised. Stalk cylindrical, hard, yellowish, spotted; 2–4 3/4 in. Late summer–fall. See *Peterson Field Guide to Mushrooms,* page 334.

Where found: Soil of spruce forests. Common in Rocky Mountains, western U.S. and Canada.

SPOTSTALK *Lactarius scrobiculatus* (Fr.) Fr.

Solitary or group-forming, medium to large mushroom *with white latex, turning yellow when exposed to air.* Cap convex, becoming funnel-shaped, *straw yellow,* with inrolled, hairy margin; *very sticky* when young; 3–10 in. wide. Gills attached or descending, close, often forked close to stalk, whitish yellow, bruising pink. Stalk cylindrical, hollow, dry-surfaced, pitted with yellowish spots; 1 1/4–2 1/2 in. Late summer–fall. See *Peterson Field Guide to Mushrooms,* p. 335.

Where found: Coniferous forests. Northern North America, south to Calif. and Colo.

POWDERPUFF MILKCAP *Lactarius torminosus* (Fr.) S.F.G.

Scattered to group-forming, large, *pink* mushroom; *latex white, color not changing when exposed to air.* Cap convex to flattish, becoming shallowly funnel-shaped, pinkish; margins hairy and strongly incurved; 2–5 in. Gills attached, close, narrow, yellowish pink. Stalk smooth, dry, hollow to filament-filled; 1 1/4–3 in. Late summer –fall. See *Peterson Field Guide to Mushrooms,* page 337.

Where found: Often under birch trees, northern North America south to Mo. and Calif.

DAMP MILKCAP *Lactarius uvidus* (Fr.) Fr.

Scattered to group-forming, medium to large, gray to brown, purple-flushed mushroom, with white or yellowish latex that gradually turns purplish, then brown when cut or bruised. Cap convex or with a knob, flattening, then becoming sunken in center; margins incurved when young; 1 1/2–4 in. Surface smooth or sticky, with various-colored spots. Gills attached, close, whitish, then yellow; bruising purplish. Stalks cylindrical, slimy to dry, yellowish at base; 1 1/4–3 in. Summer–fall. *Peterson Field Guide to Mushrooms,* p. 339.

Where found: Mixed or coniferous forests. Much of North America.

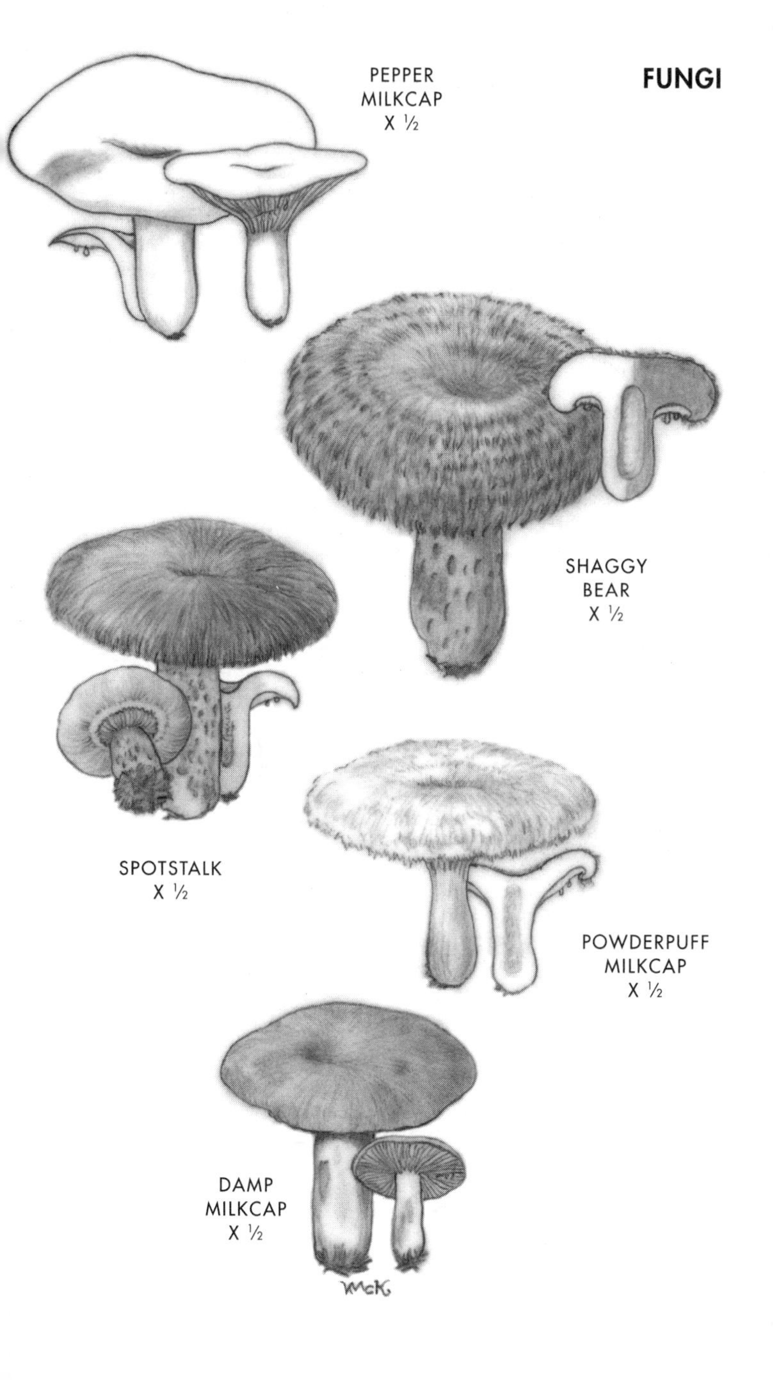
PEPPER
MILKCAP
X ½
SHAGGY
BEAR
X ½
SPOTSTALK
X ½
POWDERPUFF
MILKCAP
X ½
DAMP
MILKCAP
X ½
MCK

RICH MOIST WOODS
EASTERN DECIDUOUS FOREST: 1

WHITE BANEBERRY, DOLL'S EYE *Actaea pachypoda* **p. 76**
The white to ivory fleshy fruits, atop a reddish stem tipped with a dark dot, easily identify it as Doll's Eye. As likely to be encountered as a shade-garden ornamental as in the wild. **Flowers April–June. Fruits July–Oct.** The European Baneberry or Herb Christopher *Actaea spicata* is the source of most reports of toxicity from this genus, the bluish black berries of which have caused poisoning and death of children in Europe. The acrid-tasting fruits can cause severe stomach cramps, headache, vomiting, and dizziness. As few as 5 or 6 baneberry fruits can produce toxic effects. No deaths have been recorded from ingestion of the fruits of any American species of *Actaea*.

RED BANEBERRY *Actaea rubra* **p. 76**
Similar to White Baneberry, but the stalks are less stout. Fruits lustrous, cherry red (or sometimes white in *A. rubra* ssp. *arguta,* found in the western U.S.). **Flowers April–May. Fruits July–Oct.** Nervous system disturbances have been reported from ingesting Red Baneberry fruits.

SKUNK CABBAGE *Symplocarpus foetidus* **p. 150**
Note the large, cabbagelike, skunk-scented leaves, which emerge after the flowers in early spring. **Leafing out late April–May.** Note the broad, *hooded,* green to purple-brown flowers with a sheathed spathe, enclosing the clublike spadix. **Flowering April–early July.** Ingesting leaves may cause gastrointestinal burning and inflammation. The root, once used medicinally and described as narcotic, may contain poisonous components different from those of the leaves. A plant surprisingly little studied by chemists.

DUTCHMAN'S BREECHES *Dicentra cucullaria* **p. 74**
A delicate perennial with lacy leaves and white, yellow-tipped flowers dangling on an arched stalk. The trouserlike appearance of the flower spurs earns the name Dutchman's Breeches. Often cultivated. **Flowering April–May.** The nine North American *Dicentra* species have similarly shaped flowers. *Dicentras* are known to contain toxic isoquinoline alkaloids in the leaves and especially the tubers, including aporphine, protoberberine, and protopine. Sometimes attractive to children, given the beauty and unusual form of the flowers. Not known to have caused fatalities. Livestock have been poisoned by nibbling on the emerging spring leaves or exposed tubers in shaded pastures. Handling the plant may cause a skin rash.

WHITE BANEBERRY

RED BANEBERRY

SKUNK CABBAGE LEAVES

SKUNK CABBAGE FLOWER

DUTCHMAN'S BREECHES

RICH MOIST WOODS
EASTERN DECIDUOUS FOREST: 2

GREEN DRAGON *Arisaema dracontium* p. 140

The unusual sheathlike flowers of this member of the arum family are green with an elongated spadix, which has a fanciful resemblance to the head of a dragon. **Flowers May–July.** The root and leaves contain irritating calcium oxalates, which can severely irritate mucous membranes of the mouth and throat if eaten, especially when fresh. Toxic reactions may also be caused by proteolytic enzymes in the plant. Poisoning is usually self-limiting by virtue of the severe burning caused by the first nibble. In severe cases, irritation can cause swelling in the throat, leading to choking. Symptoms and swelling subside after a few hours.

JACK-IN-THE-PULPIT *Arisaema triphyllum* p. 150

Note the 3 leaflets, flowers green to purple–brown with whitish to green stripes in a cuplike spathe with a curved flap shading the spadix inside. **Flowers May–July.** Fruits *bright red* in clusters. As with many members of the arum family, ingesting any part of the fresh plant may cause intense burning due to irritating calcium oxalate crystals. See above.

GREEN HELLEBORE *Veratrum viride* p. 144

Veratrums are generally quite large, with clusters of pleated leaves in early spring, before deciduous trees leaf out. Flowers are yellowish 6-pointed stars in a large, pyramid-shaped, terminal cluster. **Flowers April–July.** Like the European *Veratrum album,* the 8 North American "false hellebores" are considered highly toxic because of their alkaloids that affect the heart and nervous system. Formerly used in medicine for heart ailments, which has resulted in overdoses and poisoning. Contains numerous toxic steroidal alkaloids, including germidine, germitrine, veratrin, veratradine, veratramine, and verastrosine. Ingestion of any plant part, especially the root, can result in severe slowing of heart rate and respiration, decreased arterial resistance, irregular heartbeat, excessive salivation, stomach pain, vomiting, diarrhea, spasms, and paralysis.

INDIAN PINK *Spigelia marilandica* p. 120

Note the narrow trumpetlike flowers, scarlet on the outside, with a cream-yellowish interior. **Flowers May–June.** Scattered historical reports of poisoning generally result from exceeding normal doses of medicinal preparations, once used for the treatment of intestinal worms. Nineteenth-century physicians reported narcotic effects, including dilated pupils, spasms, or convulsions. Poisoning is quite rare, though fatalities from overdose of the root, especially in children, are historically reported.

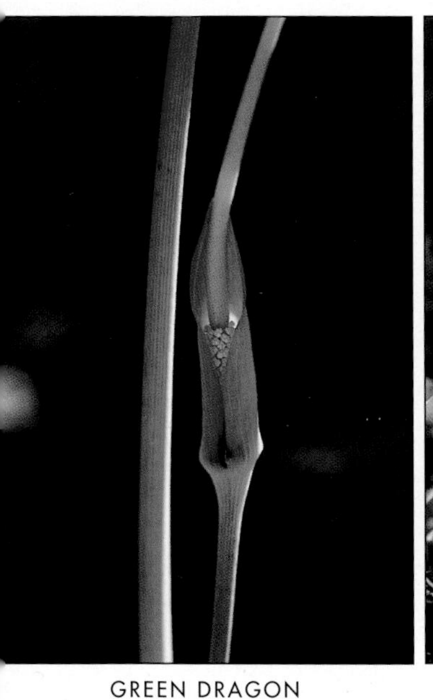

GREEN DRAGON

JACK-IN-THE-PULPIT

GREEN HELLEBORE

INDIAN PINK

RICH MOIST WOODS
EASTERN DECIDUOUS FOREST: 3

MAYAPPLE *Podophyllum peltatum* **p. 98**

Note the large, broad, umbrellalike, paired leaves with stem attached to leaf center from beneath. Flowers white, waxy, solitary, 8-petaled, nodding from division of leaf stems. **Flowers April–May.** One person's medicine is another's poison: highly toxic components of the root are used in chemotherapy for testicular and small cell lung cancer. The root, however, can be a deadly poison. It contains toxic lignans, such as podophyllotoxin and peltatins, which can cause respiratory stimulation, vomiting, catharsis, coma, or death. Produces a toxic effect on cell division. The unripe fruits have caused painful digestive disturbances, especially in children.

BLOODROOT *Sanguinaria canadensis* **p. 98**

Leaves distinctive, with deep or shallow sinuses. Orange juice in stems; roots blood-red within. **Flowers March–June.** Contains alkaloids similar to those of the Opium Poppy, including sanguinarine, which can depress the central nervous system. Overdoses cause vomiting, irritation of mucous membranes, diarrhea, fainting, shock, and coma. Most toxic reactions were reported from historical use of medicinal preparations, though persons collecting the root for commercial use have experienced dermatitis from handling it. The extremely bitter-tasting root limits ingestion to small amounts, but even minute amounts can cause toxic reactions. Levels of the active alkaloids are highly variable from region to region and population to population. Most highly concentrated in root when plant is leafed out.

BLUE COHOSH *Caulophyllum thalictroides* **p. 138**

Smooth-stemmed, blue-green compound leaves with 3 (5) leaflets with 2–3 lobes. Flowers greenish yellow (or darker). **Flowers April–June.** Fruits blue berries. **Fruits late July–Oct.** The root is used in herbal medicine, apparently without toxic results. However, the raw bluish berries have been reported to cause poisoning in children, causing irritation of the gastrointestinal tract, especially the intestinal mucosa. Fresh berries, leaves, and roots have been reported to cause contact dermatitis in sensitive individuals. Contains a number of alkaloids (with a chemical structure similar to nicotine) and saponins.

MAYAPPLE LEAVES

MAYAPPLE FLOWER

BLOODROOT LEAVES

BLOODROOT FLOWER

BLUE COHOSH

SHRUBS, VINES, FERN WOODS

POISON IVY *Toxicodendron radicans* p. 172

An extremely variable vine, trailing or climbing by aerial roots. Leaves shiny and alternate, with 3 irregularly toothed, pointed-tipped leaflets. Poison Oak, by contrast, has rounded lobes at the tip. The middle leaflet has the longest stalk. *Toxicodendron* species have hairy or smooth whitish fruit in pendulous clusters from leaf axils.

Fifty percent of the population is allergic to some degree to this plant group, which causes painful, irritating dermatitis in about two million people in the United States each year. All plant parts, especially the sap, contain irritant, nonvolatile, phenolic substances called urushiol or toxicodendrol. The oily mixture, found in resin canals, is released when the plant is bruised, even slightly. The toxins can bind to skin proteins. Washing soon after contact with the plant can remove the toxins, but molecules that have already bound to skin protein molecules cannot be removed by washing.

VIRGINIA CREEPER *Parthenocissus quinquefolia* p. 190

Leaves divided mostly into 5 (sometimes 3 –7) leaflets radiating from a central point. Fruits blue, inedible berries. **Fruits late Aug.–Nov.** Berries, attractive to children, are suspected of causing poisoning, with fatalities reported. Other symptoms have included gastric irritation such as vomiting and diarrhea, dilated pupils, weak pulse, and sweating. Sleepiness has also been reported in some individuals. Given its common occurrence in a widespread range, surprisingly little is known about the plant's chemistry or toxic compounds.

WILD HYDRANGEA *Hydrangea arborescens* p. 162

A shrub with flowers in flat-topped clusters, to 6 in. across, with papery, white, sterile, lobes along perimeter of flower clusters. **Flowers June–Aug.** Ingestion of leaves, which is rare, may cause nausea, vomiting, and bloody diarrhea.

BRACKEN FERN *Pterdium aquilinum* p. 194

Our most common fern, 3–6 ft. tall, forming large colonies. Highly variable. Leaves (fronds) are triangular in outline. Bracken fern is known to have poisoned livestock, especially in the Pacific Northwest. An enzyme in the fern, thiaminase, disturbs thiamin metabolism. Massive doses of thiamin have served as a successful treatment. Bracken also contains cancer-causing compounds. The water-soluble carcinogens can be directly imparted to cow's milk.

POISON IVY LEAFLETS

POISON IVY IN FALL

VIRGINIA CREEPER

WILD HYDRANGEA

BRACKEN FERN

WEEDS
FIELDS, WASTE PLACES

POKEWEED, POKE *Phytolacca americana* p. 78

Note large, smooth, succulent, often reddish stems. Flowers greenish white petallike sepals. **Flowers July–Sept.** Fruits purple-black. **Fruits Sept.–Nov.** All parts of the plant, especially the root, are poisonous. Young spring leaves (gathered before red coloration develops in the stalks) are considered edible after being boiled in several waters. Poke poisoning can cause severe stomach pain with cramps, vomiting, diarrhea, labored breathing, convulsions, and death. Deaths have resulted from children ingesting large amounts of the attractive fruits.

POISON-HEMLOCK *Conium maculatum* p. 82

Smooth-stemmed annual, biennial, or sometimes perennial; purple-streaked or spotted; 4–8 ft. Leaves carrot- or fernlike, lacy and delicate, overall in an equilateral triangle. Foul-scented (likened to mouse urine) when crushed. Flowers white, in umbels, in many-branched groupings. **Flowers May–Aug.** Young plant resembles wild carrot. This is the deadly poison immortalized by Socrates, who died in 399 B.C. after drinking a brew of the plant. Seeds and root, as well as leaves just before flowering, are especially toxic, containing alkaloids including coniine, N-methyl coniine, conhydrine, lambda-coniceine, and pseudoconhydrine. Many poisoning cases, some fatal, have resulted from mistaking the plant for wild carrot, or another edible member of the parsley family, then eating the root or leaves. Symptoms, developing in 1 to 3 hours, include abdominal pain, diarrhea, headache, and a rise in blood pressure, followed by reduced pulse, gradual weakening of muscle strength, loss of sight, difficulty in breathing, coma, and eventually death due to respiratory failure.

POKEWEED IN FRUIT

POKEWEED LEAVES

POISON-HEMLOCK

POISON-HEMLOCK

POISON-HEMLOCK

ALIEN WEEDS
FIELDS, WASTE PLACES, NEAR HOMES

WOODY NIGHTSHADE *Solanum dulcamara* **p. 188**

Note leaves with 1 or 2 prominent lobes at base. Flowers violet (or white) stars with back-curved petal. **Flowers May–Sept.** Fruits red. **Fruits Aug.–Oct.** Fruits and leaves contain toxic steroidal alkaloids, such as solanine, and a glycoside, dulcamarine. Alkaloid content is highest when seeds are mature but unripened. The ingestion of 10 fruits has been implicated in deaths in children, but recent German experience suggests that relatively large amounts of fruits must be ingested to produce serious symptoms (200 is a lethal dose). However, a toxic dose is not necessarily predictable, as different genetic races of the plant produce differing alkaloids in varying amounts. Ingestion of fewer than 10 fruits may cause gastric disturbances. Consumption should be avoided.

TANSY RAGWORT, TANSY BUTTERWEED *Senecio jacobaea* **p. 112**

An uncommon biennial or short-lived perennial European alien found in the Northeast, with smooth, divided leaves (terminal lobe longest), and daisylike yellow flowers to an inch across. **Flowers July–Oct.** Contains pyrrolizidine alkaloids, especially jacobine. Known to cause livestock poisoning; unlikely to be ingested by humans.

CELANDINE *Chelidonium majus* **p. 100**

Brittle-stemmed alien biennial, with orange-yellow juice within; Leaves with irregular round lobes. Flowers yellow, 4-petaled. **Flowers April–Aug.** Stem juice may irritate skin. Contains numerous alkaloids, especially chelidonine. Overdoses may cause vomiting, bloody diarrhea, circulatory disorders, etc. Such poisoning is rare and may involve medicinal preparations. In a recent case in Spain, a patient developed hemolytic anemia after ingestion of a medicinal extract of the plant.

VIPER'S BUGLOSS *Echium vulgare* **p. 128**

Bristly biennial alien with azure blue flowers on curled branches. One flower blooms at a time on curled stalks. **Flowers June–Sept.** Irritant hairs may cause dermatitis. Leaves contain pyrrolizidine alkaloids, which may cause liver disease.

COMFREY *Symphytum* spp.

Grown in gardens, sometimes escaped. **Flowers May–Sept.** Comfrey has increasingly become known for the pyrrolizidine alkaloids in its leaves and, especially, root, which cause liver disease. At least six human cases were documented in the 1980s. Those who continue to ingest the plant as an herbal remedy should know the root contains 10 times more toxic compounds than the leaves. Root ingestion should be avoided. Highest level of toxins in leaves are in first growth of year; these leaves should be discarded.

WOODY NIGHTSHADE

CELANDINE

COMFREY

TANSY RAGWORT

VIPER'S BUGLOSS

WEEDS OR CULTIVATED PLANTS
FIELDS, WASTE PLACES

MARIJUANA *Cannabis sativa*

The palmate marijuana leaf, with its 5–7 round-toothed leaflets, has been depicted on so many T-shirts and posters that this illegal plant is familiar to all. Sometimes seen by hikers in remote areas. The greatest danger the plant poses, besides the obvious mind-altering toxic effects, is that of being found where it is found.

CASTOR BEAN *Ricinus communis* **p. 144**

Note the large palmate leaves divided into 5–11 oval to lance-shaped, toothed lobes. Female flowers in burlike clusters above, reddish to green male flowers beneath. **Flowers July–Sept.** Seed contains the highly toxic lectin ricin, as well as toxic glycoproteins. Enough ricin to kill an adult (1 mg.) can be contained in just one seed. The hard, shiny seeds are attractive, inviting curiosity of children and possible ingestion. Seeds are sometimes used for jewelry, but piercing them releases toxins that can enter the body through a scratch or through the mouth, should a child suck on a necklace made from the seeds. Ricin can cause burning of the throat and mouth, vomiting, severe abdominal pain, diarrhea, thirst from loss of fluids and electrolytes, convulsions, and death.

GIANT RAGWEED *Ambrosia trifida* **p. 148**

Note the opposite, deeply 3- (5-)lobed leaves. Flowers small, greenish, upside down on conspicuous erect spikes. **Flowers July–Oct.** This and other ragweeds are probably responsible for more adverse reactions and human suffering in the U.S. than any other plant group, causing hay fever and allergies as the result of its pollen. Unpalatable ragweeds become surprisingly palatable to livestock after application of the herbicide 2,4—D; toxic levels of the herbicide can accumulate in the plants, poisoning the livestock that feed on them. Ragweeds also cause contact dermatitis.

WORMWOOD *Artemisia absinthium* **p. 148**

A fragrant garden perennial, sometimes escaped. Leaves silver-green, strongly divided into blunt segments, with silky silver hairs on both sides. Flowers small, greenish yellow. **Flowers July–Sept.** Intensely bitter herb common in herb gardens. Formerly used to flavor the alcoholic beverage absinthe; contains the toxic monoterpene thujone. Absinthe was banned in many countries soon after the turn of the 19th century (1907 in Switzerland; 1912 in the United States; 1915 in France). Vincent van Gogh (1853–1890) was believed to be addicted to absinthe. His well-known erratic behavior, resulting in self-mutilation of his left ear and finally suicide, has been attributed to congenital psychosis exacerbated by consumption of large amounts of absinthe near the end of his life.

MARIJUANA

CASTOR BEAN

GIANT RAGWEED

WORMWOOD

YELLOW COMPOSITES
SUNFLOWER FAMILY

BITTER SNEEZEWEED *Helenium amarum* **p. 112**

Note ribbed stems and stringlike leaves. Yellow flower heads to 3/4 in. across, with globe-shaped cone and triangular petals. **Flowers June–Nov.** Often blankets pastures in its range. Livestock may die from eating the plant, but they usually avoid it. Tenulin (a sesquiterpene lactone) is the primary toxic component.

ARNICA *Arnica* spp. **p. 112**

Primarily found in mountains of western North America, the genus *Arnica* includes 32 species with all-yellow flowers, opposite leaves, and hollow, minutely barbed or somewhat feathery hairs atop seeds. **Flowers June–Oct.** The best-known species is the European *Arnica montana* (pictured here), a medicinal plant. Toxic component is helenalin. May cause a toxic-allergic skin reaction. Overdoses have caused increased pulse, heart palpitations, shortness of breath, and death.

GOLDEN RAGWORT *Senecio aureus* **p. 112**

Note leaves of two shapes: basal leaves *heart-shaped,* stem leaves lance-shaped, incised. Flowers yellow, daisylike. **Flowers March–July.** *Senecio* species often contain pyrrolizidine alkaloids. Consumption of the plant, or tea, may cause acute liver lesions, which can lead to cirrhosis and, in severe cases, death.

COMMON TANSY *Tanacetum vulgare* **p. 110**

Note strongly aromatic leaves finely divided into fernlike segments. Flowers yellow buttons in flat-topped clusters. **Flowers July–Sept.** The essential oil of Tansy contains thujone, a monoterpene that is a toxic constituent of a number of other plants, including Wormwood *(Artemisia absinthium).* Some chemical races of Tansy are thujone-free. Thujone may stimulate the central nervous system partly by stimulating the reflexes of the respiratory system. Thujone may interact with the same brain receptor sites that are involved with the active chemical compounds of marijuana.

BITTER SNEEZEWEED HABITAT

BITTER SNEEZEWEED

ARNICA

GOLDEN RAGWORT

COMMON TANSY

OPEN FIELDS, MOSTLY DRY SOILS

INDIAN HEMP, DOGBANE HEMP *Apocynum cannabinum* p. 76
Stems reddish, with milky juice. Flowers terminal, small, whitish green, bell-like or urn-shaped. **Flowers June–Aug.** The milky juice in the stems and leaves is exceedingly bitter, usually discouraging ingestion. Occasionally, cattle, sheep, and horses have died after eating hay contaminated with Indian Hemp. As little as one-half ounce of the dried leaves may kill a large farm animal. A toxic cardiac glycoside found in the plant, apocynamarin, can cause cardiac arrest. Human poisonings are rare, but the plant should be considered potentially dangerous, especially to children who may be attracted to playing with the milky-juiced stalks. Poisoning, even in livestock, is rare.

DOGBANE *Apocynum androsaemifolium* p. 118
Similar to above species, but flowers are drooping pink bells with red stripes within; both in leaf axils and terminal. **Flowers June–Aug.** A toxic cardiac glycoside found in the plant, apocynamarin, can cause cardiac arrest. See above.

PRICKLY POPPY *Argemone albiflora* p. 74
Note yellowish (or white to orange-red) latex in broken stems of thistlelike, sharp-spined leaves. North American *Argemone* species number 15, mainly found in the Southwest U.S.; flower color ranges from white to yellow or lavender-tinted. Flowers 4–6 (12) petals, 2 in. wide, numerous orange stamens. **Flowers May–Sept.** Isoquinoline alkaloids are found in the leaves, stems, and seeds. Grain contaminated with the seeds has caused poisoning in humans. Symptoms include vomiting, diarrhea, fainting, and coma.

WILD LUPINE *Lupinus perennis* p. 136
Note palmate leaves with 7–9 segments lance-shaped (broadest at apex). Flowers pealike in showy raceme. **Flowers April–July.** Some lupines are poisonous; others are not. The seeds of this common garden species, widely escaped and naturalized in the Northeast, are considered toxic. Human poisoning is rare.

INDIAN HEMP

DOGBANE

PRICKLY POPPY

WILD LUPINE

OPEN FIELDS AND WOODS, GLADES MOSTLY DRY SOILS

WHITE WILD INDIGO *Baptisia alba* p. 92

[*B. lactea* (Raf.) Theriet; *B. leucantha* T. & G.]
Stems with grayish white film. Leaves divided into 3 leaflets. Pealike white flowers on a single stout raceme. **Flowers May–July.** White Wild Indigo was reported to cause the death of cows that ate hay containing the plant. May cause diarrhea and loss of appetite. Contains quinolizidine alkaloids.

BLUE FALSE INDIGO *Baptisia australis* p. 134

Smooth perennial; leaves divided into 3 leaflets, cloverlike. Flowers in erect racemes, blue to violet, 1 in. long. **Flowers April–June.** Various *Baptisia* species have been implicated in livestock poisoning. Reports of human toxicity are usually related to historical overdoses of medicinal preparations. May cause loss of appetite and diarrhea.

DELPHINIUMS, LARSKPUR Related species pp. 84, 114, 118, 132

Delphinium spp.

Delphiniums usually have deeply lobed, palmate leaves, flowers distinctly spurred in elongated clusters, often blue to violet but also in white, red, yellow, and orange. At least 70 species are native to U.S., mostly in the West. In addition, numerous species and their cultivars are grown as garden plants. Pictured here are white-flowered (also blue to violet) Dwarf Larkspur *D. tricorne* (found in open woods in the Ozarks) and a garden cultivar. Entire plant, especially young leaves and seeds, is toxic. All *Delphiniums* are considered toxic to some degree. Alkaloids include delphinine, delphineidine, ajacine, etc. May cause nervous symptoms, nausea and vomiting, abdominal pains, dry mouth, depression, restlessness, or, in large doses, death.

WHITE WILD INDIGO

BLUE FALSE INDIGO

DWARF LARKSPUR

DELPHINIUM

MILKWEEDS
FIELDS, PRAIRIES, ROADSIDES, GLADES

COMMON MILKWEED *Asclepias syriaca* **p. 122**

Downy perennial with milky latex. Leaves opposite, to 8 in. long. Flowers in showy globe-shaped clusters from leaf axils, often drooping under their own weight. **Flowers June–Aug.** Circumstantially implicated in livestock poisoning, but the bitter latex in leaves and stems would probably be distasteful to most livestock. Latex may cause dermatitis. Not known to cause poisoning in humans, probably because the fresh plant is unlikely to be ingested. Contains cardiac glycosides.

BUTTERFLY WEED, PLEURISY ROOT *Asclepias tuberosa* **p. 114**

Hairy perennial, without milky juice; to 3 ft. Leaves crowded, surrounding stem, lance-shaped to oblong-lance-shaped, up to 5 in. long. Flowers orange (rarely yellow) in showy, flat-topped, or somewhat rounded clusters. **Flowers May–Sept.** Often cultivated. Root used in folk medicine. Overdoses of the root are reportedly toxic, but historical medical literature generally refers to the safety of root in medicinal preparations rather than noting toxicity. Nevertheless, most milkweeds are considered at least potentially toxic to some degree.

GREEN MILKWEED *Asclepias viridiflora*

Flowers mostly greenish in broad, somewhat flat (as opposed to globe-shaped) clusters. **Flowers April–Aug.** Suspected of poisoning livestock. Unlikely to be consumed by humans; generally distasteful to livestock, but may be found in dry hay. Glycosides present in many milkweeds have caused weakness, bloating, difficulty in standing, rapid and weak pulse, and other symptoms in livestock. Over 20 species of *Asclepias* are native to North America, all of which are at least theoretically toxic.

SWAMP MILKWEED *Asclepias incarnata* **p. 122**

Note leaves opposite or apparently so, lance-shaped or broader at base. Flowers in small umbels, deep rose or bright pink. **Flowers June–Sept.** Sheep in Indiana are believed to have been poisoned from eating this plant. Root historically used in folk medicine, reported to be strongly laxative. Unlikely to be consumed by humans.

COMMON MILKWEED

BUTTERFLY WEED

GREEN MILKWEED

SWAMP MILKWEED

POISONOUS PLANTS CULTIVATED OR NATURALIZED

MONKSHOOD, WOLFSBANE *Aconitum napellus* p. 132

Note blue-violet flowers in many-flowered raceme; helmet-shaped sepal with visorlike beak, to 1 in. tall. **Flowers July–Sept.** More likely to be encountered in flower gardens than in wild habitats. All *Aconitum* species are poisonous. Contains toxic alkaloid aconitine. Symptoms include tingling, then numbing or burning sensation of the mouth, followed by tightening of the throat, causing difficult speech, salivation, nausea, and vomiting. May produce blurred vision, anxiety, an oppressive feeling in the chest, weakness, dizziness, and lack of coordination. Historically, numerous fatalities reported, caused by circulatory failure or asphyxiation, mostly resulting from overdoses of once widely used medicinal preparations of the European *A. napellus.* Roots and leaves are particularly toxic. Roots have also been mistaken for those of edible plants, such as horseradish, resulting in accidental fatalities.

DEADLY NIGHTSHADE, BELLADONNA *Atropa belladonna* p. 130

Note leaves paired, one larger leaf always with a smaller leaf, stalk slightly winged. Flowers single, drooping, bell-shaped, violet–yellow-brownish. **Flowers June–Aug.** Fruits glossy black berries, to 1/2 in. across, seated on star-shaped calyx. **Fruits Aug.–Sept.** Rarely seen in the U.S. but may be grown as a specimen plant in gardens. Important source of the medicinal but highly toxic alkaloids scopolamine, atropine, and *L*–hyoscyamine. Classic symptoms of poisoning include reddening of face, dry mouth, rapid pulse, and dilated pupils. In larger doses, produces hallucinations, fits of frenzies and crying, respiratory paralysis, coma, and death.

STINGING NETTLE *Urtica dioica* p. 138

Unfortunately, the easiest way to identify stinging nettles is to inadvertently brush against them. Flowers are green, inconspicuous. **Flowers June–Sept.** American plants are designated *U. dioica* subsp. *gracilis,* with 6 varieties. All the *Urticas* that occur in North America have stinging hairs like tiny syringes. The hair tip breaks off upon contact, injecting a burning chemical mixture into the skin. The burning sensation may last for up to one hour. The chemical mixture contains histamine, acetocholine, 5–hydroxytryptamine, and small amounts of formic acid.

MONKSHOOD

MONKSHOOD

DEADLY NIGHTSHADE FLOWER

STINGING NETTLE

DEADLY NIGHTSHADE FRUIT

WEEDY NIGHTSHADES

JIMSONWEED, DATURA *Datura stramonium* p. 84

Smooth-stemmed annual; 2–5 ft. Leaves coarse-toothed, angled. Flowers white to pale violet, trumpet-shaped, 3–5 in. **Flowers May–Sept.** Capsules sharp-spined. **Seed capsules Sept.–winter** (persistent when dried). A very dangerous weed. All parts are toxic, including the nectar and seeds. Most poisoning comes from intentional ingestion. Livestock avoid it. Among the toxic alkaloids found in Jimsonweed are hyoscyamine, hyoscine, and scopolamine. The alkaloids are similar to those of Monkshood *(Aconitum)* and produce similar symptoms, which can include intense thirst, dry mouth, rapid weak pulse, hallucinations, delirium, coma, and death. Human poisoning cases usually result from intentional ingestion as a hallucinogen, or from children sucking nectar from the flowers or eating the seeds. All *Datura* species are similarly toxic.

JIMSONWEED, DOWNY THORNAPPLE *Datura wrightii* p. 84

This nightshade was formerly designated *D. inoxia* or *D. meteloides* in the U.S. It has a slightly angled calyx tube and a wingless corolla. The seed capsule, pictured here, is somewhat globose in shape, with short spines, and is densely hairy.

HORSE-NETTLE, CAROLINA NIGHTSHADE p. 130
Solanum carolinense

Stem with yellow, flattish, sharp spines. Flowers violet to white; **Flowers May–Oct.** Berries yellow-orange. **Fruits Aug.–Sept.** Considered a noxious poisonous weed, often found in pastures in South Dakota. Livestock poisoning is common. In 1963, a child died from eating the berries.

DATURA STRAMONIUM CAPSULE

DATURA WRIGHTII FLOWER

DATURA WRIGHTII CAPSULE

HORSE-NETTLE

CULTIVATED OR WILD PLANTS
DRY GROUND, WOOD EDGES

FLOWERING SPURGE *Euphorbia corollata* **p. 80**

Note whorl of reduced leaves beneath flower heads. **Flowers June–Aug.** Most of the 1,600 species of spurges *(Euphorbia)* contain a milky sap with a toxic component that can cause dermatitis and, if ingested, severe internal poisoning.

CYPRESS SPURGE *Euphorbia cyparissias* **p. 142**

Flower structures (cyathia) in umbels, with yellowish oval to triangular bracts beneath, forming the most showy feature of the flowers. **Flowers May–July.** Toxins include the diterpenoids, phorbol esters, and ingenol. Contact with the skin, eyes, or mucous membranes should be avoided.

SNOW-ON-THE-MOUNTAIN *Euphorbia marginata* **p. 80**

Annual, with milky juice; the white bracts beneath the flowers are more showy than the flowers themselves. **Flowers June–Oct.** Widely cultivated as a garden plant. The milky latex is highly caustic, causing dermatitis. Most cases of dermatitis are the result of handling or trimming the plant in the garden, then touching the face or other exposed skin with the white juice, causing burning, reddening, and irritation; exposure is usually self-limiting. Symptoms may also include burning and severe irritation of the mucous membranes of the mouth, throat, and stomach.

LOBELIA *Lobelia inflata* **p. 128**

Note small pale blue flowers with upper throat of flower split; to 1/2 in. long, in raceme. **Flowers June–Oct.** Seed pods inflated. Contains the alkaloid lobeline, which was formerly used as a central nervous system stimulant for suppressed respiratory activity, and until 1993 was used in over-the-counter smoking-cessation products. Ingesting small amounts of leaves or seeds will produce burning in throat and nausea, and in larger doses, vomiting, sweating, rapid feeble pulse, and allegedly coma and death. While reports of fatalities are often cited in poisonous-plant literature, they are not well substantiated. All *Lobelia* species (48 in North America) may be similarly toxic.

WHITE SNAKEROOT *Eupatorium rugosum* **p. 80**

Flowers white, tiny, 10–30 in small rounded heads, on branched clusters; July–Oct. Poisonous to livestock; toxin is transmitted to humans via cow's milk, causing a condition known as "milk sickness," which claimed thousands of victims in the early 1800s. The most famous victim was Abraham Lincoln's mother.

FLOWERING SPURGE

SNOW-ON-THE-MOUNTAIN

WHITE SNAKEROOT

CYPRESS SPURGE

LOBELIA

PLATE 31

GARDEN FLOWERS, SOMETIMES WILD

NARCISSUS, JONQUILS, DAFFODILS p. 94
Narcissus spp.

The genus *Narcissus* contains about 26 species with hundreds of cultivated varieties (cultivars), with white, yellow, or orange blooms. Shown are *N. pseudonarcissus* L., a commonly grown, yellow-flowered species with onion-shaped bulbs, and *N. poeticus*. **Spring flowering.** Bulb and leaves are poisonous. A short time after ingesting small amounts of the plant, humans have developed symptoms such as diarrhea, severe vomiting, and sweating. Fatalities reported. Contains toxic alkaloids and needlelike calcium oxylate crystals, which may cause irritation from handling the plant. Bulbs have been mistaken for onions *(Allium cepa)*, causing poisoning. Store dormant bulbs with appropriate caution.

LANTANA *Lantana camara* p. 114

Note square stems. Flowers orange-yellow (sometimes red or white), in flat-topped clusters to 2 in. across. **Flowers May–Oct.** Naturalized in South from Fla. to Texas. Serious weed in Hawaii and tropics. In the wild, it is poisonous to grazing animals. May produce jaundice, photosensitivity, gastrointestinal disturbances, and constipation. Lantadene A and B are considered the major toxic components. Ingestion of the unripe (green) fruits has been reported to cause poisoning in two small children in Tampa, Florida.

CALLA LILY *Zantedeschia aethiopica* and other species

Note the white funnel-shaped spathe surrounding the yellow spadix. Leaves arrowlike. Native to South Africa and commonly grown as a house plant or garden flower in the U.S. Like many members of the arum family, contains calcium oxalate crystals, which can cause intense burning, swelling, and irritation of the mouth and throat if ingested.

CHRISTMAS ROSE, BLACK HELLEBORE *Helleborus niger* p. 84

Note leaves divided in 7–9 segments originating from horseshoe-shaped base. Flowers white (or pinkish). **Flowers May–July.** Poisoning is rare but may include salivation, tingling of throat and mouth, digestive disturbances, diarrhea, and vomiting.

NARCISSUS

DAFFODILS

LANTANA

LANTANA

CHRISTMAS ROSE

CALLA LILY

SHOWY GARDEN FLOWERS, SOMETIMES WILD

OPIUM POPPY *Papaver somniferum* p. 116

Annual with smooth, green-blue stems. Flowers showy, pink (white to violet). **Flowers May–Aug.** Fruits urn-shaped capsules. No other plant has relieved more pain or caused more human suffering, by means of the addictive alkaloids morphine, codeine, and the derivative heroin. Dried latex from unripe pods is crude opium. Unripe fruits can cause symptoms associated with morphine poisoning, depressing the central nervous system and the respiratory and circulatory systems.

FOXGLOVE *Digitalis purpurea* p. 128

Rough, felty biennial, flowering in the second year. First-year basal leaves resemble those of Comfrey, which has resulted in mistaken identity and subsequent deaths. Flowers violet (pink to white) spotted thimbles. **Flowers June–Aug.** Leaves intensely bitter, discouraging their ingestion. Symptoms of poisoning include dizziness, vomiting, irregular heartbeat, delirium, and hallucinations.

POINSETTIA *Euphorbia pulcherrima* p. 142

A shrub to 10 ft. in its natural range; in the U.S. it is usually seen as a Christmas-season house plant. Floral parts subtended by showy red (or yellowish green) leaflike bracts. **Winter-flowering.** Only a small percentage of cases of Poinsettia ingestion reported to poison control centers involve development of symptoms, mainly vomiting, diarrhea, and local irritation. Nevertheless, there is no good reason to ingest the plant, and it is best to keep it out of the reach of children.

OPIUM POPPY CAPSULES

OPIUM POPPY FLOWER

FOXGLOVE

POINSETTIA

PLATE 33

MISCELLANEOUS CULTIVATED PLANTS

TRUMPET VINE *Campsis radicans* **p. 186**

Weedy, high-climbing vine. Bright orange, leathery, funnel-shaped flowers in clusters. **Flowers June–Sept.** Flowers or leaves may cause dermatitis, producing inflammation and blisters. Potentially poisonous if ingested, though known cases of poisoning are rare. A common plant in its range that is poorly researched.

GROUND CHERRIES, CHINESE LANTERN *Physalis* spp. **p. 102**

The "lanterns" are the yellowish to red inflated bladderlike calyx. We have 28 native or naturalized species. Fruits or leaves may cause gastroenteritis, burning feeling in throat, fever, and diarrhea. Unripe fruits more dangerous than ripe fruits; in some species they are considered edible.

POTATO *Solanum tuberosum* **p. 86**

This familiar perennial is widely cultivated as an annual for its tubers. Flowers white (or bluish) to 1 in. wide. **Flowers May–Oct.** All green parts of the plant, including potatoes greened by exposure to sunlight, contain high concentrations of the alkaloid solanine, and are considered potentially toxic. Sprouted or green-skinned potatoes have caused most reported poisoning cases. Spoiled potatoes have also caused poisoning from bacteria and fungi. Parents should alert children to avoid accidental poisoning from ingestion of the leaves in the family vegetable garden.

LILY OF THE VALLEY *Convallaria majalis* **p. 94**

Spreading by root runners, often forming large clumps. Note leaves without teeth. Flowers white, aromatic, drooping, somewhat bell-shaped. **Flowers May–June.** Most likely to be found in gardens. Ingestion of any plant part may cause digestive irritation, diarrhea, vomiting, irregular heartbeat, coma, and possibly death due to the effects of cardiac glycosides.

PEYOTE *Lophophora williamsii* **p. 116**

Hassock-shaped, spineless cactus forming "buttons" with 7–10 ribs. Flowers pink. **Flowers March–April.** Very unlikely to be consumed except through intentional ingestion of the dried buttons, which contain the psychoactive alkaloid mescaline, as well as lophophorine; causes vomiting, stomach pain, diarrhea, vivid hallucinations, etc. Used in religious rites of Indians.

TRUMPET VINE

GROUND CHERRIES

POTATO

LILY OF THE VALLEY

PEYOTE

MISCELLANEOUS PLANTS

BALSAM PEAR *Momordica charantia* p. 188

Annual vine. Leaves deeply lobed. Flowers solitary, yellow, to 1 in. across. Fruits large, oblong to oval, yellow-orange, *warty,* splitting when mature, revealing seeds covered by red arils. When properly prepared, the bitter unripe fruits are eaten as a vegetable, especially in Asian communities. Seeds or fresh fruit pulp may cause vomiting and severe diarrhea. Plants contain toxic alkaloids, a saponin glycoside, and toxic proteins. Recovery from poisoning can take many weeks.

BLACK CHERRY, WILD CHERRY *Prunus serotina* p. 180

Leaves more than twice as long as wide, blunt-toothed; with whitish brown hairs on prominent midrib. Flowers in drooping slender racemes. **Flowers April–June.** Seeds, fresh and dried leaves, twigs, and bark contain potentially deadly cyanogenic glycosides such as amygdalin and prunasin. When ingested, these substances are transformed into highly toxic hydrocyanic acid (with an almond odor), in essence producing cyanide poisoning, which deactivates enzymes and prevents cells from taking in oxygen. Levels of toxic compounds vary greatly in different plant parts at different stages of growth. Cyanide compounds are heavily concentrated in vigorous shoots of new leaves; wilting leaves also have high concentrations. Symptoms, which can occur suddenly, include respiratory failure, loss of voice, muscle twitching or spasms, coma, and death. Toxic compounds reside in seeds; fruit with seeds removed is considered harmless. Children have been poisoned by chewing on the twigs.

AMERICAN WORMSEED *Chenopodium ambrosioides* p. 146

Leaves are wavy-toothed, rather like oak leaves, smooth or with yellow gland dots. Flowers are inconspicuous green spikes among the leaves. **Flowers and seeds Aug.–Nov.** The seed oil has traditionally been used to expel intestinal worms, but it is little used today because of its toxicity. Overdoses of the seed oil have caused deaths in humans and animals. Excessive handling of plant may cause dermatitis or dizziness from fumes of the essential oil.

EVERLASTING PEA *Lathyrus latifolius* p. 126

Note broadly winged stems. Flowers pealike; pink, blue, or white, in racemes of 4–10 flowers. **Flowers June–Sept.** Implicated in livestock poisoning from nerve degeneration, including symptoms such as hyperexcitability, convulsions, and death.

BALSAM PEAR

BLACK CHERRY

AMERICAN WORMSEED

EVERLASTING PEA

VINES

PITCHER'S CLEMATIS *Clematis pitcheri* **p. 184**

Some *Clematis* species, like Virgin's-bower (below), have split petal-like sepals at right angles; others, like this one, have urn- or vase-shaped flowers, with petallike sepals split only at the end, with up-curved segments. See Virgin's-bower for toxicity.

BITTERSWEET *Celastrus scandens* **p. 188**

Two species in our range, the native *C. scandens* and the Asian *C. orbiculatus,* are sometimes naturalized. Fruit 3-valved, scarlet-orange, splitting; seeds scarlet. **Fruits August–Oct.** If ingested, all plant parts, especially the fruits, may produce vomiting, diarrhea, chills, prostration, and convulsions. Fruits in dried arrangements may be attractive to children. However, cases of human poisoning are rare and poorly documented.

VIRGIN'S-BOWER *Clematis virginiana* **p. 184**

Note leaves divided into 3 sharp-toothed oval leaflets. Numerous flowers, with 4 narrow petallike sepals. **Flowers July–Sept.** Feathery plumes attached to seeds. The fresh sap of *Clematis* species contains protoanemonin, usually in small and variable amounts. It can be highly irritating to the skin or mucous membranes, causing blistering or ulceration. If ingested, symptoms may include gastrointestinal irritation, colic, diarrhea, or nephritis; in severe cases causes paralysis of the central nervous system. *Clematis* species are generally avoided by livestock. Though *Clematis* should be regarded as potentially toxic, cases of poisoning are very rare in the United States.

COMMON IVY *Hedera helix* **p. 190**

One of the most common evergreen vines in cultivation in the United States, grown as a ground cover and often seen climbing walls, buildings, or trees. Note alternate, leathery, oval, 3–5-lobed (or unlobed) leaves. Fruits considered toxic, but are dry and insipid, unlikely to be eaten in quantity. Fatalities of children reported in older European literature are suspicious. Can cause a burning sensation in the throat and gastrointestinal upset, with vomiting and diarrhea. Suspected of causing contact dermatitis.

WISTERIA *Wisteria* ssp. **p. 186**

Both native and introduced *Wisterias* are found in the U.S. Flowers showy, purplish (pink or white) in slender pendent racemes. **Flowers April– Sept.** All parts of Wisteria contain the glycoside wistarine and are considered potentially toxic. May cause nausea, abdominal pain, repeated vomiting, and irritation of gastrointestinal tract and mucous membranes. As few as 2–3 Wisteria seeds may be poisonous to children.

PITCHER'S CLEMATIS

BITTERSWEET

VIRGIN'S-BOWER

COMMON IVY

WISTERIA

COMMON ORNAMENTAL SHRUBS

ENGLISH YEW *Taxus baccata* **Related species** **p. 154**

Note scarlet cuplike aril surrounding seed on underside of branches; **Fruits June–Sept.** All yews are toxic to one degree or another. Yews are among the most widely planted ornamental plant groups, used extensively in landscape schemes such as hedges, around or near buildings, or in groups as a natural fence or border. English Yew *Taxus baccata* is the best known Yew in American horticulture, represented by more than 250 cultivars. Japanese Yew *Taxus cuspidata,* introduced to the United States in 1862, is also widely planted. The foliage and seeds of yews are poisonous to humans, though the fleshy red aril surrounding the seed is edible. Yews contain taxine alkaloids. Initial symptoms include dizziness, blurred vision, and dry mouth, followed by salivation, intense abdominal cramping, vomiting, and dilation of the pupils. Severe cases result in coma, cardiac or respiratory failure, then death. Chewing the leaves may also cause allergic reactions including a rash. Ingesting as few as 50 leaves has resulted in death. The Western Yew *(T. brevifolia)* and other species are the source of taxol, an important anticancer drug in chemotherapy.

OLEANDER *Nerium oleander* **p. 170**

Note the dull, dark green, smoothish, long and narrow leaves in whorls of 3 or opposite. Flowers white, yellow, pink, or purple, 1–2 in. across, trumpetlike, twisted, 5-parted; in showy clusters. Widely grown as a container plant or in the ground in milder areas, especially California. If ingested, the leaves can cause vomiting, dizziness, and heart dysfunction due to a cardiac glycoside, oleandrin. Toxic to both livestock and humans. Human fatalities are recorded. Sucking the nectar from the flowers has caused poisoning in children. Smoke from the burning branches can also cause poisoning. Severe poisoning is rare, because vomiting usually expels the toxins before they are absorbed into the system.

BOXWOOD *Buxus sempervirens* **p. 156**

Evergreen shrub; stems winged or angular. Leaves opposite, leathery, oval. A common hedge plant grown throughout our area. Leaves contain buxine and various steroidal alkaloids. May cause dermatitis. Ingesting any plant part may cause abdominal pain, vomiting, and diarrhea. Large doses can result in convulsions, respiratory failure, and death. Despite the plant's potential for harm, it is seldom ingested because of its acrid taste and peculiar fragrance.

ENGLISH YEW FRUIT

ENGLISH YEW HEDGE

OLEANDER

OLEANDER LEAVES

BOXWOOD

HEATH FAMILY SHRUBS

GREAT RHODODENDRON *Rhododendron maximum* **p. 160**

Note evergreen leathery leaves with edges rolled under. Flowers rose-pink (white), spotted, in showy clusters. **Flowers June–July.** Poisonous components and symptoms same as Sheep Laurel *Kalmia angustifolia* (see below).

SHEEP LAUREL, LAMBKILL *Kalmia angustifolia* **p. 158**

Note opposite, leathery leaves and angular, cup-shaped, pink to red-purple flowers in clusters from axils of previous year's leaves. **Flowers May–July.** Seven species of *Kalmia* are known from North American and Cuba. Two have become naturalized in Europe, causing poisoning there. The toxic constituents of *Kalmia* species are the diterpenes known as acetylandromedols (andromedotoxins). Children have been poisoned from sucking on the flowers or playing with the leaves. Symptoms produced by acetylandromedol-containing plants may include salivation, burning of the mouth, watery eyes and nose, listlessness, dizziness, vomiting, diarrhea, cramping and pain in the intestines, and itching and burning of mucous membranes and skin.

MOUNTAIN LAUREL *Kalmia latifolia* **p. 158**

Note terminal, numerous flowers, white to pink-rose with purple markings. **Flowers May–July.** In addition to symptoms listed for *K. angustifolia,* Mountain Laurel poisoning can also involve cardiac disturbance, including lowering of blood pressure and slowing of pulse. In severe cases, convulsions, paralysis, and death have been reported. As with most poisonous members of the heath family, the honey from the flowers is considered potentially toxic.

STAGGERBUSH *Lyonia* spp. **p. 156**

Flowers white to pinkish, bell-shaped, in umbel-like racemes, in clusters of old leafless branches. **Flowers April–June.** Leaves (and honey from flower nectar) contain lyoniatoxin (lyoniol A). May cause burning in mouth, then salivation, vomiting, diarrhea, and a prickling sensation of the skin. Severe poisoning can result in coma and convulsions. Poisonous to grazing animals.

FETTERBUSH *Pieris japonica* **Related species p. 160**

Note small urn-shaped white flowers in dense terminal raceme clusters. This cultivated Japanese species is more likely to be encountered than the native *P. floribunda.* **Flowers April–May.** Poisonous components and symptoms same as *Kalmia* species.

GREAT RHODODENDRON

SHEEP LAUREL

MOUNTAIN LAUREL

STAGGERBUSH

FETTERBUSH

WOODY PLANTS OF EASTERN FORESTS

COMMON ELDERBERRY *Sambucus canadensis* **p. 164**

Stem with large white pith. Flowers white-yellow, fragrant, in flat, umbrellalike clusters; to 10 in. across. **Flowers May–July.** Fruits purplish black. **Fruits July–Sept.** All plant parts of elderberries, whether purple-black–fruited species such as *S. canadensis,* the blue-fruited western North American species, *S. cerulea,* or the red-fruited western North American *S. racemosa,* should be considered potentially toxic. The berries of *S. canadensis* are considered edible *after cooking.* Avoid consumption of *raw* elderberries. The dried flowers, made into tea, have been used in folk medicine and are probably nontoxic in relatively small amounts. A strong tea will cause profuse sweating, nausea, and vomiting. The stems, with soft pith in up to 90 percent of the diameter of the stem, have been used by children for making pea shooters, whistles, and popguns. If these are placed in the mouth, irritation of mucous membranes, gastric irritation, nausea, and vomiting may result.

RED CEDAR *Juniperus virginiana* **p. 174**

Note leaves reduced to overlapping scales, twigs 4-sided. Fruits hard, roundish, pitted blue-green berries. **Fruits July–winter.** Berries can irritate the kidneys. Potential poisoning may result from excessive use of the berries in herbal teas. Children may be attracted to the fruits, but they are bitter and are unlikely to be ingested in any quantity. Rarely produces cases of poisoning.

COMMON JUNIPER *Juniperus communis* **p. 154**

Note needlelike leaves in whorls of 3, often with 2 white bands above (or 1 white band divided by a green midrib). Fruits on short stalks; globe-shaped, bluish black, with whitish film that rubs off. **Fruits May–June.** Fruits used for flavoring gin and as a diuretic. Ingestion of a few berries usually does not produce symptoms, though large or frequent doses are implicated in kidney failure. Rarely involved in cases of human poisoning.

WAHOO *Euonymus atropurpureus* **p. 162**

Note purplish, smooth (warty in related *E. americana*) fruits irregularly 4-lobed; seeds covered with scarlet aril. **Flowers and fruits May–Oct.** Historically, Wahoo has been valued for its emetic (inducing vomiting) and laxative effects—actions that point to potential toxicity if ingested in sufficient amounts. *Euonymus* species contain cardiotonic glycosides and possibly alkaloids in the leaves, stems, and fruits. Cases of poisoning are reported for the European Spindle Tree *E. europeaus,* historically producing symptoms such as colic, diarrhea, fever, circulatory problems, and collapse about 8–15 hours after ingestion. Modern poisoning cases rare.

ELDERBERRY FRUITS

RED CEDAR

ELDERBERRY FLOWERS

COMMON JUNIPER

WAHOO

BUCKEYES AND BUCKTHORNS

CALIFORNIA BUCKEYE *Aesculus californica* **p. 178**

A broad tree, with leaves divided into 5 oval-oblong leaflets. Flowers pale pink to white. Fruits smooth and pear-shaped. The seeds of this species were ground up and used as a fish poison by Indians in California. Seeds were eaten as survival food after toxins were leached out or rendered safe by roasting. Like all *Aesculus* species, this one contains the toxic glycoside aesculin. No fatalities reported in the U.S. from eating seeds, but they may cause circulatory and gastrointestinal disturbances, vomiting, diarrhea, muscular twitching, temporary paralysis, and stupor.

OHIO BUCKEYE *Aesculus glabra* **p. 178**

Leaflets 4 (rarely to 7), twigs foul-smelling when broken. Flowers yellow; April–May. Fruits with weak prickles. Toxicity same as in Horsechestnut (see below).

CASCARA SAGRADA *Rhamnus purshiana* **p. 180**

Note alternate leaves in tufts at end of branchlets, elliptical-oblong. Fruits black with 3 (rarely 2) seeds. **Fruits Aug.–Oct.** The inner bark of this small tree, aged for at least a year or heated to near boiling for over an hour, is widely used in commercial laxative products. Fresh bark or fruits are potentially toxic, causing intestinal distress and diarrhea.

HORSECHESTNUT *Aesculus hippocastanum* **p. 178**

Note whitish flowers mottled with red and yellow spots. **Flowers May–June.** Fruits warty or spiny. **Fruits Aug.–Oct.** A European native widely cultivated as a shade tree. The seed husks and seeds of Horsechestnut are considered potentially poisonous, but they are very bitter and are seldom eaten. A mixture of saponins, known as aescin, is found in the seeds. Various extracts of Horsechestnut, including aescin, are widely used in medicine in many parts of the world (but not in the U.S.) in topical products for their anti-edema and capillary-sealing effects. Aescin is poorly absorbed if ingested internally. Applied externally, it remains localized and is not systemically absorbed. If ingested, however, the compounds irritate mucous membranes, causing upset stomach and other symptoms. Rare human fatalities are reported from eating seeds.

CAROLINA BUCKTHORN *Rhamnus caroliniana* **p. 180**

Leaves elliptical to oval; untoothed or with minute teeth. Fruits black, 3-seeded. **Fruits Aug.–Nov.** Fruits and bark of all buckthorns are potentially toxic, containing purgative anthracene glycosides that act on the large intestine. We have 15 native or naturalized *Rhamnus* species. The fresh bark or berries of buckthorns can be violently laxative and may irritate the skin or mucous membranes.

CALIFORNIA BUCKEYE

CASCARA SAGRADA

CAROLINA BUCKTHORN

OHIO BUCKEYE

HORSECHESTNUT

MISCELLANEOUS WOODY PLANTS

EUROPEAN MISTLETOE *Viscum album*

Naturalized in Sonoma County, California. Toxic proteins called viscotoxins are found in the leaves but are absent from the berries, which may explain why poison control centers in Europe have not reported symptoms from ingesting berries and consider their toxicity to be slight. Berries of the American mistletoes are considered toxic.

EASTERN MISTLETOE *Phoradendron serotinum* **p. 152**

A thick-branched, parasitic, semievergreen perennial; forms spheres 1–2 ft. across. Fruits translucent white. **Fruits July–Sept.** Ingestion is most likely during the Christmas season, when children may be attracted to the berries of mistletoe decorations. The plant contains toxic amines and a toxic lectin, phoratoxin. Eating a few berries may result in vomiting, diarrhea, and moderate stomach and intestinal pain. In severe cases, symptoms may include labored breathing, dramatically lowered blood pressure, and possible heart failure. Deaths have been reported from ingesting tea made from the berries.

BLACK LOCUST *Robinia pseudoacacia* **p. 182**

Note the paired thorns to 1 in. long (though the form *inermis* is thornless). Fragrant pealike white flowers in racemes. **Flowers May–June.** Children have been poisoned by eating the seeds and inner bark or chewing on fresh twigs. Contains lectins known as toxalbumins (robin and phasin). It also contains a glycoside (robitin). Not likely to be consumed; poisoning rare.

BRAZILIAN PEPPER, FLORIDA HOLLY **p. 176**
Schinus terebinthifolius

Leaves with pepperlike scent when crushed. Fruits bright red, in clusters. Found only in warmer areas of Florida, Texas, California, Hawaii, etc. Trimming the blooming plant can result in toxic reactions including dermatitis, itching, and possible irritation of the eyes. Inhaling the odor of flowers or crushed fruits may cause respiratory irritation. Berries are considered edible, but they may cause gastroenteritis, nausea, vomiting, burning throat, and diarrhea if eaten in large amounts. In France, dried red fruits are mixed with whole black peppercorns to enhance the color and flavor of black pepper.

EUROPEAN MISTLETOE

EUROPEAN MISTLETOE

EASTERN MISTLETOE

BLACK LOCUST

BRAZILIAN PEPPER

GLOSSARY
REFERENCES
ACKNOWLEDGMENTS
INDEX

GLOSSARY

Adventive: Alien and locally escaped and naturalized, but not definitely established.
Alkaloid: Any of a large, varied group of complex nitrogen-containing compounds, usually alkaline, that react with acids to form soluble salts, many with physiological effects on humans. Includes nicotine, cocaine, caffeine, etc.

Banner petal: The upper petal (standard) of a flower, as in members of the pea family.
Basal rosette: Leaves radiating directly from the crown of the root.
Bracts: Leaflike structures just below a flower or group of flowers.

Calyx: The sepals of a flower, collectively.
Cardioactive: Affecting the heart.
Cathartic: A powerful substance that purges the bowels, or laxative.
Corymb: A flower cluster that is flat-topped or slightly convex, but whose individual flower stalks emerge at different points from the stem. Outermost flowers usually open first.
Cytotoxic: An agent that is toxic to certain organs or tissue.

Decoction: A preparation made by boiling a plant part in water.

Emetic: A substance that induces vomiting.
Entire: Without teeth; refers to leaf margins.

Fibril: A threadlike or fiberlike filament found on the cap or stalk surface of a mushroom.
Floret: A very small flower, especially the disk flowers of Aster family members.

Glaucous: Covered with a fine, white, often waxy film that rubs off.
Herbaceous: Nonwoody.
Homeopathic: Relating to homeopathy, a system of medicine founded in the late 1700s by Samuel Hahnemann. Based on the principle "like cures like": a substance that produces a set of symptoms in a well person will cure those same symptoms in a diseased individual, in minute, "potentized" doses.

Laxative: A mildly cathartic substance.

Naturalized: Established without cultivation, as an alien or introduced plant.

Obovate: Oval, but broader toward apex; refers to leaf shape.
Ovate: Oval, but broader toward base.

Palmate: With three or more leaflets or lobes radiating from a central point; hand-shaped.
Panicle: A branching flower group, the branches usually racemes (which see).
Perfect (flower): A flower that possesses a full complement of male and female parts as well as floral envelopes (petals *and* sepals).
Perfoliate: Refers to a leaf that appears to be perforated by the stem.
Pinnate: Resembling a feather; usually refers to a compound leaf with leaflets arranged on each side of a central axis.
Purgative: An agent that will cause watery evacuation of the bowels, usually with sharp pain.

Raceme: An unbranched, elongated flower group, the individual flowers on distinct stalks.
Rays: The straplike, often sterile flowers surrounding the flower head of an Aster family member; *the yellow rays of sunflowers.*
Rhizome: A creeping, underground stem.
Ring: The collar or veillike ring surrounding the stalk of a mushroom; the annulus.
Rosette (basal): Leaves radiating directly from the crown of the root.

Saponin: A glycoside, common in plants, that foams when shaken with water.

Saprophyte: A plant (usually lacking chlorophyll) that lives on dead organic matter.
Sepals: The individual divisions of the calyx.
Sessile: Without a stalk; refers to leaves.
Spadix: A thick, fleshy flower spike (usually enveloped by a spathe) as in members of the Arum family (Jack in the Pulpit, etc.).
Spasmolytic: Checking spasms or cramps.
Spathe: A modified leaflike structure surrounding a spadix, as in members of the Arum family (Jack in the Pulpit, etc.).
Spike (flower): An unbranched, elongated flower grouping, the individual flowers without stalks.
Stamens: Pollen-bearing anthers, with attached filaments (sometimes without filaments).
Stimulant: An agent that causes increased activity of another agent, organ, tissue, or organism.
Stipules: Appendages like small leaves at the base of leaf stalks of certain plants.

Tendrils: Modified leaf or branch structures, often springlike, used for climbing.
Teratogen: A substance causing the deformity of a fetus.

Umbel: A flower group with individual flower stalks, or a floral group radiating from a central axis, often flat-topped and umbrellalike (see also corymb).

Volva: A cuplike structure at the base of a mushroom stalk, consisting of the broken or lower portion of the "egg" or "button" stage of the mushroom, before expansion of the cup. Also called universal veil.

REFERENCES

Guidebooks

Borror, Donald H., and Richard E. White. 1970. *A Field Guide to the Insects of America North of Mexico.* Boston: Houghton Mifflin Co.

Chestnut, V.K. 1898. *Thirty Poisonous Plants of the United States.* Washington, D.C.: Government Printing Office.

Cobb, Boughton. 1963. *A Field Guide to the Ferns.* Boston: Houghton Mifflin. Co.

Conant, Roger, and Joseph T. Collins. 1991. *A Field Guide to Reptiles and Amphibians Eastern/Central North America.* 3d. ed. Boston: Houghton Mifflin Co.

Covell, Charles V., Jr. 1984. *A Field Guide to the Moths of Eastern North America.* Boston: Houghton Mifflin Co.

Craighead, John J., Frank C. Craighead, Jr., and Ray J. Davis. 1963. *A Field Guide to Rocky Mountain Wildflowers.* Boston: Houghton Mifflin Co.

Foster, Steven, and James A. Duke. 1990. *A Field Guide to Medicinal Plants: Eastern and Central North America.* Boston: Houghton Mifflin Co.

Hill, Steven R., and Peggy K. Duke. 1985-86. *100 Poisonous Plants of Maryland.* University of Maryland, Cooperative Extension Service Bull. No. 314.

Kingsbury, John M. 1965. *Deadly Harvest: A Guide to Common Poisonous Plants.* New York: Holt, Rinehart and Winston.

Levy, Charles, C., and Richard B. Primack. 1984. *A Field Guide to Poisonous Plants and Mushrooms of North America.* Brattleboro, Vt: The Stephen Greene Press.

Lincoff, Gary H. 1981. *The Audubon Society Field Guide to North American Mushrooms.* New York: Alfred A. Knopf.

McKnight, Kent H., and Vera B. McKnight. 1987. *A Field Guide to Mushrooms of North America.* Boston: Houghton Mifflin Co.

Millspaugh, Charles F. 1892. *American Medicinal Plants.* Reprint ed. 1974. New York: Dover Publications, Inc.

Niehaus, Theodore F., and Charles L. Ripper. 1976. *A Field Guide to Pacific States Wildflowers.* Boston: Houghton Mifflin Co.

Niehaus, Theodore F., Charles L. Ripper, and Virginia Savage. 1984. *A Field Guide to Southwestern and Texas Wildflowers.* Boston: Houghton Mifflin Co.

Peterson, Lee. 1977. *A Field Guide to Edible Wild Plants.* Boston: Houghton Mifflin Co.

Peterson, Roger Tory, and Margaret McKinney. 1968. *A Field Guide to Wildflowers.* Boston: Houghton Mifflin Co.

Petrides, George A. 1972. *A Field Guide to Trees and Shrubs.* 2nd. ed. Boston: Houghton Mifflin Co.

Schmutz, Ervin M., and Lucretia Breazeale Hamilton. 1979. *Plants that Poison: An Illustrated Guide for the American Southwest.* Flagstaff, Arizona: Northland Press.

Stebbins, Robert C. 1985. *A Field Guide to Reptiles and Amphibians of Western North America.* Boston: Houghton Mifflin Co.

White, Richard E. 1983. *A Field Guide to the Beetles of North America.* Boston: Houghton Mifflin Co.

Woodward, Lucia. 1985. *Poisonous Plants; A Color Field Guide.* New York: Hippocrene Books, Inc.

Youngken, Heber W. Jr., and Joseph S. Karas. 1973. *Typical Poisonous Plants.* Washington, D.C.: Government Printing Office.

General

Bailey, Liberty Hyde, and Ethel Zoe Bailey. Revision by L.H. Bailey Hortorium Staff. 1976. *Hortus Third.* New York: MacMillan.

Barkley, T.M., ed. 1986. *Flora of the Great Plains.* Lawrence: Univ. Press of Kansas.

Blackwell, Will. H. 1990. *Poisonous and Medicinal Plants.* Englewood Cliffs, N. J.: Prentice Hall.

Bresinky, Andreas, and Helmut Besl. 1990. *A Colour Atlas of Poisonous Fungi: A Handbook for Pharmacists, Doctors, Toxicologists, and Biologists.* London: Wolfe Publishing Ltd. **Contains treatment information for physicians.**

Campbell, Jonathan A., and William W. Lamar. 1989. *The Venomous Reptiles of Latin America.* Ithaca, N.Y.: Comstock Publishing Associates. **Contains excellent first aid and treatment information for physicians.**

Caras, Roger A. 1964. *Dangerous to Man: Wild Animals A Definitive Study of Their Reputed Dangers to Man.* New York: Chilton Books.

Correll, Donovan S., and Marshall C. Johnston. 1970. *Manual of the Vascular Plants of Texas.* Renner, Tx: Texas Research Foundation.

DerMarderosian, Ara, and Lawrence Liberti. 1988. *Natural Product Medicine: A Scientific Guide to Foods, Drugs, Cosmetics.* Philadelphia: George F. Stickley Co.

Duke, James A. 1985. *CRC Handbook of Medicinal Herbs.* Boca Raton, Fla. CRC Press.

Ernst, Carl. H. 1992. *Venomous Reptiles of North America.* Washington, D.C.: Smithsonian Institution Press. **Definitive work.**

Fernald, Merritt Lyndon. 1950. *Gray's Manual of Botany.* 8th ed. New York: Van Nostrand Co.

Frohne, Dietrich, and Hans Jürgen Pfänder. 1984. *A Colour Atlas of Poisonous Plants: A Handbook for Pharmacists, Doctors, Toxicologists, and Biologists.* London: Wolfe Publishing Ltd. **Contains treatment information for physicians.**

Fuller, Thomas C., and Elizabeth McClintock. 1986. *Poisonous Plants of California.* Berkeley: Univ. California Press.

Gibbons, Whit, Robert R. Haynes, and Joab L. Thomas. 1990. *Poisonous Plants and Venomous Animals of Alabama and Adjoining States.* Tuscaloosa: Univ. Alabama Press.

Gleason, Henry A., and Arthur Cronquist. 1991. *Manual of Vascular Plants of Northeastern United States and Adjacent Canada.* 2d ed. New York: New York Botanical Garden.

Hardin, James W., and Jay M. Arena. 1974. *Human Poisoning from Native and Cultivated Plants.* 2nd. ed. Durham: Duke Univ. Press. **Contains treatment information for physicians.**

Hickman, James, C. (ed.). 1993. *The Jepson Manual: Higher Plants of California.* Berkeley: Univ. California Press.

Hitchcock, C. Leo, and Arthur Cronquist. 1973. *Flora of the Pacific Northwest: An Illustrated Manual.* Seattle: Univ. Washington Press.

Kartesz, John T., and Rosemarie Kartesz. 1980. *A Synonymized Checklist of the Vascular Flora of the United States, Canada, and Greenland.* Chapel Hill: Univ. North Carolina Press.

Kinghorn, A. Douglas (ed.) 1979. *Toxic Plants.* New York: Columbia Univ. Press.

Kingsbury, John M. 1964. *Poisonous Plants of the United States and Canada.* Englewood Cliffs, N.J.: Prentice-Hall.

Lampe, Kenneth F., and Mary Ann McCann. 1985. *AMA Handbook of Poisonous and Injurious Plants.* Chicago: American Medical Association. **Contains treatment information for physicians.**

Lincoff, Gary, and D. H. Mitchel. 1977. *Toxic and Hallucinogenic Mushroom Poisoning.* New York: Van Nostrand Reinhold. **Contains treatment information for physicians.**

Mabberly, D.J. 1987. *The Plant Book: A Portable Dictionary of the Higher Plants.* New York: Cambridge Univ. Press.

Perkins, Kent D., and Willard W. Payne. n.d. *Guide to the Poisonous and Irritant Plants of Florida.* Gainesville, Fla: Florida Cooperative Extension Service, Univ. Florida,

Polis, Gary A. 1990. *The Biology of Scorpions.* Stanford, Calif.: Stanford Univ. Press.

Radford, Albert E., Harry E. Ahles, and C. Ritchie Bell. 1968. *Manual of the Vascular Flora of the Carolinas.* Chapel Hill: Univ. North Carolina Press.

Stephens, H.A. 1984. *Poisonous Plants of the Central United States.* Lawrence: Univ. Press of Kansas.

Turner, Nancy J., and Adam F. Szczawinski. 1991. *Common Poisonous Plants and Mushrooms of North America.* Portland, Ore.: Timber Press.

Welsh, S.L., N.D. Atwood, L.C. Higgins, and S. Goodrich. 1987. *A Utah Flora.* Provo, Ut.: Brigham Young Univ.

Westbrooks, Randy G., and James W. Preacher. 1986. *Poisonous Plants of Eastern North America.* Columbia: Univ. South Carolina Press.

ACKNOWLEDGMENTS

When a naturalist begins to acknowledge his indebtedness, he undertakes a nostalgic journey back through his lifetime of both work and play. As a boy and man I was helped and encouraged by more people than I can hope to mention. Vladimir Nobokov showed me what it was like to wonder, Ivan Sanderson taught me how to laugh, especially at myself. Laurence Klauber, although we never met, showed me through a wonderful correspondence the wonder of the incredible biological machine we call the rattlesnake.

In Africa, Joy Adamson showed me the world as she found it, full of sentimentality yet somehow precise and true. Dian Fossey showed me the meaning of strength and sacrifice and profoundly deep belief, and Jane Goodall still displays the enormously high standards and the integrity that are such special parts of her work and her person. John Williams took me birding in Kenya and taught me more about the "field" than I think even he knows; Roger Tory Peterson showed me what birding was like in Canada and the Galapagos Islands and helped me renew my enthusiasm not only for life as I live it but for all that lives. All of these people plus scores more—some well known, others obscure in the sense of celebrity but not in quality or beloved memory—help me still, in this book and all of the others, because I am the sum total of their example and encouragement.

Norman Arlott, who did the fine illustrations of animals in this book, was a birding companion in Africa, too, and is as fine a field companion as he is a painter and illustrator. He has added immeasurably to this undertaking.

Thanks to Harry Foster, who stayed the course and was endlessly encouraging. Special thanks to Steven Foster, who stepped in and took over not only the botanical part of this book but rounded the project off and helped it come into being in more ways than I can acknowledge here.

And through it all, of course, there has been Jill: my wife of 39 years, the mother of two children who joined us in the field, and grandmother of four who have now begun to do so. I am ever grateful.

Finally, there are all of the researchers, field biologists, naturalists, zoo directors, and all the other "professional animal people" I have known in my life, personally or through their work. I am in their debt.

Roger Caras

No book can be completed without the help of others—friends, families, contemporaries one has or has not met, and the many researchers, writers, and observers of the natural world who have come before. As always, these individuals are far too numerous to enumerate in the confines of an acknowledgment. This book took me on a journey into new realms of knowledge, literature, and new ways of looking at plants. The classic works on North American poisonous plants by L.H. Pammel, Walter Conrad Muenscher, and John M. Kingsbury proved particularly useful and inspiring.

Without the help and encouragement over the years of teachers including Les Eastman, Dr. James A. Duke, Dr. Shiu Ying Hu, and Prof. Yue Chongxi, my work would not have been possible. I must also thank Dr. Dennis Awang, Mark Blumenthal, Dr. Norman Farnsworth, Dr. Albert Y. Leung, and Dr. Varro Tyler for supplying information on plants and their toxic interactions with humans.

I am indebted to Amy Eisenberg for producing the fine black and white plant illustrations that could not be found in other field guides. Numerous Peterson Field Guide authors and illustrators graciously and generously provided access to and permission to use their artwork in this volume. My warm thanks go to Roger Tory Peterson, Lee Peterson, Kent and Vera B. McKnight, Charles Ripper, and Jim Blackfeather Rose.

Kent McKnight provided extremely useful comments and corrections on the mushroom section. I am grateful to you for your careful eye.

I would like to thank Susan Kunhardt and Lisa White of Houghton Mifflin for their attention to editorial and production details, without which this book would not be possible. And thank you, Harry Foster, for inviting my involvement with this project and for keeping it on track.

I am indebted to Roger Caras for initiating this book and seeing it through with charm and grace.

A special thanks to my assistant, Mary Pat Boian, who kept my papers shuffled in the right piles.

My wife, Jude Farar, graciously gave me the time and encouragement I needed to complete this book. My son Colin and daughter Abbey served as useful sounding boards of curiosity and inquisitiveness into the realms of venomous animals and poisonous plants. From them I learned much. Thank you.

Steven Foster

Photo credits

Plant photographs: © Steven Foster.

Animal photographs except those listed below: © Suzanne L. Collins and Joseph T. Collins.

Eastern Cottonmouth, Io Moth caterpillar, Saddleback Caterpillar, Bumblebee, Bald-faced Hornet, Paper Wasp, Yellowjacket, Red Velvet Ant: © E. R. Degginger.

Shorttail Shrew, Blister Beetle: © Bill Ivy/Tony Stone Images.

INDEX

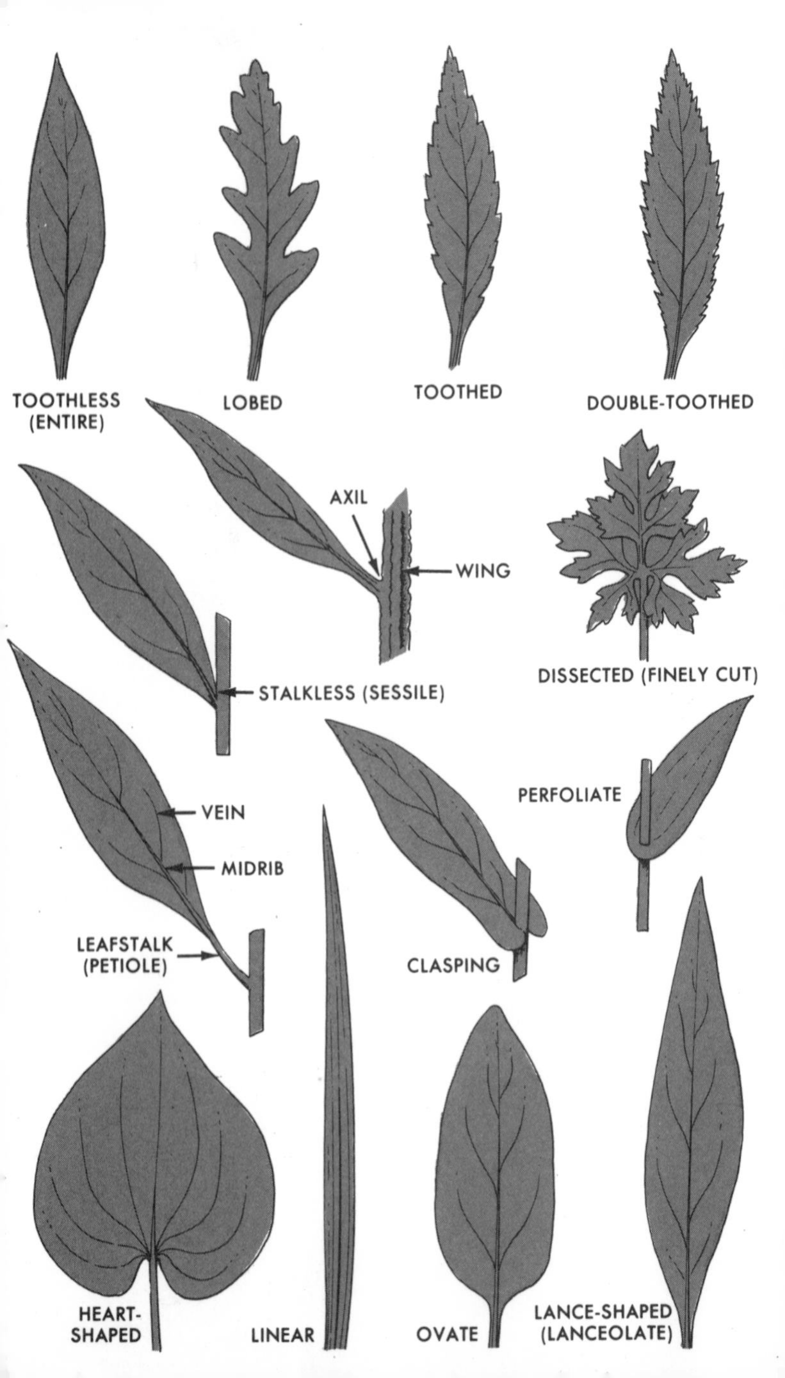
TOOTHLESS
(ENTIRE)
LOBED
TOOTHED
DOUBLE-TOOTHED
AXIL
WING
DISSECTED (FINELY CUT)
STALKLESS (SESSILE)
PERFOLIATE
VEIN
MIDRIB
LEAFSTALK
(PETIOLE)
CLASPING
HEART-
SHAPED
LINEAR
OVATE
LANCE-SHAPED
(LANCEOLATE)